NUTRIENT ELEMENTS IN GRASSLAND
Soil–Plant–Animal Relationships

David C. Whitehead

Department of Soil Science
University of Reading
UK

Formerly of the Grassland Research Institute, Hurley, Berkshire,
and the Institute for Grassland and Environmental Research,
North Wyke Research Station, Devon, UK

CABI *Publishing*

CABI *Publishing* is a division of CAB *International*

CABI Publishing
CAB International
Wallingford
Oxon OX10 8DE
UK

Tel: +44 (0)1491 832111
Fax: +44 (0)1491 833508
Email: cabi@cabi.org
Web site: http://www.cabi.org

CABI Publishing
10 E 40th Street
Suite 3203
New York, NY 10016
USA

Tel: +1 212 481 7018
Fax: +1 212 686 7993
Email: cabi-nao@cabi.org

A catalogue record for this book is available from the British Library, London, UK.

Library of Congress Cataloging-in-Publication Data
Whitehead, D. C. (David Charles)
    Nutrient elements in grassland : soil-plant-animal relationships / D.C. Whitehead.
       p. cm.
    Includes bibliographical references (p.   ).
    ISBN 0-85199-437-7 (alk. paper)
    1. Trace elements in animal nutrition. 2. Ruminants--Nutrition. 3. Trace elements in plant nutrition. 4. Forage plants--Composition. 5. Soils--Trace element content. I. Title.

S592.6.T7 W55 2000
636. 2′0845--dc21   00-028927

ISBN 0 85199 437 7

Typeset by AMA DataSet Ltd, UK.
Printed and bound in the UK at the University Press, Cambridge.

Contents

Preface

This book is concerned with the various chemical elements that are nutrients for either plants or animals. Its primary objective is to bring together information on the concentrations and main transformations of these elements in soils, in grassland plants and in ruminant animals. The data are restricted to those relevant to the grassland areas of temperate regions: data from tropical grasslands are not included. For each element, attention is given to its forms and availability in soils, its uptake and distribution in grassland plants, its role in animal nutrition and the amounts and forms excreted by grazing animals. The influences of soil, plant, weather and management factors on the concentrations of the elements in grassland herbage are described and the concentrations related to the needs of ruminant animals, particularly cattle and sheep. In addition, typical annual balances of the inputs and outputs of each element, on a per hectare basis, are estimated for both intensively managed and extensively managed grassland.

The book will probably be used mainly as a source of reference, and the way in which the chapters are subdivided and the extensive list of references are intended to facilitate this type of use. Readers are assumed to have a basic knowledge of soil science and plant and animal physiology, but a glossary of some of the more specialized terms is included. For the sake of brevity, chemical symbols of the elements are used in the text and, although all the elements often occur as ions, the electrical charge (+ or –) is indicated only where there is specific reference to the ionic form.

In compiling this book, I have drawn partly on information gathered during a period of more than 25 years spent at the Grassland Research Institute, Hurley, UK, until its closure in 1992. Since that time, secondment from the North Wyke Research Station of the Institute for Grassland and Environmental Research and a subsequent Fellowship in the Department of Soil Science at the University of Reading have provided access to the University Library, which has been a major source of additional information.

I would particularly like to thank, for their encouragement and the provision of office accommodation, the current Head of the Department of Soil Science, Professor B.J. Alloway, and his predecessors, Professor P.J. Gregory and Professor R.S. Swift.

David C. Whitehead
Reading
January 2000

Notes

Units

Data for herbage yields and rates of fertilizer application are stated in kg ha^{-1}, some having been converted from lb acre^{-1}. For conversion:

1.0 kg = 2.21 lb
1.0 ha = 2.46 acres
1.0 kg ha^{-1} = 0.89 lb acre^{-1}
1.0 lb acre^{-1} = 1.12 kg ha^{-1}

Terminology

Chemical symbols are used for the nutrient elements. Quantities and concentrations of nutrient elements are stated in terms of the actual elements, not their oxides. The concentrations cited for soil and herbage refer to concentrations in the dry matter, unless otherwise stated; concentrations for animal tissues refer to concentrations in live weight (the usual practice), unless otherwise stated.

Abbreviations

DM	dry matter
DTPA	diethylenetrinitrilopenta-acetic acid
EDTA	ethylenediamine tetra-acetic acid
M	molar (concentration in solution)
MAFF	Ministry of Agriculture, Fisheries and Food (UK)
OM	organic matter

Chapter 1

Introduction

Nutrient Elements Essential for Plants and Animals

Plants and animals are composed primarily of carbon (C), hydrogen (H) and oxygen (O), but a number of other elements, the nutrient elements, are necessary as components of their structural tissues or as participants in biochemical reactions. The nutrient elements known to be essential for plants are nitrogen (N), phosphorus (P), sulphur (S), potassium (K), calcium (Ca), magnesium (Mg), chlorine (Cl), iron (Fe), manganese (Mn), zinc (Zn), copper (Cu), boron (B) and molybdenum (Mo). Plants, such as legumes, that rely directly on N_2 fixation for their N supply also require cobalt (Co) as a nutrient for their rhizobial bacteria. In addition, nickel (Ni) has been shown to be essential for some plant species (Marschner, 1995); and sodium (Na), selenium (Se) and silicon (Si) have been shown to be beneficial in some situations, without satisfying the criteria for essentiality. An element is regarded as essential for plants if: (i) a deficiency makes it impossible for the plant to complete the vegetative or reproductive stage of its life cycle; (ii) the deficiency is specific to the element in question and can be prevented or corrected only by supplying this element; and (iii) the element is directly involved in the nutrition of the plant and is not simply correcting some unfavourable condition of the soil or culture medium (Wild and Jones, 1988).

All the elements essential for plants, with the possible exception of B, are essential for animals and, in addition, animals require iodine (I), Na, Se, chromium (Cr) and Ni. A few other elements, including fluorine (F) and Si, have also been shown to be beneficial when added in minute amounts to purified diets under rigorously controlled laboratory conditions (Underwood and Mertz, 1987; Mills, 1992; Underwood and Suttle, 1999). However, there is no evidence of Cr, Ni, F or Si being deficient in agricultural practice or in natural ecosystems, and they are not discussed further in this book.

The quantity of a nutrient element required by either plants or animals depends on its metabolic functions and varies widely from element to element. There is a marked contrast in plants, for example, between N and

Mo. Herbage plants are deficient in N when the concentration of N in young leaf tissue is less than about 2.5% of dry weight, whereas Mo is deficient only when its concentration is less than about 0.00005% (0.5 mg kg^{-1}). The large requirement for N is due to its being a major constituent of proteins and nucleic acids, while the much smaller requirement for Mo reflects its function as an essential, though quantitatively minor, constituent of several enzymes. The nutrient elements are often categorized into macronutrients and micro-nutrients (or major and minor elements) but the categories differ to some extent between plants and animals. The macronutrients for both plants and animals are N, P, K, Ca, Mg and S. Two elements, Na and Cl, are micro-nutrients for plants but macronutrients for animals; Fe is a micronutrient for plants and on the borderline between micronutrient and macronutrient for animals, while Mn, Zn, Cu, Co and Mo are micronutrients for both plants and animals. Boron is a micronutrient for plants, and I and Se are micro-nutrients for animals. The differences between plants and animals, together with the high concentrations of Fe in many soils, and uncertainties regarding the essentiality of elements such as F and Si, make it difficult to adopt a consistent terminology in this area. The term 'trace element' is sometimes used as an alternative to 'micronutrient', but is often ambiguous, due to uncertainty as to whether other non-essential elements are included and whether the 'trace' applies to soil, plant or animal. In general, the terms 'macronutrient' and 'micronutrient' are used in this book, with Na being included as a macronutrient and Fe as a micronutrient.

Despite their essentiality, it is possible for several of the nutrient elements to be present in soils in concentrations that result in toxic effects in plants and animals. Excessively high concentrations in the soil of potentially plant-available forms sometimes occur naturally through geochemical processes, sometimes through the addition of fertilizers and sometimes as a result of pollution. With some elements, e.g. Cu and Mo, the amounts taken up by plants may be insufficient to damage the plants themselves but sufficient to be toxic to grazing animals. In localized areas, such as those contaminated by mining or industrial wastes, other non-nutrient elements, particularly cadmium (Cd), lead (Pb) and mercury (Hg), may also occur in the soil in sufficiently high concentrations to cause toxic effects in plants or grazing animals.

The concentrations of most nutrient elements in soils normally reflect the nature of the soil parent material, together with the long-term effects of climate and vegetation. However, in some instances, the effects of large fertilizer inputs or of serious pollution may dominate. Although, for most elements, only a small proportion of the total soil content is immediately available for plant uptake, there is sometimes a clear relationship between the concentration of a nutrient element in the soil and the growth of plants or the health of livestock. For example, a deficiency of Mo is responsible for the restricted growth of plants, especially legumes, in parts of Australia, New Zealand and the USA; and a deficiency of Se in the soil causes animal health problems in some areas of various countries, particularly the USA, Canada and China. During the past 30 years, a number of regions have been mapped on the basis of geochemical characteristics, including the concentrations of nutrient elements in soils or sediments, and these maps provide an indication of the possible incidence of problems in plants or animals. For example, in England and Wales, maps have been prepared showing the concentrations in soils of P, K, Na, Ca, Mg, Fe, Mn, Zn, Cu, Co and Cr, though not the other nutrient elements (McGrath and Loveland, 1992). In the USA, maps have been prepared showing areas that are deficient in Zn, Cu, Co, Mo and Se, or that have toxic concentrations of Mo, but based on the analysis of plant material rather than soil (Welch *et al.*, 1991).

Types of Grassland in Temperate Regions

Some areas of temperate grassland occur naturally but others are man-made. In most of the areas where grassland is the natural vegetation (e.g. the steppes of central Asia and the prairies of North America), the climate is semi-arid and the growth of trees is prevented by the lack of rainfall. However, the lack of rainfall also restricts the productivity of grassland, and these areas are increasingly used for arable cultivation rather than livestock production. In the more humid temperate regions (e.g. north-western and central Europe, New Zealand, Japan, parts of Australia and parts of North and South America), most grassland is man-made and would eventually revert to forest if not managed for agriculture. In general, the man-made areas of grassland are more productive and are managed more intensively for milk or meat production from ruminant animals than are the natural grasslands. However, some previously forested areas of grassland are managed extensively and are maintained as grassland by light grazing or occasional cutting: these are regarded as semi-natural grassland. In many such areas, the grassland is important in protecting the soil against erosion.

In natural and semi-natural grassland, plants derive their nutrients almost entirely from recycling processes, mainly the death and decay of plant material, with small inputs from the atmosphere and the weathering of soil minerals. When grassland is managed more intensively, fertilizers are usually applied to increase the supplies of N, P and/or K, and to replace the losses of these elements caused by removing plant and animal products. There is also more recycling of nutrients through animal excreta than in natural and semi-natural grasslands. Sometimes, if maximum productivity is to be achieved, there is a need for nutrient elements other than N, P and K to be applied as fertilizers. However, intensively managed grassland differs from arable agriculture in its need for nutrient elements, due mainly to the involvement of grazing animals and their additional requirements. Although for some nutrient elements (e.g. P) the needs of plants and animals are similar in terms of the concentration in the plant tissue, for other elements (e.g. Na, I, B, Se) the needs differ considerably. In relation to plant growth, the needs of grass-land for fertilizer nutrients are influenced by the intensity of management, by whether the sward is harvested by cutting or grazing and by the botanical composition of the sward. The intensity of management is normally reflected in the amount of herbage produced per unit area per year, and this can vary from less than 1000 kg to more than 15,000 kg DM ha^{-1}. In the absence of legumes, the amount of herbage produced by agricultural grassland often depends on the rate of fertilizer N applied, which may range up to 400 kg N ha^{-1} year^{-1} or more, but such high rates of fertilizer N are effective only if the growth of grass is not severely restricted by other factors, in particular by a lack of water. Variation in rainfall and/or irrigation is therefore an important factor in the amount of herbage produced and in the amount of fertilizer nutrients that can be utilized effectively. Where clover grows vigorously in association with grass and fixes N_2 from the atmosphere, fertilizer N has little effect, though the addition of P, K and/or S may be required for maximum yield. In general, the need for fertilizer is greater when the herbage is harvested by cutting rather than grazing, since nutrients are removed in the hay or silage and the nutrients in excreta are not necessarily returned to the same area.

Some grassland swards are sown to contain only a single grass species, often perennial ryegrass, timothy or brome-grass, depending on geographical region. Other swards may contain one or more grass species, together with a legume, such as white clover or lucerne. Yet others may contain several grass species, possibly with one or more legumes and possibly with non-legume dicotyledonous species (forbs). Recently sown swards usually contain either one species or a mixture of a few species but, as the number of years since sowing increases, the number of species tends to increase. Natural and semi-natural grasslands usually contain a wide variety of species, though a single species, or perhaps two or three, may dominate. However, the application of fertilizers, particularly the repeated application of fertilizer N, to swards containing a mixture of species tends to reduce their species diversity (Heddle, 1967; Smith, 1994; Wedin and Tilman, 1996; Schellberg et al., 1999). This may not be important in terms of productivity in intensively managed systems, but there is evidence that, when the supply of N is limited, a greater diversity of species increases productivity and reduces the loss of nutrients by leaching (Tilman et al., 1996). And the loss of species diversity inevitably reduces the wildlife interest of an area. After many years of receiving fertilizer N, the lack of species diversity cannot be overcome simply by ceasing the application of fertilizer because, by that time, there is a lack of seeds of the appropriate species (Pegtel et al., 1996).

In many countries, and particularly in western Europe, there are conflicting pressures to promote different types of grassland management. On the one hand, there are pressures to maximize animal production per unit area by intensive management. This generally involves the use of single-species swards receiving high inputs of fertilizer, and often with a substantial proportion of the herbage being converted into silage to be fed to livestock in yards. On the other hand, there are pressures to increase the area of grassland that is managed more extensively, in order to minimize the pollution of rivers, lakes and aquifers and to maintain habitats for wildlife. These objectives are favoured by swards containing a variety of species, with little input of fertilizer nutrients and with the herbage utilized as far as possible by grazing. Utilization by grazing avoids the sources of pollution that are liable to result from the production of silage and the disposal of excreta in the form of slurry. One outcome of the conflicting pressures is that, in humid temperate regions, grassland is tending to become increasingly differentiated according to its intensity of management (Wilkins, 1987; Parente, 1996). Three main types can be distinguished:

1.   Intensive production from single-species grass swards receiving fertilizer N at rates of 250–400 kg N ha^{-1} year^{-1}, with appropriate P and K; associated mainly with milk production, and usually involving the production of silage for use during the winter.

2.   Moderate production from grass/clover swards receiving limited inputs of fertilizer nutrients; associated mainly with meat production, with relatively little silage, and occurring predominantly on larger farms.

3.   Grassland managed extensively, partly for agricultural production and partly for the protection of wildlife and landscapes and/or to prevent rural depopulation.

The most effective means of combining adequate food production with environmental objectives is likely to be achieved by maintaining a clear distinction between fields, or larger areas, managed primarily for agricultural production and those managed primarily for their wildlife and landscape value (Wilkins and Harvey, 1994). In some regions, such as much of New Zealand, where clover grows well and winters are relatively mild, the pattern is rather different from that described above, as intensive production can be achieved from grass–clover swards without fertilizer N and with little need for silage.

With grassland in semi-arid regions, attempts to maximize production are liable to result in overgrazing, which curtails plant growth and may result eventually in desertification. This sequence has occurred, for example, in parts of the Middle East and in some grassland areas of China (Hu *et al.*, 1992). Long-term productivity in these regions is dependent almost entirely on the supply of water, through rainfall or irrigation, and on restricting the numbers of animals grazing per unit area.

The Significance of Legume Species

In some regions, legume species, such as clovers, are important components of grassland swards, their importance being due to their ability to utilize atmospheric N_2. Of all living organisms, only a few genera of bacteria are able to fix atmospheric N_2, and these produce the nitrogenase enzyme system that converts N_2 to ammonia (NH_3):

$$N \equiv N + 8H^+ + 8e^- \rightarrow 2NH_3 + H_2$$

The ammonia is then metabolized either by the bacteria themselves, or by a higher plant with which the bacteria have a symbiotic relationship. Some N_2-fixing bacteria (e.g. *Azotobacter*) live freely in the soil, but the most effective are the rhizobia that live symbiotically in the roots of legume species, such as clovers and lucerne. Although the fixation process has a high requirement for energy, the legume species appear to be able to provide this energy from photosynthesis with little, if any, curtailment of growth (Ledgard and Giller, 1995). Consequently, with favourable soil and weather conditions, the amount of herbage produced from grass–clover swards based on N_2 fixation may equal that from all-grass swards receiving high rates of fertilizer N.

In Europe and New Zealand, white clover is the most important legume species in grassland whereas, in much of North America, lucerne, red clover and birdsfoot trefoil are more commonly grown. In temperate areas with a pronounced dry season, such as parts of Australia, the most successful forage legumes are annual clovers and medicks. White clover is particularly important in many areas because it grows well in association with grasses and is tolerant of grazing. Its stolons grow laterally, thus ensuring that most of its growing points are below grazing height and enabling it to colonize gaps in the sward. Of the other cultivated legumes, lucerne, red clover and sainfoin are not stoloniferous; they are usually cut for hay or silage rather than grazed, and are often grown alone rather than in mixtures with grass. However, despite the advantages of including white clover as a component of grassland swards in the more humid temperate regions, it does have a number of disadvantages. First, in comparison with grass, it grows slowly in the cool conditions of early spring. Secondly, it has a relatively small root system, and so competes poorly with grass for nutrients other than N. Thirdly, white clover is more susceptible than grass to cold wet conditions, and so its proportion in grass–clover swards tends to decline during the winter. Fourthly, in some grass–clover combinations, the clover may be shaded by the grass to an extent that reduces photosynthesis or the branching of stolons. A fifth potential disadvantage of white clover is that it is more susceptible than grass to attack by pests, such as nematodes. These various factors provide some explanation for the decline in clover growth that sometimes occurs unexpectedly in well-managed grass–clover swards, and they also explain the reluctance of many farmers to rely on clover rather than fertilizer N.

All the legume species show a wide variation in the amount of N_2 fixed from the atmosphere. For example, with white clover in grass–clover swards, the amount fixed may range from nil to more than 500 kg N ha^{-1} year^{-1}. The amount often reflects the vigour of clover growth, and this is usually limited by factors such as temperature, soil acidity and the supply of water and by competition from the grass. However, fixation is curtailed by the presence of available N in the soil, and therefore by the application of fertilizer N and organic manures and by the N deposited in the urine and, to a lesser extent, the dung of grazing animals. In some situations, a lack of appropriate rhizobia in the soil may limit the amount of N_2 fixed by legumes.

Nutrient Deficiencies in Temperate Grassland Systems

Limitations to grass growth caused by inadequate supplies of N, P or K are widespread and are often overcome by the routine application of fertilizers. Deficiencies of nutrients other than N, P and K occur less widely, though areas of S deficiency are increasing. Deficiencies of Ca and Mg affecting plant growth are rare, but in many areas there is a need for lime ($Ca(OH)_2$ or $CaCO_3$) to overcome soil acidity, and the supply of Mg is sometimes insufficient for grazing livestock. With the micronutrient elements, deficiencies are often localized, though, with some of the elements, there are widespread deficiencies in some regions. Micronutrient deficiencies can be dealt with either by the addition of the nutrient to the soil or, if they affect animals but not plant growth, by direct provision to the livestock.

Although nutrient deficiencies in plants sometimes produce visible symptoms, it is rare for the symptoms in grasses to be characteristic. Diagnosis of a specific deficiency therefore requires additional information, such as the herbage concentration of those elements suspected of being deficient. In agricultural areas, nutrient deficiencies that affect the health, or rate of growth, of ruminant animals are generally more widespread than those affecting grass growth. Thus, with Ca, Mg and several of the micronutrients, the supply from the soil may be sufficient for the maximum growth of plants, but nevertheless insufficient to produce a concentration in the herbage that is adequate for ruminant animals. However, as with plants, a mild deficiency often results in no more than a failure of the animals to thrive, and is therefore difficult to diagnose. In general, clear symptoms in animals are produced only when a deficiency is severe, and this is so with P, Ca, Mg, Zn, Cu, Co, Se and I. When the supply is marginal, the incidence of deficiency in animals is sometimes influenced by the intensity of grassland management. For example, P deficiency occurs most frequently under extensive management with little or no fertilizer input, while deficiencies of Ca and Mg are most likely with large fertilizer inputs, which are liable to alter herbage composition in a way that impairs the absorption of these elements by animals. Also, when the intensity of management is increased, any marginal deficiency of a nutrient other than N, P and K is likely to be accentuated, due to the increased herbage growth causing an increased demand from the soil. In addition, if the soil pH is raised as a result of liming, the availability of some elements, such as Mn and Co, will be reduced. On the other hand, the availability of Mo will be increased, with the possibility of toxic effects on livestock. If the effects of more intensive management include a change in the botanical composition of the sward, with an increase in the proportion of the more productive grass species, these species may have lower concentrations than the original species of some nutrient elements.

Nutrient Cycling in Grassland Systems

The cycling of nutrient elements between soil, plant and animal is particularly important in grassland, because ruminant animals often rely entirely on grassland herbage, including hay and silage, for their nutrient supply, and because the animals return, in their excreta, large proportions of the elements that they consume. Due to their reliance on herbage, the animals need each of the nutrient elements to be present in adequate amounts, and certain combinations of elements to be in an appropriate balance (Underwood and Suttle, 1999). Numerous processes are involved in the cycling of any nutrient through the soil, plant and animal components of a grassland system. They include the conversion of insoluble to soluble forms in the soil, the uptake of the nutrient by plant roots, its translocation from roots to shoots, its utilization by plants, its consumption and subsequent excretion by animals and its return to the soil through the death and decomposition of plant and animal residues. Other processes involve the reactions of the recycled forms with inorganic and organic constituents of the soil. In addition, there may be inputs to the system from the atmosphere or through fertilizers or manures, and losses from the system through leaching and volatilization from the soil. The main processes are illustrated in Fig. 1.1. Clearly, they do not follow a simple sequence but, at various points, provide the possibility of alternative pathways. Overall, the processes combine the effects of a localized cycle, involving a relatively rapid turnover between biological components, with a geochemical cycle, in which the transformations occur much more slowly. However, human activities, such as the production of fertilizers and the burning of fossil fuels, increase the rate at which some nutrients are transferred from the geochemical cycle to the biological cycle. In natural and semi-natural grasslands, geochemical inputs are negligible on an annual basis, and plant uptake is met largely by the biological recycling of nutrients that remain within the system from year to year. However, when the intensity of grassland management is increased through the application of fertilizers and through higher stocking density, inputs of nutrients from external sources and their removal in animal products become increasingly important. The external sources are mainly fertilizers but, in dairy farming, they often include supplementary feeds, provided in addition to grassland herbage, hay and silage.

Although ruminant animals may consume a large proportion of the herbage material produced each year, the remainder, plus all dead root material, undergoes decomposition *in situ*, with the return of the constituent

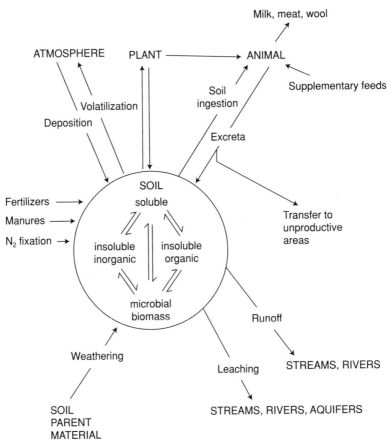

Fig. 1.1. Major transformations involved in the cycling of nutrient elements in grassland systems.

nutrient elements to the soil. Also, since the animals excrete a large proportion of each of the nutrient elements that they consume, substantial amounts are returned to the soil, either directly during grazing or, sometimes, by the application of slurry or manure. However, with most of the nutrients, large proportions of the amounts returned in plant residues and animal excreta are not immediately available for plant uptake, and they undergo various transformations that influence the extent to which they are reutilized. Soil microorganisms are involved in many of these transformations, and especially in the release of nutrients in soluble and plant-available forms. In addition, the soil fauna, such as earthworms and dung beetles, contribute to nutrient cycling, partly by mixing the plant residues and animal excreta more thoroughly with the soil, and partly by consuming and partially digesting such residues so that they are decomposed more rapidly by the microorganisms.

The concentrations of the various nutrient elements in grassland herbage are central to any consideration of soil–plant–animal relationships and of the adequacy of supplies. For any particular nutrient, the concentration in herbage can vary widely, and it is therefore important to understand the factors that control the concentration in a given situation. The supply of the nutrient from soil or fertilizer is clearly a major influence, but many other factors are involved. For example, the application of fertilizer may accentuate a marginal deficiency by increasing yield and thus diluting the herbage concentration of any nutrient not added in the fertilizer. In addition, both positive and negative interactions between nutrient ions can occur, with effects on their solubility in the soil and/or on their uptake by the plant. Other factors that may have a substantial effect on the concentrations of nutrient elements in herbage include the stage of maturity of the herbage, the season of the year and weather conditions, and the plant species present. Consequently, a single value for the herbage concentration of a nutrient, presented as being typical or average for a particular species or type of grassland, may conceal a wide range of variation.

When herbage is grazed, the return of nutrients in excreta is generally beneficial for subsequent plant growth, but not all the grazed area is affected. Nutrients are returned in a concentrated form to the patches affected by dung and urine and, in any one year, these usually represent considerably less than half the area being grazed. The resulting spatial variability in the nutrient status of the soil is most marked with N and K, and it is common for grazed swards to contain patches with excessive amounts of available N and K, alongside patches where N and K are deficient. Grazing animals, as well as returning nutrients in their excreta, also affect plant growth in other ways, for example by repeated defoliation, by compaction of the soil and by contamination of the herbage with dung. These effects are mainly negative and tend to offset the benefits of the return of nutrients in excreta. The return of nutrients in slurry, from the excreta of housed animals fed silage or hay, results in a more uniform distribution of nutrients and this generally favours their reutilization, though it may increase the loss of N through the volatilization of ammonia. The application of slurry also increases the risk that nutrients will be lost in runoff, if heavy rainfall occurs soon after the application.

In general terms, the management of nutrient cycling can be used to increase the productivity of grassland through: (i) increasing the amount of a limiting nutrient in circulation; (ii) increasing the rate of transfer between soil, plant and animal components of the system; and/or (iii) decreasing losses through leaching, runoff and volatilization. Within the soil, the rates at which insoluble forms of the various nutrients are converted to soluble and plant-available forms are increased by a high degree of microbiological activity, a feature usually associated with a plentiful supply of nutrient-rich OM and with a soil pH in the range 5–7. The transfer of nutrients from plant to animal is increased when the stocking density of grazing animals is increased, and an increase in stocking density also tends to increase the rate at which nutrients in the plant component of the system are returned in available forms to the soil. However, increases in the intensity of management and in stocking density also tend to increase losses from the system, especially losses of N. Measures to reduce losses, as well as improving the efficiency of milk or meat production, are important in the context of the wider environment, since nutrients lost from the grassland system are liable to have adverse effects on the composition of water in rivers, lakes and aquifers and on the composition of the atmosphere. For example, enhanced concentrations of N and P in rivers and lakes, due to leaching and runoff, tend to increase eutrophication; and enhanced concentrations of nitrate in aquifers may result in the presence of undesirable amounts of nitrate in drinking-water. The effects on the composition of the atmosphere are due mainly to N, with adverse effects arising partly through increases in the amounts of acidifying compounds, such as ammonium aerosols, and partly through increases in 'greenhouse' gases, such as the nitrous oxide that results from denitrification. In response to concerns about the adverse environmental effects of losses of nutrients from intensive agriculture, various management practices have been proposed to reduce them and also to reduce the consumption of energy in agriculture as a whole. For grassland, such practices include the tactical adjustment of rates of fertilizer N, the injection rather than the surface application of slurry, the greater use of white clover and the partial replacement of grass by maize silage (Jarvis *et al.*, 1996b). In the context of making greater use of clover, it has been calculated that the expenditure of fossil fuels for a fertilized grass sward receiving 200 kg N ha^{-1} year^{-1} is typically 20 times greater than that for a grass–clover sward fixing 200 kg N ha^{-1} year^{-1} (Wood, 1995). Although the measures outlined above are generally effective in achieving a reduction in losses of N and in overall energy consumption, they often entail a reduction in profitability for farmers in current economic circumstances. However, as environmental concerns become more dominant in political terms, it is likely that measures to minimize the impact of agricultural practices on the wider environment will become increasingly cost-effective.

Problems of Sampling and Analysis

Although there is a large amount of published information on concentrations of many of the nutrient elements in soil, plant and animal materials, some of the data have limited value because the analytical procedures that were used were less than satisfactory. In general, inaccurate data are most likely with the micronutrients, which, by definition, are present in extremely low concentrations and for which the analytical methods may not be sufficiently sensitive. The micronutrients are also more likely than the macronutrients to be subject to errors due to contamination of the sample. However, for all elements, errors may arise at various stages of the analytical process. Appropriate procedures for sampling the bulk material are essential in order to achieve accurate analytical results, and are particularly important with soil, due to the marked spatial variability that occurs, especially in grassland soils. Soil samples are usually obtained by taking a number of cores to a specific depth, often 15 cm, and then combining them to provide a composite sample, from which subsamples are taken for analysis. At least 25 cores per sample should be taken, at intervals, from a zigzag path across the field or experimental plot (MAFF, 1994). Inaccuracies due to contamination of the sample are more likely with plant material than with soil or animal material. Grassland herbage may be contaminated by soil before or during collection, or by dust during the drying and grinding operations, and such contamination will increase the apparent concentration of those micronutrient elements, especially Fe, Co and I, whose concentration is much greater in soil than in plant material. For example, the contamination of herbage by soil (containing 3.5% Fe) to the extent of only 0.1% would increase the apparent concentration of Fe by 35 mg kg^{-1}, in comparison with a typical herbage concentration of about 100 mg kg^{-1}. Contamination by soil can be reduced by washing the herbage with water; washing for 5 min in flowing water has been reported to remove almost all soil contamination without causing any measurable loss of N, P, S, K, Na, Ca or Mg from the herbage (Gillingham *et al.*, 1987). Nevertheless, washing does entail the risk of some loss of soluble elements, such as K and Cl, and is probably best avoided unless contamination is visible (Jones, 1991) or unless attention is centred on Fe, Co and/or I. Plant material should always be dried as soon as possible after collection (or after temporary storage at < 10°C) in order to minimize respiration and decomposition, which result in a loss of dry weight and therefore an increase in the apparent concentration of non-volatile elements. Drying is best carried out at 60–80°C, with the herbage being loosely packed to minimize the time available for respiration and decomposition. Samples, particularly of plant material, may be subject to contamination from the packaging used during collection or storage, as various types of plastic and paper contain appreciable amounts of some micronutrient elements, especially Zn and B (Scott, 1970). In addition, soil, plant and animal tissue samples can all be contaminated during grinding, through abrasion of the grinding mechanism. Steel components may contaminate the sample with Fe, while brass components may add Cu and Zn. Rubber fittings may add Zn (Jones, 1991).

   With all types of material, it is important to ensure that the subsamples of any sample to be analysed are homogeneous and representative of the bulk sample. Variation between subsamples in their distribution of different particle sizes can introduce substantial errors into the analytical results. For example, when various forage species were milled and then separated into five particle-size fractions, ranging from > 0.84 mm to < 0.15 mm in diameter, concentrations of N, P, Ca and Mg generally increased with decreasing particle size, often by more than 25% (Smith *et al.*, 1968). Similarly, with grass that had been dried, milled and then separated into fractions (> 0.350 mm, 0.075–0.350 mm and < 0.075 mm), the concentration of I in the finest fraction was more than twice that in the coarsest fraction (Norman and

Mackey, 1997). The problems arising from analytical procedures, as distinct from sample preparation, are most serious for the micronutrients and for S. This is illustrated by a comparison of the results obtained from the analysis of identical samples of plant material in laboratories in 54 countries during the period 1981–1985, a comparison in which the analytical methods used were those in normal use at each laboratory. Coefficients of variation were assessed for a range of elements, and were > 20% for most of the micronutrients and for S, but were considerably less for N, P and K (Houba *et al.*, 1986). For a laboratory to produce reliable analytical data, all stages in the process should be subject to 'quality control' to minimize errors. Consideration should be given to sampling, pretreatment of the samples, storage, sampling and the method of dissolution of the sample, if this is necessary, as well as the choice of analytical method; and, if possible, certified reference samples of soil, plant or animal material should be analysed at the same time in order to check that reliable results are being obtained (Jones, 1991; Houba *et al.*, 1997).

In some reports involving analytical data, ambiguity results from the use of the term 'content'. Sometimes 'content' is used synonymously with 'concentration'; sometimes it refers to the total amount of an element, i.e. concentration × weight of material; and sometimes it is used in a general sense, embracing both concentration and amount. It is possible for content, defined as total amount, to increase with time or with a particular experimental treatment, while the actual concentration of the elements shows a decrease. This type of ambiguity is most likely with plant material, since the yield of DM ha^{-1} is important in quantifying the component processes of nutrient cycling, and it is then essential to distinguish concentration from total amount.

Mass Balances of Nutrient Elements in Grassland Systems

The preparation of the mass balance of a nutrient element, on an annual basis, can provide useful evidence of the extent to which the requirement for a specific output of herbage or livestock product is being met. It may also suggest how to improve the efficiency of use of the element and how to minimize losses. In this context, annual balances for N, P and K, on individual farms in The Netherlands, are being used as part of a programme to limit the amounts of these nutrients lost to the wider environment (Breenbroek *et al.*, 1996). The preparation of a mass balance can also assist in modelling the cycling of a nutrient element, particularly if balance data are available for a range of situations.

A mass balance for a nutrient in a particular grassland system is obtained by summating the various inputs and outputs, to and from the system. The inputs that need to be taken into account are those from atmospheric deposition, from fertilizers and, possibly, sewage sludge and/or from feedingstuffs imported to the system. With N, there may be an additional input from the fixation of atmospheric N_2 by legumes. The outputs are the amounts removed in animal products, the amounts removed in silage or hay (with allowance for returns in slurry) and the amounts lost from the soil by leaching, runoff and volatilization. It is usual to express all the items in the balance in terms of g or kg ha^{-1} year^{-1}. The soil component of the balance may be expressed in terms of either the total amount or the plant-available amount of the nutrient element, the latter requiring estimates of the amounts of the nutrient released through the decomposition of OM and the weathering of parent material and of the amounts immobilized in non-available forms in the soil. The complexity of the balance varies with the element, and is greatest for N and least for the micronutrient cations, whose loss from the soil is usually negligible. Information on the total amounts of a nutrient present in various components of the grassland system at a specific time is not necessary for the preparation of a mass balance, though, if available, it adds to an understanding of the processes involved. Nevertheless, obtaining measurements for all the relevant inputs and outputs at a particular site is complex and time-consuming, and relatively few complete balances have been prepared in this way. Where actual measurements are missing, it is often possible to make reasonable estimates from other published information, but this approach inevitably increases the uncertainty attached to the various items of the balance (Oenema and Heinen, 1999). However, in preparing many of the balances included in Chapters 5–11, the absence of a full set of measurements for any specific location has led, necessarily, to their being based on data from several different sources.

Chapter 2

Nutrient Elements in Soils

Origin of Major Soil Constituents

The nature of the soil parent material is usually the main influence on the amounts of the nutrient elements, other than N, present in a soil. The influence is due partly to the amounts originally present in the parent material and partly to the ability of the soil constituents to retain soluble forms of the elements against loss by leaching. With the large majority of soils, the parent material is inorganic and consists of fragments of weathered rock, together with the products of weathering, such as clay minerals, and oxides of Fe and aluminium (Al). However, with a small proportion of soils, the parent material is mainly organic, in the form of peat derived from past vegetation.

The rocks from which most soil parent materials are formed may be igneous, metamorphic or sedimentary. Igneous rocks are of two main types, depending mainly on the temperature at which they solidified. Those that were formed at relatively low temperatures and pressures, such as granite, are generally acidic and are composed of a mixture of quartz, feldspars and muscovite micas (Table 2.1). On the other hand, igneous rocks that were formed at high temperatures and pressures are generally basic and contain a relatively high proportion of the ferromagnesian minerals, such as pyroxenes, amphiboles, olivines and biotite mica. The feldspars and muscovite micas have relatively high concentrations of Na and K, whereas the ferromagnesian minerals are relatively rich in Fe, Mg and Ca.

The chemical structure of a mineral has a major influence on its behaviour during the processes of weathering. Thus quartz, which is composed entirely of linked silica tetrahedra, is stable and contains no elements other than silicon and oxygen. Feldspars, micas and the ferromagnesian minerals have more complex, less rigid structures and are less stable than quartz. Feldspars are basically Al silicates forming a three-dimensional framework and incorporating varying amounts of Na, K and Ca. Micas vary in composition but consist basically of a layer of Al–O octahedra, linked on each flat surface to a layer of Si–O tetrahedra, an arrangement often referred to as a 2:1

Table 2.1. Some primary and secondary minerals present in soils (from Paul and Huang, 1980).

Mineral group	Name	Chemical formula of basic unit
Primary minerals		
Silica	Quartz	SiO_2
Feldspars	Orthoclase	$KAlSi_3O_8$
	Plagioclase	$NaAlSi_3O_8$
Micas	Muscovite	$K(Si_3Al)Al_2O_{10}(OH)_2$
	Biotite	$K(Si_3Al)(Mg,Fe)_3O_{10}(OH)_2$
Pyroxenes	Augite	$Ca(Mg,Fe,Al)(Si,Al)_2O_6$
Amphiboles	Hornblende	$(Ca,Na)_2(Mg,Fe,Al)_5(Si,Al)_8O_{22}(OH)_2$
Olivines	Forsterite	Mg_2SiO_4
Phosphates	Apatite	$Ca_{10}(F,OH,Cl)_2(PO_4)_6$
Oxides	Rutile	TiO_2
Secondary minerals		
Clay minerals	Kaolinite	$Si_2Al_2O_5(OH)_4$
	Chlorite	$(AlSi_3)AlMg_5O_{10}(OH)_8$
	Smectite/montmorillonite	$M^*_{0.4}(Al_{0.15}Si_{3.85})(Al_{1.3}Fe_{0.45}Mg_{0.25})O_{10}(OH)_2.nH_2O$
	Vermiculite	$M^*_{0.55}(Al_{1.15}Si_{2.85})(Al_{0.25}Fe_{0.35}Mg_{2.4})O_{10}(OH)_2.nH_2O$
Oxides, hydroxides	Goethite	$FeOOH$
	Haematite	Fe_2O_3
	Gibbsite	$Al(OH)_3$
	Allophane	$Al_2O_3.SiO_2.nH_2O$
	Pyrolusite	MnO_2
Carbonates	Calcite	$CaCO_3$
	Dolomite	$CaMg(CO_3)_2$
Sulphur-containing	Gypsum	$CaSO_4.2H_2O$
	Pyrite	FeS_2

M^*, variable metallic cation, such as K, Na, Ca.

lattice. Both feldspars and micas are negatively charged, and this is due largely to isomorphous replacement (i.e. the partial replacement of one element by another, during the formation of the mineral structure). For isomorphous replacement to occur, the difference between the elements in ionic radii must be less than about 15% and the difference in ionic charge no greater than 1. The most common type of isomorphous replacement is of Al^{3+} by Mg^{2+} and/or Fe^{2+}. However, in some minerals, especially the ferromagnesian minerals, Fe^{2+} is often partially replaced by other metallic ions, including Co^{2+}, Zn^{2+} and Cu^{2+} (Davies, 1994). In addition to any negative charge resulting from isomorphous replacement, a further negative charge may result from the dissociation of H^+ ions at the edges of the lattice structure, a dissociation that varies with the ambient pH. The combined negative charge from both sources is balanced by cations, which are adsorbed on to the surface of the lattice structure and, in general, are exchangeable with other cations.

When rocks are exposed to conditions at the earth's surface, they are subject to both physical and chemical weathering. Physical weathering is due mainly to the expansion and contraction induced by changes in temperature, and results in the gradual disintegration of the rocks into smaller fragments. Chemical weathering is due to the interaction of rock material with water, oxygen and carbon dioxide (CO_2). It increases with increasing acidity and therefore with increasing concentration of CO_2. The rate of chemical weathering also tends to increase with increasing temperature, and is greater where new mineral surfaces have been exposed by physical weathering. The resulting changes in composition of the rock material include the partial loss of soluble constituents and the formation of new secondary minerals, such as clays, as shown in Fig. 2.1.

Hydrolysis is a major component of chemical weathering and is involved, for example, in the transformation of feldspars and micas to clay minerals. Potassium feldspar is converted largely to the clay minerals, kaolinite and halloysite, with the release of silicic acid and free K^+ ions. The weathering of muscovite involves, in addition to hydrolysis, the replacement of interlayer K^+ by other cations, such as Mg^{2+}, Ca^{2+} and Al^{3+}, resulting in illite. Similarly, the weathering of biotite yields chlorite or vermiculite. As chemical weathering becomes more prolonged or intense, illite and vermiculite are transformed further to produce smectite, kaolinite and halloysite (see Fig. 2.1).

In general, the clay minerals are composed of layers of Si–O tetrahedra and Al–O octahedra, linked in the ratio of either 1 : 1 or 2 : 1. Typically, there is some isomorphous replacement of Si^{4+} by Al^{3+} and of Al^{3+} by Mg^{2+} or

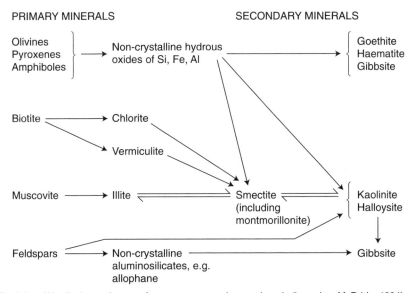

Fig. 2.1. Weathering pathways of some common primary minerals (based on McBride, 1994).

Fe^{2+}, especially in the 2 : 1 lattice structures. Moderately weathered soils are often dominated by clay minerals that have a 2 : 1 lattice structure (e.g. smectite and illite), whereas more strongly weathered soils are dominated by clays with a 1 : 1 structure (e.g. kaolinite). Illite has the characteristic that it contains K^+ ions between successive 2 : 1 layers, in positions in which the K is 'fixed' and therefore not readily exchangeable. Some clay minerals also incorporate small amounts of other metallic elements, such as Mn^{2+} and Zn^{2+}, through the isomorphous replacement of Al^{3+} (West, 1981). Clay minerals, like micas, carry negative charges, due partly to isomorphous replacement and partly to the dissociation of H^+ from the edges of the lattice structure, and these are balanced by adsorbed cations. The number of charges per unit weight of clay mineral, i.e. its cation exchange capacity, is an important property in relation to soil fertility and is often expressed in terms of milliequivalents (mEq) 100 g^{-1}. In addition to clay minerals, the products of chemical weathering may include the insoluble oxides or hydroxides of Fe, Al and Mn, the carbonates of Ca and Mg, and various insoluble phosphates and sulphides (Davies, 1994). Although there is little N in primary minerals, some 2 : 1 clay minerals contain small amounts of N as 'fixed' ammonium within the crystal lattice.

The quantity of any element that remains after weathering is influenced by a wide range of factors, including its abundance in the original rock, whether it behaves as a cation or anion, its ionic potential (valency/radius), and conditions such as pH and redox potential. During chemical weathering, there is normally some loss by leaching, and different constituents of the rock are lost differentially. As a result, the concentrations of nutrient elements in weathered and transported materials may differ markedly from the concentrations in the original rock. The monovalent cations, Na^+ and K^+, are lost more readily than the divalent ions, Mg^{2+} and Ca^{2+} and, amongst the common anions, chloride is lost more readily than sulphate, and sulphate more readily than phosphate.

Past weathering sequences have resulted in the formation of sedimentary rocks, some of them consisting largely of clay minerals and others of fragments of igneous rock plus a cementing material, such as gypsum, calcite, silica and/or iron oxide. Shales are the most widespread type of sedimentary rock, while sandstones and limestones are dominant in some areas. The shales, although consisting mainly of clays, vary considerably in composition, especially in their proportions of clay- and sand-size particles and of OM. Limestones also vary, ranging from almost pure $CaCO_3$ to dolomite limestone, which contains a mixture of $MgCO_3$ and $CaCO_3$, and with both types containing variable amounts of non-carbonate material. Sedimentary rock material has often been transported over long distances by water or ice, and this movement is partly responsible for the regional variation in soil parent material. Thus, much of the topography and the distribution of sediments in the northern hemisphere is due to the effects of the Pleistocene glaciation and the subsequent action of water. Wind erosion has also contributed to the distribution of sediments, resulting, for example, in the deposition of loess in the plains of central Europe and Asia. During the time span of geological activity, sedimentary rocks may have been subject to a number of cycles of weathering and deposition, or they may have been converted to metamorphic rocks by the extreme heat and pressure at depth below the earth's surface. Slate, for example, is a metamorphic rock derived in this way from shale.

As a result of differences in the extent of weathering and transport, the parent materials of soils range from slightly weathered igneous rock to sediments that have been weathered and transported repeatedly, and these differences result in wide variations in particle size distribution and in mineralogical composition. Within soils, the sand, silt and clay size fractions differ markedly in mineralogical composition. The sand and silt fractions (> 0.002 mm diameter) are both dominated by residual primary minerals, such as quartz, feldspars and micas, though the silt fraction may also contain a small proportion of secondary minerals. In contrast, the clay fraction (< 0.002 mm diameter) consists almost entirely of secondary minerals, such as clays and the hydrous oxides of Fe and Al (Paul and Huang, 1980). In temperate regions, hydrous Fe oxides are usually much more abundant than aluminium oxides, and they often occur as coatings on or interlayers between the crystalline clay minerals. Other elements are often associated with these hydrous oxides, due to the coprecipitation or adsorption of cations, such as Mn^{2+}, Zn^{2+}, Cu^{2+} and Co^{2+}. There may also be some coprecipitation or adsorption of anions, such as phosphate, particularly under acid conditions. Chemical weathering continues after the formation of soil and often at an increasing rate, as microorganisms and plants release organic compounds, including acids, into the soil. Organic acids that form chelates with divalent and trivalent cations are particularly effective in releasing these ions from insoluble inorganic minerals. Although the nutrient elements released by weathering in soils are subject to uptake by plants, they are also susceptible to leaching, and this process tends, over many years, to deplete the surface soil of some nutrient elements.

Influence of Parent Material on Nutrient Elements in Soils

The average concentrations of the nutrient elements in two igneous and three sedimentary rocks are shown in Table 2.2, and these differences are reflected to some extent in the composition of soils derived from the rocks. For example, soils derived from basalt tend to have higher concentrations of Mg and of the micronutrient cations than do soils derived from granite. However, although soils derive their initial supply of nutrient elements from their parent material, this influence tends to decline with time. The processes of leaching and gleying, and soil properties, such as pH and redox potential, all affect the extent to which the nutrient elements are retained in soils and the

Table 2.2. Average concentrations (mg kg^{-1}) of nutrient elements in two igneous and three sedimentary rocks (data from Mason and Moore, 1982; Krauskopf and Bird, 1995).

	Granite	Basalt	Shale	Sandstone	Limestone
N	59	52	60	–	–
P	390	610	700	170	400
S	58	123	2,400	240	1,200
K	45,100	5,300	26,600	10,700	2,700
Na	24,600	16,000	9,600	3,300	400
Ca	9,900	78,300	22,100	39,100	302,300
Mg	2,400	39,900	15,000	7,000	47,000
Fe	13,700	77,600	47,200	9,800	3,800
Mn	195	1,280	850	10–100	1,100
Zn	45	86	95	16	20
Cu	13	110	45	1–10	4
Co	2.4	47	19	0.3	0.1
Cl	70	200	180	10	150
I	< 0.03	< 0.03	2.2	1.7	1.2
B	1.7	15	100	35	20
Mo	6.5	0.6	2.6	0.2	0.4
Se	0.007	0.3	0.6	0.05	0.08

forms in which they occur. Biological processes, such as the production of organic acids by plants and soil microorganisms, are also important, through their effects on soil pH and the ability of some organic acids to form soluble complexes or chelates with metal ions.

Although the concentrations of many nutrient elements in soils are influenced by the nature of the inorganic parent material, N is an exception. Apart from the small amount of ammonium which may be fixed by clay minerals, almost all soil N is present in the OM and is derived ultimately from N_2 in the atmosphere, by either biological or chemical fixation. Other elements, particularly P and S, are also present in soil OM, and their total concentrations in soils are therefore influenced by the amount of OM present. In general, the concentrations of most nutrient elements are higher in clay soils than in sandy soils, with loam and silty soils being intermediate: often the concentrations are inversely related to the proportion of quartz in the soil. Typical concentration ranges of 16 nutrient elements in the soils of temperate regions are shown in Table 2.3, together with the concentrations reported for three soils differing in parent material. Soils derived from peat are usually strongly leached and tend to have low concentrations of K, Ca, Mg and the micronutrient cations, but relatively high concentrations of N (McGrath and Loveland, 1992).

Usually only a small proportion of the soil content of any nutrient element is available for uptake by plants at any one time, and this fraction is subject to continued depletion by uptake and leaching. However, it is

Table 2.3. Typical concentration ranges of 17 nutrient elements in soils of temperate regions (based on Kilmer, 1979; Brady and Weil, 1999; plus data in subsequent chapters) mean concentrations of > 500 soils in USA (from Helmke, 1999) and concentrations in three agricultural soils from Scotland differing in parent material (from Ure *et al.*, 1979).

	Soils of temperate regions (range values)	Soils from USA (mean values)	Soil from granite	Soil from shale	Soil from sandstone
N (%)	0.08–0.5	–	–	–	–
P (%)	0.04–0.4	0.026	0.09	0.12	0.04
S (%)	0.02–0.2	0.120	0.06	0.10	0.02
K (%)	0.3–2.5	1.50	3.90	1.50	1.70
Na (%)	0.02–1.0	0.59	1.60	1.20	0.71
Ca (%)	0.2–25.0	0.92	0.36	0.13	0.53
Mg (%)	0.1–1.5	0.44	0.14	0.96	0.39
Fe (mg kg^{-1})	5,000–50,000	18,000	10,000	35,000	20,000
Mn (mg kg^{-1})	50–3,000	330	450	920	440
Zn (mg kg^{-1})	20–300	48	36	47	57
Cu (mg kg^{-1})	2–50	17	7.0	6.0	8.5
Co (mg kg^{-1})	2–40	6.7	1.2	19.0	4.0
Cl (mg kg^{-1})	10–1,000	–	34	160	500
I (mg kg^{-1})	0.5–50	0.75	< 0.1	2.5	0.9
B (mg kg^{-1})	2–100	26	3.6	22.0	9.9
Mo (mg kg^{-1})	0.2–5.0	0.59	1.8	0.7	1.1
Se (mg kg^{-1})	0.1–2.0	0.26	0.09	0.12	0.24

constantly but slowly renewed by chemical weathering and, in some instances, by the mineralization of nutrients from OM and inputs from the atmosphere.

Inputs from the Atmosphere

Several processes result in nutrient elements being transferred from the earth's surface to the atmosphere and, as a consequence, being moved from one area to another before being deposited. Some of the transfers to the atmosphere occur naturally and some are the result of human activities. The natural processes include volcanic and geothermal activity, the formation of volatile compounds through chemical and biochemical reactions and the transport of dust by wind. Some elements are also transferred to the atmosphere from the sea surface, through the suspension and evaporation of droplets of sea spray. The processes that are due to human activity include the combustion of fossil fuels, the emission of pollutants from industrial processes and the increase in wind erosion caused by cultivation. Once in the atmosphere, gaseous molecules and solid particles are transported laterally for distances that may range from less than a millimetre to many thousands of kilometres. Eventually, almost all such molecules and particles are deposited on the land or sea surface, though possibly after chemical interaction with other molecules or particles in the atmosphere. Some of the deposition is 'wet', in rain or snow, and some is 'dry', occurring through the sorption of gaseous molecules by vegetation or soil or through the settling of aerosol particles by gravity.

Wet deposition often supplies significant amounts of plant-available N, S and Cl, and sometimes Na and Mg, to soils. The amounts of Na, Cl and Mg deposited from the atmosphere are particularly influenced by transfers from the sea and therefore by proximity to the coast (Lag, 1987). In general, inputs by wet deposition are usually greatest in areas of high rainfall, while dry deposition is relatively more important in areas of low rainfall. The input of a nutrient by wet deposition reflects the amount of rainfall and the concentration of the nutrient in the rain, both of which can be measured fairly easily. However, the input by dry deposition is more difficult to assess, because both the concentration in the atmosphere and the average deposition velocity fluctuate with time and are influenced by the characteristics of a particular location. Thus, deposition velocity is influenced by particle size, by weather conditions and by the nature of the ground cover, including the type of vegetation (Sposito and Page, 1984).

Assessments of the amounts of 16 nutrient elements deposited from the atmosphere at three rural locations in the UK are shown in Table 2.4. In industrial areas, much larger amounts of some elements, especially Zn and Cu (and several other non-nutrient elements, e.g. Cd), are deposited from the atmosphere and, if continued for many years, these may have a substantial

Table 2.4. Deposition of nutrient elements at three rural sites in the UK: average annual amounts (kg or g ha^{-1} year^{-1}) during the period 1972–1981 (Cawse, 1987).

	Range for three sites	Mean
N (nitrate only) (kg)	36–44	40
S (sulphate only) (kg)	43–58	52
K (kg)	9–20	15
Na (kg)	21–41	29
Ca (kg)	15–19	17
Mg (kg)	5–11	7.2
Cl (kg)	42–74	54
Fe (g)	1400–5700	3100
Mn (g)	90–200	135
Zn (g)	480–1000	660
Cu (g)	170–250	220
Co (g)	1.6–6.0	3.3
I (g)	< 30–< 70	< 50
Mo (g)	< 10	< 10
Se (g)	2.8–5.2	4.0

effect on the concentrations of the elements in soils (Sposito and Page, 1984). Even in the absence of marked industrial pollution, the total deposition of some elements, especially Fe, Zn, Cu and Co, may exceed the amounts removed by crops of silage or hay, as shown by comparing Table 2.4 with Table 3.5 (p. 55). Although, for most elements, much of the atmospheric deposition is not immediately plant-available, the addition to soil reserves may be important, especially where there is little release of nutrient elements by weathering.

Inputs from Fertilizers and Liming Materials

In general, fertilizers are applied to grassland and other crops to meet the needs for N, P and/or K. However, it is unusual for the amount added in fertilizer to be taken up completely, and the remainder adds to the soil reserves, though there may be some loss by leaching and, with N, by volatilization. Also, a proportion of the fertilizer nutrient that is taken up is subsequently returned to the soil in dead plant residues and contributes to the soil reserves in this way. The repeated application of fertilizers, year after year, can therefore add appreciably to the total amounts of N, P and K in the soil, though the effect is generally greatest with P, which is least susceptible to loss. Nutrient elements other than N, P and K may also be supplied in fertilizers, as either intentional or incidental constituents. Sulphur is sometimes included deliberately in fertilizers and, in some regions, one or more of the micronutrient elements may be incorporated in order to overcome a potential deficiency. However, these elements and others, such as Ca, also occur as incidental constituents of fertilizers. For example, superphosphate contains approximately 11% S in the form of incidental calcium sulphate. Some fertilizers contain appreciable amounts of micronutrients, usually derived from the phosphate rock and/or potassium minerals used in their manufacture. The amounts vary widely, as illustrated in Table 2.5, largely because different sources of phosphate rock and potassium minerals have markedly different contents of the micronutrients. Liming materials, which are predominantly calcium carbonate or hydroxide, also contain small but variable amounts of nutrient elements, other than Ca, as incidental constituents (Table 2.6; Reith and Mitchell, 1964).

Recycling through the Decomposition of Plant Material

The stems and roots of grassland plants have a maximum lifespan of a few years, and individual leaves and roots often survive for only a few weeks or months. Even when some of the herbage material is removed by cutting or grazing, a substantial proportion of the herbage, plus all the root material, is eventually returned to the soil through death and decomposition. The

Table 2.5. Reported concentrations (mg kg^{-1}) of some micronutrient elements in ammonium nitrate, superphosphate and potassium chloride fertilizers: (i) Verloo and Willaert, 1990; (ii) Adriano, 1986; (iii) Raven and Loeppert, 1997.

	Ammonium nitrate	Superphosphate		Potassium chloride		
	(i)	(i)	(ii)	(i)	(ii)	(iii)
Fe	192	1150	–	145	–	440
Mn	39	9	890	< 0.7	3.5	2.8
Zn	11.7	244	165	6.1	10	4.6
Cu	3.6	23	15	1.7	3.1	2.7
Co	3.3	15	77	22.5	22	< 0.07
B	–	–	132	–	16	–
Mo	–	–	335	–	26	–

Table 2.6. Concentrations of 13 nutrient elements in agricultural limestones from the USA (from Adriano, 1986).

	Mean	Range for 194 samples
P (%)	0.02	0.001–0.56
S (%)	0.11	< 0.01–1.35
K (%)	0.23	< 0.001–1.80
Na (%)	0.03	< 0.001–0.15
Ca (%)	30.2	18.0–39.8
Mg (%)	4.9	0.04–12.9
Fe (mg kg^{-1})	4,300	100–31,000
Mn (mg kg^{-1})	330	20–3,000
Zn (mg kg^{-1})	31	< 1–425
Cu (mg kg^{-1})	2.7	< 0.3–89
Co (mg kg^{-1})	< 1	1–6
B (mg kg^{-1})	4	< 1–21
Mo (mg kg^{-1})	1.1	< 0.1–92

nutrient elements in the dead herbage and root material are thus recycled and, if there is little input from external sources, this recycling provides the main source for new plant growth. The decomposition of plant residues depends mainly on microbial activity in the soil, and recycling is therefore more rapid in warm moist conditions and when the soil pH is in the range 5–7. Decomposition is also increased by the activities of soil fauna, such as earthworms and insects, which transfer some of the dead plant material to below the surface of the soil, thus making it more accessible to the soil microorganisms. Earthworms and insects also consume dead material, which, after excretion, is more susceptible to further decomposition by micro-organisms. Increasing the productivity of a grassland sward – for example, by promoting the growth of clover or by the use of fertilizer N – tends to increase the population of earthworms and, in highly productive grassland, the weight of earthworms per hectare may equal the weight of grazing livestock (Haynes and Williams, 1993).

During the decomposition of plant residues, those nutrient elements that are constituents of organic molecules are converted to inorganic forms. This process, known as mineralization, is due mainly to extracellular enzymes released by the soil microorganisms, and results in the release of N as ammonium, P as phosphate and S as sulphate ions. At the same time, because soil microorganisms have relatively high concentrations of these nutrient elements, a proportion of the amount mineralized is immobilized in the microbial biomass. The balance between mineralization and immobilization depends mainly on whether the size of the microbial biomass is increasing or decreasing: when it is increasing, there is inevitably some increase in immobilization but, when the microbial biomass is decreasing, there are increases in the amounts of N, P and S mineralized, due to release from the dead microbial tissue. The mineralization and immobilization of the nutrient elements in plant residues do not necessarily match the mineralization and immobilization of carbon, which may undergo net mineralization to CO_2 at the same time that N, for example, is undergoing net immobilization. Such differences are due largely to differences in composition between the plant material and microbial biomass. Plant residues typically have a C : N ratio of > 20 : 1, whereas the microbial biomass has a ratio of about 10 : 1. Similarly, the C : P ratio of plant litter is often > 250 : 1, while that of microbial biomass is about 30 : 1 (Clark and Woodmansee, 1992).

The amounts of herbage material that decompose *in situ* in grassland swards are difficult to measure, partly because some of the material is consumed during the senescent stage by soil fauna and partly because the rate of decomposition varies widely with factors such as temperature and rainfall. In a semi-natural chalk grassland in the UK, not fertilized and not grazed by farm livestock, grass leaves were estimated to live for periods varying from 6 weeks to 6 months, depending on the species and the time of year (Williamson, 1976). With white clover, little of the stolon material appears to survive for more than one growing season (Marriott, 1988). Physiological measurements carried out on intensively managed grass swards indicated that the amount of herbage decomposing *in situ* each year was broadly similar to the amount consumed by grazing animals or harvested for hay or silage (Parsons, 1988). With swards that are grazed lightly or are cut infrequently, the proportion of the herbage decomposing *in situ* is higher. Leaves senesce more rapidly during periods of extreme temperatures and of drought, but some leaf material dies throughout the year. Damage caused by grazing animals also increases the death of leaves and, in general, the proportion of the herbage that decomposes *in situ* is probably about 20% greater when a sward is managed by grazing rather than cutting (Hassink and Neeteson, 1991). Overall, depending on the intensity of management and other factors, the total amount of herbage material decomposing *in situ* varies from less than 1000 DM ha^{-1} year^{-1} in some natural and semi-natural grasslands to more than 10,000 kg DM ha^{-1} year^{-1} in heavily fertilized swards. In some natural grasslands, particularly those on acid and/or poorly drained soils, there is a tendency for dead herbage and stem material to accumulate as a partially decomposed mat at the soil surface. Such an accumulation inevitably reduces the rate at which the nutrient elements are recycled.

Assessing the amount of root material that decomposes in grass swards is also difficult. A newly sown grass sward may produce about 3000 kg root OM ha^{-1} in 6 months, increasing to 10,000 kg ha^{-1}, including both living and dead roots, after 3 or 4 years. However, in dry conditions, the amount of roots may be only 5000 kg or less after several years (Haas, 1958). Under long-term grassland, the total amount of root material is often between 10,000 and 20,000 kg ha^{-1} and, in prairie grassland, it may be as much as 25,000 kg ha^{-1} (Black and Wight, 1979). Although there is a lack of information on the length of life of grass roots, there is some evidence that, when swards are defoliated at intervals of 3–4 weeks, the average length of life is about 5–6 months (Garwood, 1967a; Troughton, 1981). Certainly, the death of roots is increased by defoliation (Evans, 1973; Eason and Newman, 1990). When defoliation is infrequent, roots may live for more than a year, and the average length of life of grass roots in prairie grassland has been estimated at between 2 and 5 years (Dahlman and Kucera, 1965; Clark, 1977; Redmann, 1992). Soil temperature may also have an effect, as illustrated by the finding, with young perennial ryegrass, that the length of life of the roots declined as temperature increased over the range 15–27°C (Forbes *et al.*, 1997). In grass–clover swards, the clovers produce much less root material than do the grasses (Young, 1958), but clover roots decompose more quickly than grass roots (Laidlaw *et al.*, 1996). From these various observations, it appears that the annual turnover of root material probably ranges from less than 1000 kg DM ha^{-1} in some natural and semi-natural grasslands to more than 8000 kg DM ha^{-1} in some intensively managed swards. Evidence from radiocarbon dating indicated that the annual turnover of OM, mainly roots, in an old grassland sward in lowland UK was 4000–5000 kg ha^{-1} (Jenkinson, 1969). Even where the grass species are mainly annual, as in parts of California, the amount of root material undergoing decomposition each year may be as much as 4500–7500 kg ha^{-1} (Jones and Woodmansee, 1979).

In general, the amounts of both herbage and root material undergoing decomposition each year increase with increasing intensity of management, and this implies a corresponding increase in the amounts of nutrient elements being recycled. However, in most grassland soils there is a tendency for the content of OM to increase, and this trend continues for many years if the sward remains undisturbed. Consequently, with N, P and S, which are inherent constituents of soil OM, a proportion of the amount returned each year accumulates in the OM and is not released for renewed plant uptake. Only when the equilibrium content of soil OM is attained does the annual amount of N, P or S returned in dead plant residues (plus animal excreta if the sward is grazed) equal the amount that is potentially available for uptake. However, whether or not there is an accumulation of OM, the uptake of nutrients by plant roots and their return in dead herbage and roots are important in counteracting the tendency towards nutrient depletion caused by leaching. Where there is an accumulation of soil OM, the amounts of nutrient elements in circulation may increase slowly from year to year, due to small inputs from the weathering of soil parent material and/or from the atmosphere. On the other hand, if a grassland sward is ploughed and resown, the OM content of the soil decreases, due to an increased rate of decomposition, and, consequently, there is an increase in the release of nutrients in soluble forms. These soluble nutrients may be taken up by plants but, if not, are susceptible to loss by leaching.

Recycling through the Excreta of Grazing Animals

The presence of grazing animals increases the rate at which nutrient elements are recycled, because the animals utilize for live-weight gain or milk production only a small proportion of each of the nutrients that they consume, and nutrients that are not utilized are returned to the soil in dung and urine. The nutrients in dung are mainly insoluble and become available slowly, whereas those in urine are water-soluble and are therefore potentially available for plant uptake immediately after deposition. The extent to which the recycling of nutrients in grassland is increased by the presence of grazing animals is, of course, influenced by the density of stocking, by the proportion of their time that the animals spend in the field and by the concentrations of nutrients in the herbage consumed. Dairy cattle are usually taken from the field twice per day for milking, and some of their excreta are deposited on farm tracks and in buildings, with the result that the recycling of nutrients is incomplete. Often, 10–35% of the excreta are deposited in milking sheds or on yards and roadways (Nguyen and Goh, 1994a). However, dairy cattle are often given supplementary feed in addition to grazed herbage, and there may then be a net addition of nutrients to the soil through the return of excreta.

The proportions of the grazed area that are affected by dung and urine during any 1 year will depend on the density of stocking, the type of stock and the daily production of dung and urine. However, for dairy cattle grazing at an intensity of 700 cow-days ha^{-1} year^{-1}, assuming an average of 14 faecal excretions per day with 85% in the field being grazed, each covering an area of 0.07 m^2 (see Table 4.12, p. 93), then about 6% of the grazed area would be actually covered by dung each year. However, a larger area, perhaps 20–30% of the total, would be affected by its proximity. With urine, assuming a urination frequency of ten times per day and 85% of the urinations in the field being grazed, plus an average area per urination of 0.35 m² (see Table 4.12) with no overlapping, then urine would be deposited on 21% of the area each year. However, the area of the sward actually covered by a single urination is influenced by a number of factors, including soil moisture content, the presence of macropores that open to the soil surface and the amount of herbage present (Williams and Haynes, 1994). Whatever the area covered by urine, a larger area is affected due to the lateral diffusion of nutrients and the lateral component of root growth, and this may extend to two to five times the area actually covered (see Table 4.12; Lantinga *et al.*, 1987). The recycling of nutrients in excreta, and especially in urine, differs to some extent between cattle and sheep, even though the concentration of nutrients in urine is similar for sheep and cattle if they consume the same pasture (Williams and Haynes, 1993). Cattle produce larger volumes of urine but usually at less frequent intervals (see Table 4.12), and one effect of the larger volume is that the urine penetrates to a greater depth in the soil and tends to result in a greater concentration of nutrient ions, such as NH_4^+ and K^+, in the soil solution of the urine patch. With sheep, the urine patches are smaller and more evenly spread, resulting in a lower concentration of the nutrient ions in the soil solution. Studies with artificially applied urine showed that, typically, cattle urine moved to a depth of 400 mm by flow through macropores, whereas, with a sheep urination, the depth was only about 150 mm (Williams and Haynes, 1994). Sometimes urine scorches the grass on which it is deposited and the reduction in growth, though temporary, results in a reduced uptake of the nutrients contained in the urine. Scorching is most likely to occur in dry conditions and especially when no rain falls in the days following deposition, and when there is little if any accumulation of dead leaf, stubble and root material at the soil surface (Williams *et al.*, 1999).

The visible effect of excreta on grass growth is due mainly to the N and K in urine, and usually this is most apparent about 2–3 months after deposition, though the effect of K in particular may continue for longer (Wolton, 1979). There is often no visible effect of the other nutrient elements in either dung or urine, though there may be a significant impact in the long term. With moderately intensive management, the areas actually covered by urine typically receive the equivalent of about 500 kg N ha⁻¹ and 500 kg K ha⁻¹ (see Table 4.13, p. 93) and, in some situations, the rates of return are considerably higher. A rate equivalent to about 1000 kg K ha⁻¹ was thought to be typical of productive pastures in New Zealand (Saunders, 1984). Since both the N and K are in readily available forms, the inputs to a urine patch exceed the amounts that would be taken up by grass during a period of several months, and they are therefore susceptible to loss. Depending on rainfall, substantial losses of both N and K can occur from grazed swards by leaching and, whatever the rainfall, appreciable amounts of N can be lost as gases through the volatilization of ammonia, denitrification and nitrification. Losses of nutrients through leaching are reduced when the soluble forms in the soil are distributed as evenly as possible over the area and, with a grazed sward, a more even distribution of nutrients is achieved by a high density of relatively small animals than by a lower density of large animals. The effect of urine on uptake by plants is often more long-lasting with K than with N, since the N is lost more readily through leaching and volatilization. In a sward that was intensively grazed by sheep, the proportion of the grazed area affected at any one time by K was estimated to be about three times greater than that affected by N (Morton and Baird, 1990).

With moderately intensive management, the patches actually covered by dung might well receive the equivalent of about 1000 kg N, 350 kg P and 380 kg K ha^{-1} (see Table 4.13). Most of the N and P in dung is in relatively unavailable forms, and some of the N that is potentially available is lost through the volatilization of ammonia. The herbage actually covered by dung may be smothered temporarily, but growth is often stimulated in the area immediately surrounding the patch (MacDiarmid and Watkin, 1971; Weeda, 1977), and the area affected is often about three to five times the area covered (MacDiarmid and Watkin, 1972; During and Weeda, 1973) or up to 15 cm from the edge of the dung patch (Deenen and Middelkoop, 1992). The increase in growth is usually most apparent 2–3 months after the deposition, but the visible effect is accentuated because herbage in the affected area is rejected by grazing animals while visible dung is present (Marsh and Campling, 1970). Sometimes a positive effect on growth may continue for up to 2 years (Haynes and Williams, 1993). There is often a rapid increase in the amount of microbial biomass in dung during the first 5 days after deposition and, during this period, there is a substantial mineralization of organic N (Yokoyama *et al.*, 1991). The decomposition of dung occurs most rapidly in moist conditions (Dickinson and Craig, 1990) and is accelerated by the presence of soil fauna, such as earthworms and dung beetles. Earthworms are particularly important in cool temperate regions, at least when their presence is not curtailed by a lack of moisture (Holter, 1979), whereas dung beetles are most prevalent in warm climatic conditions, such as those in Australia (Tyndale-Biscoe, 1994). The effect of soil fauna in burying and fragmenting dung is reflected in a greater reutilization of the nutrients by grassland plants when such fauna are present (Fincher *et al.*, 1981). Rainfall after the deposition of dung also increases its rate of decomposition (MacDiarmid and Watkin, 1972), as do other causes of fragmentation, such as birds searching for insects. In general, the rates of fragmentation and decomposition are greater for sheep dung than for cattle dung. In a study in New Zealand, sheep dung deposited in spring remained visible on the sward surface for only a few weeks, whereas cattle dung remained for about 12 months (Williams and Haynes, 1993b).

As dung and urine are distributed unevenly over grazed grassland, there is always a large spatial variation in the nutrient status of the underlying soil, variation which resembles a mosaic whose pattern is slowly changing. The change in the pattern is largely due to the decline in the amounts of nutrients under an individual dung or urine patch, resulting partly from plant uptake and partly from losses through leaching or volatilization. Sometimes the dung and urine patches are distributed fairly evenly within the grazed area (MacDiarmid and Watkin, 1972), but there is often spatial variation on a larger scale, due to the tendency of animals to excrete in certain areas, such as near water troughs or in shaded or sheltered areas (West *et al.*, 1989; Wilkinson *et al.*, 1989; Mathews *et al.*, 1994). Sheep have a greater tendency than cattle to 'camp' and excrete in such areas (Wilkinson and Lowrey, 1973; Nguyen and Goh, 1994a). Whatever the pattern of distribution of the excreta, the non-uniform return often results in some parts of a grazed sward being deficient in one or more nutrients, while other parts have an excess (West *et al.*, 1989). However, there is little information on the magnitude of this problem under different grazing systems or on the lateral and vertical distribution of excreted nutrients in the soil and the changes that occur with time. In general, the spatial variation in nutrient status tends to be minimized by grazing with small rather than large animals and by measures that discourage the animals from congregating repeatedly in certain areas. Spatial variation is also reduced when animals produce relatively dilute urine at relatively frequent intervals. With dung, the effects on spatial variation tend to be less when there is an active population of earthworms or dung beetles, and variation can also be reduced by harrowing.

When livestock are housed, their excreta are normally collected and stored as slurries or manures before being applied to the land. Slurries contain both dung and urine, mixed with a variable amount of water, mainly from washing the floors of the housing. Farmyard manure consists largely of dung, with a variable amount of urine and bedding material, such as straw. Some average concentrations of nutrient elements in cattle manures and slurries are shown in Table 2.7. Losses of N can occur from both slurry and farmyard manure by volatilization and, partly for this reason, the ratios of N : P and N : K in slurries and manures are usually lower than those required by plants. Consequently, when slurries are applied to grassland at rates chosen to supply adequate N, the amounts of P and K may be excessive. Thus, in a study in which three rates of cattle slurry were applied annually to grassland for 16 years, with no application of inorganic fertilizers, there was a substantial accumulation of P and K in the surface soil, though little effect below 10 cm (Christie, 1987). In order to achieve the maximum utilization of nutrients when slurry is applied in conjunction with fertilizers, the rate and timing of each application should be adjusted to take into account the effects of the others (MAFF, 1994).

Inputs from the Disposal of Sewage Sludges

Sewage sludge, which is a major waste product of urban communities, often presents a disposal problem, but it does contain appreciable amounts of nutrient elements, especially N and P. The amounts of K are relatively small,

Table 2.7. Some reported average concentrations of nutrient elements in cattle manures and slurries (% or mg kg^{-1} in DM): (i) 400 samples, Maryland (Brady and Weil, 1999); (ii) monthly samples, 1 year, seven farms, North Carolina (Safley *et al.*, 1984); (iii) samples from six farms (Levi-Minzi *et al.*, 1986); (iv) average values (Van Dijk and Sturm, 1983); and (v) 13 farms, north-west Spain (Diaz-Fierros *et al.*, 1987).

	(i) Manure, dairy cow, USA	(ii) Manure, dairy cow, USA	(iii) Manure, beef cattle, Italy	(iv) Slurry, cattle, The Netherlands	(v) Slurry, dairy cattle, Spain
N (%)	2.4	4.7	2.70	5.00	4.00
P (%)	0.7	0.95	0.55	1.05	0.70
S (%)	0.3	–	–	–	–
K (%)	2.1	2.18	3.37	5.67	4.15
Na (%)	–	0.41	–	–	0.98
Ca (%)	1.4	1.17	1.4	1.86	1.71
Mg (%)	0.8	0.60	0.6	0.90	0.70
Cl (%)	–	1.00	–	–	–
Fe (mg kg^{-1})	1800	1400	1220	–	–
Mn (mg kg^{-1})	165	190	150	–	–
Zn (mg kg^{-1})	165	200	143	–	–
Cu (mg kg^{-1})	30	40	49	–	–
B (mg kg^{-1})	20	–	–	–	–

as most of the K remains in the liquid effluent. However, sludges vary in their composition, depending on factors such as the source of the sewage, the method of treatment and how they are stored before application to the soil (Oberle and Keeney, 1994). The simplest method of sludge treatment involves screening to remove fragments of metal and glass, followed by settlement of the solid particles in water to produce raw or primary sludge. Raw sludge is often treated by the activated sludge process, producing a secondary sludge which contains a large proportion of bacterial material. Although this secondary sludge can be applied to land, it is often treated further, in order to reduce the bulk of the sludge and to reduce odour and the population of pathogenic organisms. Anaerobic digestion at about 35°C for 5–35 days is widely used. Until recently, most sewage sludge was disposed of either at sea, in landfill sites or by incineration. However, it is now increasingly being disposed of in ways that are less damaging to the wider environment and, in particular, by application to agricultural land. This method of disposal has the advantage that nutrient elements are returned to the soil, but there are potential disadvantages if the sludge contains toxic materials or pathogens, or if it has to be transported over long distances. Grassland is often more suitable than arable land for the disposal of sludge, because grassland areas are more readily available throughout the year and because the presence of the grass reduces the compaction of the soil caused by vehicles applying the sludge. There is also less risk of surface runoff from grassland. In contrast, the use of arable land is restricted to those times of the year when crops are absent and when there is little risk of soil compaction (Smith, 1996).

Some data on the concentrations of nutrient elements in sewage sludges are shown in Table 2.8. In comparison with typical concentrations in soils (Table 2.3), sewage sludges usually contain much higher concentrations of Cu and Zn, and appreciably higher concentrations of N, P, S, Ca, Mo and Se. Of the toxic materials that may be present in sewage sludge, the most important are the metallic elements that are derived from industrial wastes, particularly Zn, Cu, Cd, Cr and Ni. The actual amounts clearly depend on the proportions of industrial and domestic wastes in the sludge and on the types of industry in the locality. A high proportion of each of these metallic elements is bound initially to the OM of the sludge and, although the metals are released slowly in available forms as the sludge decomposes, they tend to be retained in the soil. Repeated applications may therefore result in an accumulation of metallic elements in the soil, which may ultimately have toxic effects on plants or on livestock consuming the plants. Some metallic elements may also have a harmful effect on microbial processes, such as nitrification, but, in general, the soil microbial population appears to adapt to concentrations in the soil that are less than those that are toxic to plants or livestock (Smith, 1991). However, since the metallic elements in sewage sludge tend to accumulate in the surface layers of the soil, there may

Table 2.8. Mean and range concentrations of nutrient elements in sewage sludges as reported in four investigations (% or mg kg^{-1} DM): (i) 16 samples, cities, USA (Furr *et al.*, 1976); (ii) > 250 samples, north-east USA (Sommers, 1977); (iii) 42 samples, UK (Berrow and Webber, 1972; and, for I, Whitehead, 1979); and (iv) 555 samples, UK survey (Smith, 1996).

	(i)	(ii)	(iii)	(iv)
N (%)	2.9 (1.5–5.8)	3.9 (0.1–17.6)	–	–
P (%)	1.6 (1.0–2.7)	2.5 (0.1–14.3)	–	–
S (%)	–	1.1 (0.6–1.5)	–	–
K (%)	1.2 (0.3–3.9)	0.4 (0.02–2.6)	–	–
Na (%)	0.4 (0.08–1.5)	0.6 (0.01–3.1)	–	–
Ca (%)	3.6 (0.8–11.6)	4.9 (0.1–25.0)	–	–
Mg (%)	0.6 (0.2–1.13)	0.5 (0.03–2.0)	–	–
Cl (%)	0.4 (0.05–1.02)	–	–	–
Fe (%)	3.1 (0.9–8.3)	1.3 (0.1–15.3)	2.4 (0.6–6.2)	1.6 (0.2–10.7)
Mn (mg kg^{-1})	190 (30–520)	380 (20–7,100)	500 (150–2,500)	380 (55–1,390)
Zn (mg kg^{-1})	2,130 (560–6,890)	2,790 (100–27,800)	4,100 (700–49,000)	1,140 (280–27,600)
Cu (mg kg^{-1})	1,350 (580–2,890)	1,210 (80–10,400)	970 (200–8,000)	590 (70–6,140)
Co (mg kg^{-1})	10 (3.7–17.6)	5.3 (1–18)	24 (2–260)	10 (< 2–600)
I (mg kg^{-1})	7.8 (1.0–17.1)	–	5.7 (0.9–17.4)	–
B (mg kg^{-1})	37 (16–90)	77 (4–760)	70 (15–1,000)	30 (15–1,000)
Mo (mg kg^{-1})	14 (1.2–40)	28 (5–39)	7 (2–30)	5 (< 2–154)
Se (mg kg^{-1})	3.1 (1.7–8.7)	–	–	3 (< 2–15)

be a risk to grazing animals, not only through the consumption of herbage but also through their direct ingestion of soil. Many countries have introduced restrictions to prevent the excessive accumulation of metallic elements in the soil – for example, by specifying: (i) minimum periods between the application of sludge and the grazing or harvesting of grassland; (ii) maximum permissible concentrations in soils; and/or (iii) maximum permissible average rates of addition over a 10-year period (MAFF, 1993; Kabata-Pendais and Adriano, 1995). In the UK, it is permissible for livestock to be grazed on sludge-treated grassland after a minimum period of 3 weeks, but this restriction still entails some risk of the animals consuming sludge adhering to the herbage or present in ingested soil (Smith, 1996).

Availability of Nutrient Elements to Plants in Relation to Chemical Form and Soil Processes

With any nutrient element, the fraction immediately available for plant uptake is that present in the soil solution. For most elements, this soluble fraction is largely in simple ionic form, but a variable proportion may occur in small molecular organic compounds. For example, some N may occur in amino acids, some P in inositol phosphates and a proportion of the divalent metallic ions in various organic complexes. Many nutrient elements are able to occur in the soil solution in more than one chemical form, i.e. they have a mixed speciation, and the speciation may vary with factors such as pH, redox conditions and the presence of organic complexing agents. However, the simple ions are generally taken up more readily than the organic compounds. The fraction of any nutrient present in the soil solution is usually only a small proportion of the total, but, when this fraction is depleted by plant uptake and leaching, replenishment tends to occur through a number of natural processes. These include the slow weathering of soil minerals, the mineralization of OM, inputs from the atmosphere, ion exchange and equilibria between soluble and insoluble inorganic compounds. With some elements, particularly N and S, the mineralization of OM is the process that has the greatest influence on the available supply. However, with other elements, such as K and Mg, the available supply is strongly influenced by the cation exchange capacity of the soil and hence by the amount and nature of the clay minerals. With a number of elements, particularly the metallic micronutrients, the formation of soluble complexes with organic molecules is important. Not all soil processes tend to increase the solubility of nutrient elements; some have the opposite effect. Thus, an element released in soluble form in the soil solution may be adsorbed or occluded in a non-exchangeable form by soil constituents such as hydrous Fe oxides; it may become insoluble by reacting with other soluble compounds, or it may be immobilized in organic material, particularly the microbial biomass. Nutrients in these insoluble forms are currently unavailable, though they may become more soluble and available again over a period of time. With the divalent and trivalent cations that form soluble organic complexes, the formation of a complex tends to increase the amount in solution, but it increases plant uptake only if the complex is absorbed intact, or if it is dissociated readily at the root surface.

The nutrient elements in grassland soils, and especially their available forms, are usually present in highest concentration in the top few centimetres. This is largely because leaf litter and animal excreta are deposited and fertilizers are applied on the surface of grassland, and the nutrients are moved downwards only by water or by soil fauna. Within the soil, the accessibility of nutrient elements to plant roots is influenced by the extent to which the soil particles form aggregates, by the size and distribution of pores between the aggregates and by the relative amounts of soil solution and air in the pore space. Fertile soils generally have a large proportion of their aggregates with a diameter of between 1 and 5 mm, and roots grow and take up water and nutrients mainly from the pores and fissures between them. However, micropores within aggregates may also be accessible to root hairs (see p. 43). In addition to their effects on root growth, the amount of pore space and the size distribution of the pores affect the rate at which soluble nutrients can move in the soil. The rate of movement is also influenced by the proportion of the pore space that is occupied by the soil solution: as this diminishes during the drying of the soil, the rate of movement decreases and uptake by plant roots tends to decline.

Influence of Cation Exchange Capacity of the Soil on Availability

The cation exchange capacity of soils is due partly to the negatively charged sites of clay minerals and partly to OM. The contribution of OM is due to the dissociation of H^+ ions from carboxyl and phenolic hydroxyl groups, and increases with increasing pH. In general, the cations that are adsorbed to balance the various negative charges can be replaced by other cations in solution, i.e. by cation exchange. The cation exchange capacity of most soils is within the range 5–50 mEq 100 g^{-1}, though, with peat soils, it may be more than 200 mEq 100 g^{-1}. The greater the cation exchange capacity of a soil, the greater its ability to retain nutrient cations in a form that is potentially available but not readily susceptible to leaching. The replacement of one cation by another depends partly on their relative concentrations and partly on the strength with which they are bound. In general, trivalent cations are bound more strongly than divalent cations, which, in turn, are bound more strongly than monovalent cations. However, the extent to which an ion is hydrated also has an effect, as the size of the hydrated ion determines how close it can come to the exchange sites. When not hydrated, the Na^+ ion is smaller than K^+, and Mg^{2+} is smaller than Ca^{2+}, but when hydrated, K^+ is smaller than

Na^+, and Ca^{2+} is smaller than Mg^{2+} (Mengel and Kirkby, 1987). Those cations which are bound least strongly are the most easily leached. In general, in non-acid soils, Ca^{2+} and Mg^{2+} are the dominant cations on the exchange sites, often accounting for more than 90% of the total cations. Potassium accounts for up to 5% of the total, while Na^+ is often negligible, except in arid saline soils, where it may account for as much as 50% of the total cations. In general, the ratio of $Ca^{2+} : Mg^{2+}$ varies between 5 : 1 and 1 : 2. However, in acid soils, the relative amounts of the nutrient cations are modified by the presence of H^+ ions, together with Al^{3+} ions released from hydrous oxides or clay minerals. In highly acid soils, Ca^{2+} and Mg^{2+} may account for less than 10% of the total exchangeable cations, while H^+ and Al^{3+}, which are negligible in non-acid soils, may account for more than 90% of the total.

Influence of Soil pH on Availability

There is a continuing trend for soils to become more acid, due to the effects of rain containing dissolved CO_2 and other acidic compounds, and to the acidity arising from nitrification and the accumulation of OM. Where soils are subject to leaching, H^+ ions from these various sources displace other cations, which are then lost in the percolating water. Soil acidification tends to occur to a greater extent under grassland than under arable crops, as there is generally more nitrification (of ammonium from the urea in animal urine) and more accumulation of OM. Nevertheless, soil pH is strongly buffered by various reactions, including cation exchange, the dissociation of H^+ and OH^- from soil particles, and the equilibrium between calcium carbonate and CO_2. The presence of free calcium carbonate is particularly effective as a buffer, because it tends to dissolve in the presence of acids, releasing CO_2 but with little, if any, effect on soil pH. In non-calcareous soils with a pH > 5, most buffering is due to the dissociation of H^+ (or OH^-) from clay minerals, hydrous oxides of Fe and Al and OM. In mineral soils with a pH < 5, any further acidity causes Al^{3+} ions to dissolve from clay minerals and occupy the cation exchange sites. If the pH of highly acid soils is increased by liming, the Al then precipitates as $Al(OH)_3$ and Ca is adsorbed by the exchange sites. These various reactions maintain the pH of most soils within the range 4.0–8.5, though sandy soils, which are poorly buffered, may have a pH of < 4.0.

Although active plant growth can occur throughout the pH range of 4.0–8.5, individual plant species often require a more restricted range of pH. Some plant species are susceptible to low pH and others to high pH. The harmful effects of low pH are often due not to low pH *per se*, but to high concentrations of Al^{3+} and Mn^{2+} in the soil solution having toxic effects on plants, or to a reduced availability of P. The harmful effects of high pH may also be due to a reduced availability of P, as well as to a reduced availability of Fe. In general, the availability of the micronutrients occurring as cations, especially Fe, Mn and Co, tends to increase with increasing soil acidity, whereas the availability of the micronutrients occurring as anions (B, Mo and Se) tends to decrease with increasing acidity. In highly acid soils with a pH lower than about 5, there is a reduction in the activity of the microbial population and also in the rate of root growth, and these factors may curtail the uptake of nutrients.

Influence of Soil Organic Matter on Availability

Almost all the N and much of the P and S in grassland soils occurs in the soil OM, and each of these three elements is released in plant-available forms by the mineralization brought about by microbial enzymes. Organically bound N is released as NH_4^+, organically bound P as $H_2PO_4^-$, and organically bound S as SO_4^{2-}. The greater the amount of OM in the soil, and the more rapid its decomposition, the greater the supply of plant-available N, P and S. As it relies on microbiological activity, mineralization does not occur at a uniform rate during the year. It is usually greatest during late spring, when the soil is becoming warmer but is still moist, and is probably enhanced when freezing and thawing episodes have occurred during the winter. The rate of mineralization is slow during the winter and is zero when the soil is frozen.

Soil OM has only a small effect on the available supply of the macronutrient cations, K, Na, Ca and Mg, though it does contribute to the soil's cation exchange capacity, often to the extent of 20–70% (Stevenson, 1994). OM has a greater effect on the availability of the micronutrient cations (Fe, Mn, Zn, Cu, Co), which, in addition to being retained by cation exchange, form complexes with carboxyl, phenolic and other organic groups. Depending on the rate of decomposition of the OM, and on the amounts and types of complexing compounds produced, the availability of the micronutrient cations may be either increased or decreased. Complexation by small and soluble organic molecules will tend to increase the amount of a micronutrient cation in the soil solution and therefore increase its mobility and availability, whereas complexation by an insoluble component of OM will reduce the amount in solution and hence reduce availability. There may also be interactions between organic compounds and hydrous Fe, Al and Mn oxides, which may sometimes increase and sometimes decrease their capacity to adsorb other cations (Harter and Naidu, 1995). The presence of large amounts of OM in a soil was found to increase the uptake of Cu by roots, probably by the soluble OM increasing the amount of Cu in solution and thus promoting movement to the root (McBride, 1994).

One component of the soil OM which has a major influence on nutrient availability is the microbial biomass. In temperate grassland soils, estimates of the amount of microbial biomass have ranged between 500 and

9000 kg ha^{-1}, with typical values of 2000–3000 kg ha^{-1} (Úlehlová, 1992). Its effect on nutrient availability is due partly to the fluctuations that occur in the amount of the biomass and therefore in the balance between mineralization and immobilization. The microbial biomass also has an effect through the production of organic acids and other complexing agents, which influence the solubility of metallic elements and of phosphate, in the inorganic fraction of the soil. Organic acids produced by soil microorganisms include citric, tartaric, malic and α-ketogluconic (Stevenson, 1991), and each of these is capable of forming complexes with metallic cations in the soil. As well as having a direct effect on the availability of the cations and of phosphate associated with them, the formation of organic complexes can result in the movement of cations in the soil profile. While in a complex, a cation has increased mobility, but subsequent degradation of the complex may result in its being reprecipitated or sorbed in a different position, usually at greater depth.

Influence of Soil Redox Conditions on Availability

In general, the redox potential of a soil (i.e. its oxidizing or reducing capacity) depends on the supply of oxygen in gaseous or dissolved form, in relation to the amount of readily decomposable OM. The supply of oxygen is influenced by the ease with which gases can move in the soil, and thus by the texture and degree of aggregation of the soil, and by the amount of water present. Oxidizing or aerobic conditions are associated with sandy soils, low rainfall and good drainage, while reducing conditions are associated with clay soils, high rainfall and poor drainage. The greater the amount of readily decomposable OM in the soil, the less oxygen is present, since decomposition is accompanied by microbial respiration and results in oxygen being used and carbon dioxide being produced. Redox conditions in soils may show marked differences within distances of a few millimetres, due to differences in gaseous diffusion and the localized presence of fragments of OM. Even in soils that appear to be well aerated, the interior of soil aggregates may be anaerobic while the exterior is aerobic.

Redox conditions influence the availability of many of the nutrient elements, including N, S, Fe, Mn, Zn, Cu and Co. Some of the effects are purely chemical, while others are due to changes brought about by microbial activity. With N, aerobic conditions favour the nitrification of ammonium to nitrate, which may then be susceptible to loss from the soil through leaching. Anaerobic conditions favour the denitrification of nitrate with the loss of gaseous N$_2$ and N$_2$O. With S, severely anaerobic conditions may result in sulphates being converted to insoluble metal sulphides or, in some circumstances, to volatile organic sulphides. Since the metal sulphides are insoluble, severely anaerobic conditions may result in elements such as Zn, Cu and Co becoming less available for plant uptake. On the other hand, anaerobic conditions may increase the availability of Fe through changing its oxidation state from Fe^{3+} to Fe^{2+}, and moderately anaerobic conditions may increase the solubility of other elements, such as Mn, Cu, Co and Mo, associated with the hydrous Fe oxides. In a comparison of herbage from freely drained and poorly drained soils in Scotland, concentrations of Mn and Co were much higher from the poorly drained soils, though there was only a small difference with Cu and no difference with Zn (West, 1981). Conversely, an increase in aeration, resulting, for example, from improved drainage may reduce the availability of Fe, Mn and Co.

Redox conditions may also have indirect effects on nutrient uptake through an influence on the growth and activity of plant roots. Widespread anaerobic conditions in a soil tend to inhibit root growth and damage existing roots, mainly due to the production of toxic compounds, such as ethylene and, possibly, sulphides.

Assessment of Nutrient Availability in Soils

The ability of a soil to supply a particular nutrient in plant-available forms can be assessed through soil analysis, plant analysis and/or pot or field experiments. Both soil analysis and plant analysis depend on a relationship having been established between plant growth and the concentration of the nutrient in the soil or plant material. Both types of procedure have advantages and disadvantages, but they may produce rather different results. Soil analysis indicates the amount of a nutrient that is potentially available for uptake under conditions that are favourable for root growth, whereas plant analysis reflects the actual status of the plant in relation to the nutrient, and this may be influenced by other factors (Marschner, 1995). Although pot and field experiments usually produce clear results for the specific conditions involved, they are time-consuming and the results do not necessarily apply to a different set of conditions.

Soil analysis is normally based on the extraction of a soil sample with an appropriate aqueous reagent. For a reagent to be effective for a particular nutrient, there has to be a good correlation between plant uptake and the amount extracted by the reagent from a range of different soils, sampled in different years and weather conditions (Sims and Johnson, 1991). Mild reagents, such as water alone or a dilute salt solution, extract little more than the amount of the nutrient already in the soil solution, but they are often effective for elements such as B and Cl, which have a relatively high solubility in the soil. Stronger reagents, such as solutions of an acid, an alkali or a chelating agent, extract a larger proportion of the soil nutrient and, ideally, indicate the capacity of the soil to replenish the amount in the soil solution (Marschner, 1995). For most nutrient elements, such reagents are more effective than water or dilute salt solutions, but it is important, for any reagent, that the concentration selected should not extract too large a proportion of the total soil nutrient. Examples of methods that have been adopted widely include the determination of 'available' P using a solution of 0.5 M NaHCO$_3$ at pH 8.5, and the determination of 'available' micronutrient cations using EDTA.

The use of plant analysis for assessing whether or not a particular nutrient is deficient is described in Chapter 3.

Losses of Nutrient Elements from Grassland Soils

Losses of nutrient elements from grassland soils can take place through a number of pathways. The most important are: (i) the leaching of soluble nutrients into the subsoil or into a drainage system; (ii) the erosion of particulate material from the soil surface by water or wind; (iii) volatilization to the atmosphere; and (iv) removal in harvested plant material, such as silage, or in animal products, such as milk. Usually, the losses occur preferentially from the plant-available forms of the nutrients. Although the different loss pathways vary in importance, depending on the particular nutrient and on environmental conditions, leaching is usually the most important in humid temperate regions. Leaching from grassland soils tends to be increased by the high concentration of some nutrients in urine patches, and is enhanced when there is a rapid flow of soil solution through macropores, as this curtails plant uptake and the reactions that lead to the retention of nutrients in the soil. In some situations, leaching from grazed grassland may amount to as much as 200 kg N ha^{-1} and 50 kg K ha^{-1} year^{-1}. With the nutrient cations, leaching tends to be increased by soil processes that result in the production of acidity and the displacement of the cations from cation exchange sites.

The erosion of particulate material by water or wind is generally much less from grassland than from arable soils, due to vegetative cover being present throughout the year. However, surface runoff can result in an appreciable loss of nutrients from sloping fields, particularly when heavy rain or the melting of snow follows soon after the application of slurry or fertilizer. Loss by surface runoff is usually more important for P than for N and K, because P is associated to a greater extent with insoluble particulate material and is less susceptible to leaching. Not all runoff results in a loss of nutrients from the field, as the material washed from a slope may be deposited in low-lying areas within the field, a factor that contributes to spatial variability in nutrient status.

The loss of nutrients by volatilization is of most importance with N, which can be lost in the form of ammonia (NH$_3$), and as N$_2$, N$_2$O and NO through denitrification and nitrification (see p. 101). Loss by volatilization is normally negligible for S, and there are no volatile losses of P or of the macro-nutrient or micronutrient cations. However, it is possible that appreciable amounts of I and Se volatilize from soils, though there is a lack of quantitative measurements in field situations. As with leaching, losses of nutrients by volatilization often occur predominantly from localized patches and at times when soil and weather conditions are particularly favourable for the process.

The removal of nutrient elements in silage, hay or animal products and the extent to which any such losses are offset by the application of slurry or manure clearly depend on the farm management system. Estimated amounts of the nutrient elements removed in typical annual yields of silage and hay from intensively managed and extensively managed grassland are shown in Table 3.5.

Chapter 3

Uptake and Concentrations of Nutrient Elements in Grassland Herbage

Growth and Characteristics of Roots in Grassland Plants

The density of roots per unit volume of soil is normally much higher in grassland than under arable crops or trees; and prairie grassland has a particularly high root density (see p. 26). Although rooting depth in grassland is often less than 1 m, in prairie grassland roots may extend to a depth of 1.5–2 m (Weaver, 1961). The actual amount of root material in a particular area of grassland, and the spatial distribution of the roots in the soil, are influenced by the plant species present, as well as by soil, weather and management factors (Deinum, 1985), but, in general, the roots become finer and more widely spaced with depth. There are major differences in root development between monocots (e.g. grasses) and dicots (e.g. legumes and forbs). With grasses, each seedling produces a seminal root, which then develops lateral roots and, as the plant matures, the lateral roots produce numerous fine branches. In contrast, a dicot seedling forms a tap root and a more limited number of branch roots. As a result, the total amount of root material produced by dicots is generally smaller than that produced by grasses. However, some species of both monocots and dicots are stoloniferous, and these develop an extensive, though rather shallow, secondary root system from nodes of the stolons. Grasses of *Agrostis* spp. and white clover provide examples of this type of rooting.

Although the density of plant roots, per unit volume of soil, is normally greatest near the soil surface, where supplies of nutrients are highest, deeper roots are important for the acquisition of moisture during periods of drought. The growth and branching of roots are influenced by both soil and weather conditions and occur most readily in well-aggregated moist soils and in zones of the soil that are relatively rich in nutrients (Robinson, 1994).

Nevertheless, with an increasing supply of nutrients, especially of N, there is an increase in the ratio of shoot weight to root weight and sometimes an actual reduction in root growth. Conversely, if the supply of nutrients declines beyond a certain point, there is often an increase in root length but a decrease in root diameter, and also a decrease in leaf area per unit of plant weight (Ryser and Lambers, 1995). Root growth is restricted by a low percentage of pore space and a high bulk density in the soil, though to an extent that is influenced by soil texture. In general, the restriction of root growth begins when the bulk density of the soil exceeds 1.55 g cm^{-3} in clay loams, 1.65 g cm^{-3} in silt loams and 1.80 g cm^{-3} in fine sandy loams (Bowen, 1981). In clay soils in which there are few aggregates in the size range of 1–20 mm diameter, most of the roots are present in cracks, worm-holes or old root channels, as only the finest roots and root hairs are able to penetrate into blocks of clay.

The depth of rooting may also be influenced by the supply of water, as illustrated in a study of prairie grassland in Kansas: after a period of moist years, the roots of most species extended to below 1.5 m, many roots reached 2.7 m and a few reached 6 m, whereas, after a period of dry years, most roots were in the top 0.6 m and only a few roots of dicot species penetrated below 1.5 m (Albertson and Weaver, 1943, cited by Gregory, 1988). However, in the less dry conditions of south-east England, the depth of rooting of ryegrass was not affected by summer irrigation, provided in addition to the normal rainfall (Garwood, 1967b). Although root growth is generally more extensive in moist soil than in dry soil, the depth of rooting is sometimes curtailed by anaerobic and chemically reducing conditions. These conditions are most likely to occur in areas of high rainfall and where the subsoil is poorly drained, perhaps due to the presence of a compacted layer. Where there is a compacted layer, root growth can often be improved by subsoiling.

Another factor that may restrict root growth is acidity, especially when the soil pH is lower than about 4.5, thus inducing a potentially toxic concentration of Al^{3+} or Mn^{2+} in the soil solution. However, species vary in their susceptibility to acidity, and legumes are generally more susceptible than grasses. Acidity is often more severe in the subsoil than in the surface soil, a difference that may cause shallow rooting and consequently result in the plants being more susceptible to drought. In grassland, the management of the sward can also affect root growth, as a result of differences in the severity of defoliation. Root growth depends on the production of carbohydrate by photosynthesis, and the frequent removal of leaf material, by intensive grazing or cutting, therefore tends to cause shallow rooting (Schuster, 1964). The growth of roots may cease for several weeks after severe defoliation.

Species differ in their pattern of root development and this influences their ability to compete for water and nutrients. In general, the roots of clovers are thicker, shorter and less branched than those of grasses (Table 3.1) and, partly for this reason, the clovers compete less effectively for nutrients (Evans, 1977). In a comparison of ryegrass and red clover, the greater uptake

Table 3.1. Some root characteristics of perennial ryegrass, timothy, white clover and red clover: mean data for plants grown in soil and sand culture (from Evans, 1977), and mean root lengths of ryegrass and white clover grown separately in field plots (from Evans, 1978).

	Ryegrass	Timothy	White clover	Red clover
Length/weight ratio (cm mg^{-1})	30.6	36.7	27.6	26.8
Root diameter (mm)	0.19	0.17	0.26	0.27
Surface area/weight ratio (mm^2 mg^{-1})	181	199	227	230
Length of root hairs (mm)	0.55	0.77	0.23	0.20
Percentage of roots with root hairs	95	97	68	58
Area of root hair cylinder (m^2 mg^{-1})	1,230	1,980	490	420
Root length (cm kg^{-1}) in top 20 cm	14,490	–	3,310	–
Root length (cm kg^{-1}) in 20–40 cm horizon	2,180	–	630	–

of K by the ryegrass was attributed mainly to the greater length and surface area of the ryegrass roots (Mengel and Steffens, 1985). There are also differences between species in the amount of root biomass produced (Fransen *et al.*, 1999) and in the extent to which they are able to maintain their rate of root growth when nutrients are deficient, and thus explore deeper layers of the soil. In New Zealand, browntop (*Agrostis tenuis*) was found to be particularly effective, in comparison with other grassland species, in obtaining P from the soil through increasing its growth of roots (Mouat, 1983). Species also differ in the rooting depth that they require for optimal growth. For example, *A. tenuis* grows well on relatively shallow soils, whereas *Dactylis*, *Lolium* and *Trifolium* spp. require deeper soils. These differences were reflected in the distribution of the species in a semi-natural grassland in Wales, with *A. tenuis* growing more densely on the shallower soils and the other species being mainly restricted to areas of deeper soil (Kershaw, 1963).

The morphology of individual roots has an influence on nutrient uptake and, in this context, roots have three fairly distinct regions. Close to the root tip, there is a small region where the xylem and phloem are not differentiated from the cortex and where the root surface is covered by mucigel. Next is an intermediate region, where xylem and phloem are formed but the endodermis is not fully developed. This merges into the third region, where the vascular stele, containing mature xylem and phloem, is separated from the cortex by the endodermis, a layer of cells that contains the suberized Casparian strip (Fig. 3.1). Water and nutrient ions are absorbed mainly, but not only, in the second region, which often comprises that part of the root between about 5 mm and 50 mm from the root tip (Morrison and Cohen, 1980). Roots often develop root hairs in this region, the root hairs being protrusions from cells of the epidermis about 10 μm in diameter and usually between 0.2 and 1 mm long. The root hairs, which are able to grow into micropores within soil aggregates, therefore increase the effective radius of the root and its absorbing surface. In general, plants with numerous fine roots and with relatively long root hairs appear to be the most effective in

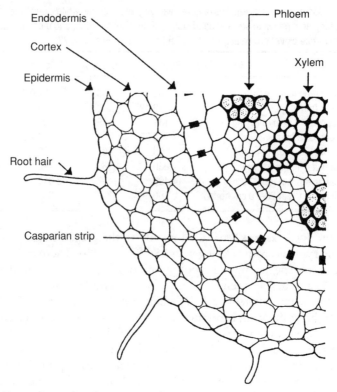

Fig. 3.1.   Transverse section of a young root showing the location of the main types of cell (based on a diagram from Nobel, 1974).

taking up nutrients from soils when supplies are limited (Clarkson, 1985). This pattern, especially the network of fine roots, ensures that the supply of nutrient ions close enough to the root surface to move to it by mass flow or diffusion is as large as possible. As well as differing in the pattern of root growth, plant species differ in their length of root hairs; the root hairs of grasses are normally longer and more numerous than those of legumes (Table 3.1; Mengel and Steffens, 1985).

Contact between Nutrient Ions and the Root Surface

Grassland plants derive their nutrient elements almost entirely from the soil via uptake through the roots, though they also absorb small amounts of some nutrients from the atmosphere through the leaves. The nutrients derived from the soil are taken up predominantly as soluble ions, and the rates of uptake are influenced by both soil and plant processes. The amount of an ion that comes into contact with a plant root depends partly on the growth of the root and partly on the rate at which the ion moves to the root surface. Although there is some contact between nutrient ions and the root surface as a direct result of root growth, most contact involves the transport of ions in the soil solution by mass flow and diffusion. Mass flow occurs when water and soluble ions move towards a root as a result of transpiration; the amount of an ion coming into contact with the root surface in this way therefore depends on the plant's consumption of water and on the concentration of the ion in the soil solution. For some nutrient ions, such as Ca^{2+} and, to a lesser extent, Mg^{2+} and SO_4^{2-}, the concentration in the soil solution is normally high enough for mass flow to satisfy plant needs and, for these ions, there may even be some accumulation at the root surface (Jungk and Claassen, 1997). For many other ions, movement by diffusion makes an important contribution to the amount taken up. Diffusion occurs when an ion moves randomly in solution from a relatively high concentration to a lower concentration, and diffusion is therefore necessarily involved when the uptake of an ion is greater than the amount provided by mass flow. Ions such as $H_2PO_4^-$, K^+, Zn^{2+} and Cu^{2+}, which are strongly adsorbed by soil constituents and whose concentration in the soil solution is therefore low, tend to become depleted in the zone close to the root and, for these ions, diffusion is generally more important than mass flow (Hinsinger, 1998). Although diffusion results in a net movement towards the root, there may still be a concentration gradient that extends into the soil to between 0.1 mm and 15 mm from the root surface, depending on the ion and the rates of diffusion and uptake (Barber, 1984). Soil texture and the moisture content of the soil influence the extent to which the ions move by both mass flow and diffusion. The higher the clay content of the soil, the slower the movement of ions, since there is more adsorption by soil constituents and the pores between soil particles are smaller. The drier the soil, the thinner the water films around soil particles, and therefore the slower the movement of water and ions by mass flow, and the slower the rate of diffusion.

Processes of Nutrient Uptake by Plant Roots

The uptake of nutrient ions into roots occurs by both passive and active processes. The passive processes include mass flow in transpired water, diffusion and the adsorption of ions by components of root cells. Active processes are involved when the needs for nutrient ions are not met by passive uptake, and the active processes require specialized carrier molecules and the expenditure of metabolic energy. However, the presence of a nutrient ion at a greater concentration in the plant than in the soil solution does not necessarily indicate that active uptake has occurred: it may reflect physical adsorption to cell components. Much of the physical adsorption is due to cation exchange, and the cation exchange capacity of roots influences their uptake of nutrient ions and also the ratio between monovalent and divalent cations (Haynes,

1980). In general, the cation exchange capacity of roots is higher in legume than in grass species. As a consequence, legumes normally have a higher content of divalent than of monovalent cations, whereas in grasses this pattern is reversed. In a comparison of more than 20 grass species, the higher the cation exchange capacity of the roots, the higher the proportion of divalent cations in the content of total cations (Wacquant, 1977). With many plants, including grasses, the cation exchange capacity of the roots increases with increasing N supply, and the application of fertilizer N therefore tends to increase the ratio of divalent to monovalent ions (Mouat, 1983). The rate at which any nutrient ion is absorbed also varies with the supply of the nutrient, with plant species and with factors such as ambient pH, temperature and moisture, which influence the growth rate of the plant. For an individual ion, the rate of absorption generally increases with increasing concentration of the ion in solution, until a maximum rate is attained. It is generally greatest at a pH of about 5–6. Uptake may also be influenced by an antagonism or synergism between ions. Antagonism occurs between ions of the same charge, and a high concentration of one cation (or anion) in the soil solution tends to depress the uptake of other cations (or anions). Synergism occurs between ions of opposite charge, and a large uptake of one or more cations increases the uptake of accompanying anions, and vice versa. Nevertheless, the cations and anions are often absorbed at rather different rates, and any resulting imbalance in the cytoplasm is matched by the synthesis or degradation of organic acids, particularly malic acid, to maintain a pH of about 7.0 (Jacoby, 1995; Jones, 1998). This maintenance of a near-neutral pH is reflected in the exudation from the roots of either H^+ or OH^- (or HCO_3^-) ions.

The movement of ions from the soil solution into the cell wall portion, or 'apoplasm', of the root tissue is always passive and results from mass flow and diffusion. In most plants, the cell walls constitute about 10–15% of the root volume. They have a complex structure of microfibrils, consisting mainly of cellulose and hemicellulose, together with some glycoprotein, and, in grasses, they also contain silica. As plant cells mature, the cell walls become increasingly impregnated with lignin and associated phenolic compounds. Following the movement of nutrient ions into the cell walls, they are susceptible to interactions, which may influence their onward movement towards the xylem. For example, carboxylic acid groups in the galacturonic acid component of hemicellulose tend to adsorb cations and repel anions and, although some of the adsorbed cations are readily exchangeable, there is some retention of the divalent and trivalent cations, partly in the form of complexes that are relatively stable.

Ions that are in soluble or exchangeable forms in the cell walls can move readily through the root cortex towards the vascular stele, but, in order to enter the stele, they have to cross two membranes, at which their transfer can be controlled. Whether or not they encounter the Casparian strip of the endodermis, which is largely impermeable to ions in solution, the ions must move into the symplasm, i.e. the cytoplasm of the living cells. The symplasm is contained within the plasma membrane (plasmalemma) of each cell, though it is linked between adjacent cells through the plasmodesmata (Fig. 3.2). Ions moving towards the stele may either enter the symplasm within the cortex and then move via the plasmodesmata, or they may enter the symplasm at the endodermis itself. Then, in order to enter the stele, ions must cross a second plasma membrane at the junction between the endodermis and the stele or, in the young region of the root, between the cortex and the stele. In terms of molecular structure, the plasma membrane consists of a lipid bilayer with globular proteins embedded in it, and this enables it to exert some metabolic control over the movement of ions, partly by selectivity and partly by providing a mechanism for the 'active' transport of ions moving against an electrochemical gradient. Selectivity and 'active' transport involve carrier molecules, which can be activated by phosphorylation and are then able to form complexes with appropriate ions. Such a complex is then able to move from the outside of the plasma membrane towards the inside, where a phosphatase splits the complex and the ion is released into the cytoplasm. The activity of the carrier molecule can be restored subsequently by phosphorylation, a process that requires energy from photosynthesis. As plant cells have an overall negative charge, active transport occurs to a greater extent with anions than with cations, and nitrate, phosphate, sulphate and chloride all appear to be accumulated against an electrochemical gradient (Blevins, 1994). In addition to the presence of ion carriers, the plasma membrane appears to contain a limited number of ion channels, which are formed from proteins, are hydrophilic and allow some movement of soluble ions through

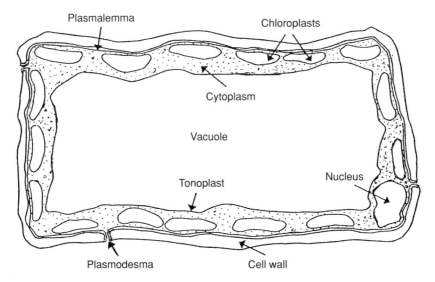

Fig. 3.2. Diagram of a mesophyll cell showing major components and membranes (based on a diagram from Nobel, 1974).

the membrane by diffusion. However, as ion channels are not open continuously and there are relatively few per cell, the extent to which they contribute to the movement of nutrient ions is uncertain (Marschner, 1995).

As the absorption of cations is largely passive and is less specific than that of anions, it depends to a large extent on the concentrations of the various cations in the soil solution. The total amount of all cations in a particular type of plant tissue is fairly constant, and so an increase in the supply of one cation tends to reduce the concentrations of the others. Of the macronutrient cations, only K appears to show active uptake, and then only at low concentrations (Mengel and Kirkby, 1987). With the other cations, including the micronutrient cations, uptake is thought to result mainly from the electrochemical potential gradient of the ion that exists across the plasma membrane. The gradient is composed of two components, the gradient in concentration of the ion and the electrical potential gradient of the plasma membrane, which results from the overall negative charge of plant cells. However, there is some selectivity in the uptake of cations, possibly brought about by the ion channels of the plasma membrane, which may vary in their radius and in the electrostatic properties at the mouth of the channel (Kochian, 1991).

Once cations and anions have reached the vascular stele, they move to the shoots in the xylem, dissolved in the transpiration stream (Kochian, 1991). In the shoots, the ions are transferred from the xylem to the symplasm and vacuoles of the leaf cells; any subsequent redistribution within the shoots or from shoots to roots occurs in the phloem.

The Importance of the Rhizosphere and of Mycorrhizal Associations

The rhizosphere is the zone of soil closely surrounding a root where the presence of the root influences soil properties. One such effect is a change in the concentration of ions in the soil solution, due partly to the uptake of water and partly to the uptake of ions. A second effect arises when there is a difference between the amounts of cations and anions taken up, since roots exude either H^+ or OH^- (or HCO_3^-) ions in an amount that is equivalent to the difference, and the exudation tends to change the pH of the rhizosphere. The difference between the amounts of cations and anions taken up is influenced to a large extent by the form in which N is absorbed: when ammonium is the main form, plants take up more cations (including ammonium) than anions, and there is a net exudation of H^+ ions, whereas when nitrate is the main form, plants take up more anions (including nitrate) than cations, and there is a net exudation of OH^- (or HCO_3^-) ions. The pH of the rhizosphere may be up to 1 unit lower than in the bulk soil for plants supplied with ammonium, and up to 1 unit higher than the bulk soil for plants supplied with nitrate (Mengel and Kirkby, 1987). Leguminous plants fixing N_2 also induce soil acidity, probably because such plants absorb more cations than anions, with a resultant exudation of H^+ ions (Jarvis and Hatch, 1985; Tang *et al.*, 1998). The more acid the pH of the rhizosphere, the greater the availability of some nutrient ions, particularly Fe, Mn and Co, but the lower the availability of others, particularly Mo.

Soil in the rhizosphere is also influenced by organic compounds exuded from plant roots, compounds which include mucilage and a range of smaller molecules, such as aliphatic acids and amino acids. In addition, there is a continual loss of cells from the root surface. The amounts of these organic exudates and decomposing cells are difficult to measure with plants growing in soil, but it is thought that, together, they often amount to 5–20% of total dry matter production by the plant, and sometimes more (Tate, 1995). Some of the exudate compounds, particularly di- and tricarboxylic acids, influence the availability of nutrient ions through the formation of complexes, as well as through an effect on rhizosphere pH (Hinsinger, 1998; Jones, 1998). Plant species differ in the amount and type of organic acids that they produce: calcicole plants exude more of the di- and tricarboxylic acids than do calcifuge plants, a feature that enables them to obtain Fe and possibly other cations more readily from calcareous soils (Tyler and Strom, 1995; Jones, 1998). Malic acid, which is dicarboxylic, and citric acid, which is tricarboxylic, both form stable complexes with metal cations and therefore tend to dissolve these cations from insoluble soil constituents. There is some evidence that plants increase their exudation of malic and citric acids when they are deficient in micronutrient cations (Jones and Darrah, 1994). Some organic acids, and particularly oxalic acid, which is exuded by calcicole plants, also tend to increase the availability of P in calcareous soils by forming stable complexes with the Ca of apatite (Tyler and Strom, 1995). However, although the various organic acids undoubtedly have some influence on the availability of nutrient ions, their quantitative importance, especially in relation to the micronutrient cations, is uncertain. In non-acid soils, the cations most likely to be dissolved from soil minerals are Ca and Mg and there may be little effect on the micronutrient cations, whereas, in acid soils, where the mobilization of micronutrients by malate and citrate is greater, these nutrients are unlikely to be deficient (Jones and Darrah, 1994). Also, the organic acids are themselves subject to degradation by soil microorganisms and, in one study, had an average half life of only 2–3 h (Jones, 1998). A more specific effect on the availability of nutrients in the rhizosphere results from the secretion by plant roots of larger complexing molecules, known as phytosiderophores, which have an extremely strong affinity for micronutrient cations, especially Fe. Mugineic acid (see p. 229) is an important example. As well as increasing the solubility of micronutrient cations in the soil, the complexes formed by these siderophores are capable, in some instances at least, of direct absorption by the root (Jones and Darrah, 1994).

The various root exudates and dead root cells encourage the growth of microorganisms in the rhizosphere and these microorganisms, as well as degrading organic acids, may also produce organic acids from other sources of OM. Bacteria are generally more important than fungi in the rhizosphere, and their production of organic acids may accentuate the change in rhizosphere pH brought about by exudates from roots. Depending on soil factors such as the initial pH of the soil and its buffering capacity, together with the form of N taken up by the plant, the rhizosphere may have a pH as much as 2 units lower than that of the bulk soil (Marschner, 1995). Another effect of the decomposition of root exudates and dead cells by soil micro-organisms is a change in the redox potential of the rhizosphere towards more chemically reducing conditions, a change that tends to increase the solubility of Fe and Mn (see p. 226).

Mycorrhizae, in which fungal hyphae are associated with plant roots, are potentially important in nutrient uptake but, as with the release of organic acids, their quantitative importance is uncertain. Their ability to increase nutrient uptake is due to the hyphae extending several centimetres away from the root, and thus increasing the volume of soil from which uptake can occur. The hyphae are typically about 3 μm in diameter, and thus less than half the diameter of root hairs (Jungk and Claassen, 1997). There are several types of mycorrhiza but the most abundant are the arbuscular mycorrhizae, which form branched haustorial structures (arbuscules) within the cortex of a plant root. Although the arbuscules typically have a lifespan of only about 10–12 days, they are the main sites for the transfer of nutrients from the fungal hyphae to the higher plant (Marschner, 1995). Mycorrhizae occur commonly in grassland species, including both grasses and clovers (Hetrick *et al.*, 1988; McGrath and Jarvis, 1993). They appear to have most effect on the uptake of the less mobile nutrients, such as P and some micronutrients, especially when soil supplies of these nutrients are low. Often the rate of P uptake, per unit root length, is two to three times higher in mycorrhizal plants than in non-mycorrhizal plants (Marschner, 1995). Mycorrhizae have also been shown to increase the uptake of Zn and Cu, but there is less evidence that they enhance the uptake of other nutrient elements (Tinker and Gildon, 1983; Marschner and Dell, 1994; Marschner, 1995).

Uptake of Nutrient Elements through Leaves

In addition to being taken up by roots, small amounts of some nutrient elements are absorbed through plant leaves. This process is of more significance for N and S than for the other macronutrients, as these two elements are present in the atmosphere in gaseous forms, such as ammonia and sulphur dioxide. There is also evidence that plants readily absorb gaseous forms of I and Se through their leaves, though the quantitative importance of the process for these elements is uncertain. As well as being absorbed through leaves as gaseous compounds, N, S and various other elements may be absorbed from particulate aerosol material, and possibly to a small extent from dust. Although much of the particulate material deposited on leaves remains on the surface, water-soluble forms of the nutrient elements have the potential to enter the plant tissue. Solubility varies considerably with the element, as shown in a study of atmospheric deposition at several sites in the UK (Cawse, 1980). Solubility was high, at more than 80%, for deposited K, Na, Ca, Cl, Zn and Se; it was intermediate, at 50–80%, for Mg, Mn, Cu and Co, and was < 25% for Fe. However, the amounts actually absorbed by leaves from particulate material will be influenced by factors such as the timing of rainfall and the formation of dew.

Transport and Metabolism of Nutrient Elements in Plants

Nutrient ions are transported from roots to shoots mainly in the xylem, whereas transport within the shoots and from shoots to roots occurs in the phloem. The macronutrient elements are transported from roots to shoots mainly as inorganic ions, N as nitrate, P as phosphate, S as sulphate and K, Na, Ca and Mg as the free cations. However, a small proportion of some of these elements in the xylem may be in the form of organic compounds – for example, N in amino acids and Ca and Mg in organic complexes. Of the micronutrient elements, Fe, Zn and Cu, though not Mn, are transported mainly as complexes with organic compounds, such as citrate (Mehra and Farago, 1994) and peptides (Welch, 1993).

The metabolic requirement for most nutrient elements is located mainly in the leaves, where processes such as photosynthesis and protein synthesis are most active. Many nutrient elements contribute either to the structure or to the activity of the various enzymes involved in these processes. As leaves senesce, their metabolic activity declines, and nutrient elements may be remobilized and translocated to other tissues or leached from the leaves by rain. Remobilization is of most benefit when nutrient supply from the soil is poor or is temporarily interrupted, for example by dry soil conditions (Marschner, 1995). The nutrient elements differ in the ease with which they are remobilized (Table 3.2), though the pattern varies to some extent with species and with the nutritional status of the plant (Marschner, 1995). Those elements that are remobilized most readily are generally present at a higher concentration in young, actively growing leaves than in older leaves. Conversely, elements that are not readily remobilized are often at a higher concentration in old leaves than in young leaves.

Table 3.2. Nutrient elements categorized according to their mobility in the phloem (from Kochian, 1991; Marschner, 1995).

High mobility	Intermediate mobility	Low mobility
N, P, S, K, Mg, Na, Cl	Fe, Mn, Zn, Cu, Mo	Ca, B

The leaching of nutrient elements from leaves by rain occurs most readily when the nutrients are present in excess, and/or when the leaves are beginning to senesce. However, at least with K, there may be some leaching even from actively growing leaves that do not contain excessive amounts. Thus, young ryegrass grown in soil in pots and containing 2.5% K in the herbage was found to lose the equivalent of about 1.5 kg K ha^{-1}, representing more than 2% of its K content, in 1 day of light simulated rain or during a single heavy storm (Clement *et al.*, 1972). In this investigation, there was less leaching from red clover and from lucerne than from ryegrass. Another factor that influences the extent of leaching from leaves is the pH of the rain and, in general, the leaching of cations increases with increasing acidity of the rainfall (Marschner, 1995).

Distribution of Nutrient Elements in Grasses and Legumes

With some, but not all, nutrient elements, there are marked differences in concentration between shoots and roots. The difference depends partly on the functions of the element and partly on whether the element is susceptible to being bound by the root tissue. Some elements are generally in higher concentration in shoots than in roots, some are generally higher in the roots and others show no consistent difference between shoots and roots. In grasses, N is normally at a higher concentration in shoots than in roots but, in legumes, there is sometimes a higher concentration in roots than in shoots, due to the high concentration of N in the root nodules. Phosphorus and S are also normally at a higher concentration in shoots than in roots in grasses, though not necessarily in legumes. The concentrations of K, Na, Ca and Mg are usually, though not always, higher in shoots than in roots in both grasses and legumes. However, with the micronutrient cations, Fe, Mn, Zn, Cu and Co, concentrations are often greater in the roots than in the shoots.

Within the shoots, the concentrations of most nutrient elements are generally higher in the leaves than in the stems and leaf sheaths. Often, the difference is particularly marked with Ca (Table 3.3). However, with K, the concentration is sometimes higher in the stems than in the leaves (Thomas, 1967). The flowering heads of grasses tend to accumulate P and Zn and have higher concentrations than the leaves of these elements (Table 3.3; Fleming, 1965; Smith and Rominger, 1974). Phosphorus and Zn also accumulate in the flowering heads of white clover (Wilkinson and Gross, 1967). Relatively high concentrations of P have also been reported in seeds, compared with leaf material, in a range of herbaceous plants, but, with Zn, the average concentration was lower in the seeds than in the leaves (Tyler and Zohlen, 1998).

Although the concentrations of many nutrient elements are lower in roots than in shoots, as much as 90% of the plant weight of grassland, at any one time, may be below the soil surface, and there is therefore a considerable

Table 3.3. Concentrations of 15 nutrient elements in the leaf blade, leaf sheath, stem and flowering spikelet fractions of cocksfoot (from Davey and Mitchell, 1968).

	Leaf blade	Leaf sheath	Stem	Flowering spikelet
N (%)	1.80	0.90	0.50	1.70
P (%)	0.20	0.14	0.17	0.33
S (%)	0.35	0.15	0.14	0.30
K (%)	1.40	1.20	1.00	0.60
Na (%)	0.37	0.20	0.12	0.08
Ca (%)	1.20	0.31	0.10	0.24
Mg (%)	0.25	0.11	0.08	0.11
Fe (mg kg^{-1})	177	68	15	96
Mn (mg kg^{-1})	106	93	48	76
Zn (mg kg^{-1})	45	15	9	50
Cu (mg kg^{-1})	5.5	2.7	3.5	4.9
Co (mg kg^{-1})	0.15	0.07	0.09	0.14
B (mg kg^{-1})	29	11	7	16
Mo (mg kg^{-1})	0.65	0.18	0.20	0.34
Se (mg kg^{-1})	0.06	0.07	0.08	0.12

storage of nutrients in roots. A proportion of this store may then be transferred to the shoots when required for the growth of new leaf material (Clark and Woodmansee, 1992). In some perennial species, there is a transfer of nutrients from leaf material to roots during the autumn and remobilization from the roots during the next growing season. This type of transfer appears to be most important in regions with cold winters: it was clearly demonstrated in a short-grass prairie in Colorado, USA, where a single application of ^{15}N-labelled KNO_3 was monitored over a 5-year period. In each autumn, N was transferred to the roots in amounts that represented, on average, one-third of the N present in the shoot material in June (Clark, 1977).

Ranges of Concentration of Nutrient Elements in Grassland Herbage

The concentrations of all the nutrient elements in grassland herbage show wide variations, even within individual species and within limited geographical areas. The means and ranges of concentration reported for a number of nutrient elements in > 9000 samples of grass and legume hay and silage from Pennsylvania and New York State, USA, are shown in Table 3.4, together with data for > 2000 samples of timothy in Finland and mean values for > 1400 samples of grassland herbage in the UK. In both of the two sets of data that include ranges of concentration, the highest concentration of many elements was 50–100 times greater than the lowest. Although some variation

Table 3.4. Concentration ranges and mean concentrations of some nutrient elements reported for > 9000 samples of grass and legume herbage from Pennsylvania and New York State, USA (Adams, 1975), and of > 2000 samples of timothy from Finland (Kähäri and Nissinen, 1978; Paasikallio, 1978), together with mean concentrations for > 1400 samples of grass herbage from England and Wales (MAFF, 1975).

	Pennsylvania and New York, USA		Finland		England and Wales (mean)
	Range	Mean	Range	Mean	
N (%)	1.0–5.3	–	–	–	–
P (%)	0.05–0.81	0.26	0.12–0.98	0.28	0.35
S (%)	0.04–0.43	0.23	–	–	–
K (%)	0.21–4.93	2.06	0.76–4.26	2.36	2.59
Na (%)	0.00–0.39	0.02	0.001–0.03	0.00	0.25
Ca (%)	0.03–2.73	0.86	0.10–0.67	0.26	0.60
Mg (%)	0.03–0.79	0.20	0.03–0.45	0.13	0.17
Fe (mg kg^{-1})	10–2600	208	8–800	44	–
Mn (mg kg^{-1})	6–1200	53	8–500	67	99
Zn (mg kg^{-1})	3–300	27	15–120	32	45
Cu (mg kg^{-1})	2–214	13	0.4–20	4	8
Co (mg kg^{-1})	–	–	0.0–0.6	0.06	0.13
B (mg kg^{-1})	1–94	19	1.4–34	4.9	–
Mo (mg kg^{-1})	–	–	< 0.02–17	0.5	1.22
I (mg kg^{-1})	–	–	–	–	0.22

may have been due to analytical error and/or to contamination of the sample, much is undoubtedly genuine. The factors that contribute to variation in herbage concentrations are discussed below (pp. 58–68), and include the stage of maturity of the herbage, species differences, seasonal and weather factors, soil type and the application of fertilizers.

Removal of Nutrient Elements in Harvested Herbage

When grassland herbage is converted into silage or hay, the amounts of the nutrient elements that are removed from the field depend partly on the yield of silage or hay and partly on the herbage concentrations of the elements. Table 3.5 shows the approximate amounts that may be removed by: (i) a relatively small yield of herbage with rather low concentrations of nutrients; and (ii) a large yield of herbage with rather high concentrations of nutrients. If the same yields of herbage are consumed by grazing animals, rather than being removed in silage or hay, the amounts of the nutrient elements shown in Table 3.5 correspond to the amounts consumed per hectare by the grazing animals.

Table 3.5. Amounts of nutrient elements removed in cut herbage or consumed by grazing animals: (i) with a herbage yield of 3000 kg DM ha^{-1} and rather low concentrations of nutrients; and (ii) with a herbage yield of 10,000 kg DM ha^{-1} and rather high concentrations of nutrients (kg or g ha^{-1} year^{-1}).

	(i)		(ii)	
	Assumed nutrient concentration	Removal in 3000 kg herbage DM	Assumed nutrient concentration	Removal in 10,000 kg herbage DM
N	2.0%	60 kg	4.0%	400 kg
P	0.2%	6 kg	0.6%	60 kg
S	0.2%	6 kg	0.4%	40 kg
K	1.5%	45 kg	3.5%	350 kg
Na	0.05%	1.5 kg	0.5%	50 kg
Ca	0.5%	15 kg	1.2%	120 kg
Mg	0.1%	3 kg	0.3%	30 kg
Cl	0.2%	6 kg	2.5%	250 kg
Fe	50 mg kg^{-1}	150 g	300 mg kg^{-1}	3000 g
Mn	20 mg kg^{-1}	60 g	150 mg kg^{-1}	1500 g
Zn	15 mg kg^{-1}	45 g	60 mg kg^{-1}	600 g
Cu	5 mg kg^{-1}	15 g	15 mg kg^{-1}	150 g
Co	0.05 mg kg^{-1}	0.15 g	0.25 mg kg^{-1}	2.5 g
I	0.1 mg kg^{-1}	0.3 g	0.5 mg kg^{-1}	5.0 g
B	3 mg kg^{-1}	9 g	15 mg kg^{-1}	150 g
Mo	0.1 mg kg^{-1}	0.3 g	4 mg kg^{-1}	40 g
Se	0.05 mg kg^{-1}	0.15 g	1 mg kg^{-1}	10 g

Plant : Soil Concentration Ratios

When the concentrations of nutrient elements in grassland plants are compared with those in soils, both on a DM basis, it is clear that there are large differences between the elements in the extent to which they are accumulated by the plants. Actual plant : soil concentration ratios in a specific situation are influenced by numerous soil and environmental factors, but ratios based on typical values for the soils and plants of temperate grassland (Table 3.6) show a wide variation between the nutrient elements. Nitrogen is the element showing the greatest degree of accumulation, with a plant : soil ratio of about 10 : 1, followed by Cl, with a ratio of about 7 : 1, whereas Fe shows the greatest degree of exclusion, with a plant : soil ratio of about 0.0004 : 1.

The Concept of the Critical Concentration

The critical concentration of a particular nutrient element in plant tissue is defined as the concentration that occurs when the supply of the nutrient is

Table 3.6. Typical plant : soil concentration ratios of nutrient elements for grassland in temperate regions.

	Typical concentration in soil DM	Typical concentration in herbage DM	Plant : soil concentration ratio (x : 1)
N (%)	0.28	2.8	10.0
P (%)	0.2	0.4	2.0
S (%)	0.10	0.35	3.5
K (%)	1.5	2.5	1.7
Na (%)	0.25	0.25	1.0
Ca (%)	1.8	0.6	0.33
Mg (%)	0.8	0.2	0.25
Fe (mg kg^{-1})	35,000	150	0.0004
Mn (mg kg^{-1})	1,600	165	0.10
Zn (mg kg^{-1})	150	37	0.25
Cu (mg kg^{-1})	30	9	0.3
Co (mg kg^{-1})	20	0.1	0.005
Cl (mg kg^{-1})	500	3,500	7.0
I (mg kg^{-1})	5	0.2	0.04
B (mg kg^{-1})	50	5	0.10
Mo (mg kg^{-1})	2.6	0.9	0.35
Se (mg kg^{-1})	0.4	0.05	0.12

slightly less than that needed for maximum growth, when all other nutrients are adequate. The concept of the critical concentration assumes that the growth, or DM yield of plant material, and the concentration of the nutrient in the plant tissue are related in the way shown in Fig. 3.3. As the supply of the nutrient is increased, there is an increase in both yield and nutrient concentration, until the maximum yield is attained, after which there is no further increase in yield but a continuing increase in concentration. Eventually, a toxic concentration is reached and there is a decline in yield. Sometimes, when a nutrient is extremely deficient, the addition of a small amount of the nutrient will increase yield but actually cause a small decrease in concentration, though larger additions will increase both yield and concentration. The critical concentration is often taken to be the concentration associated with a yield of 5% or 10% below the maximum, though sometimes with the maximum yield itself (Kelling and Matocha, 1990), and this difference in definition is undoubtedly responsible for some of the variation in reported critical concentrations.

In assessing the validity or usefulness of a particular critical concentration, it is important to take into account factors such as possible differences between plant species, the changes that occur with increasing maturity (see below) and the differences in concentration between different parts of the

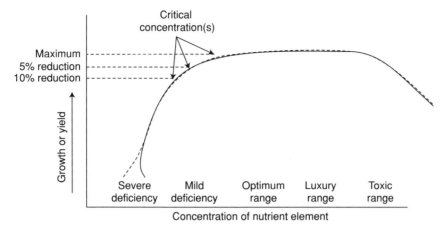

Fig. 3.3. General relationship between growth or yield and the concentration of a nutrient element in the plant tissue.

plant. An additional complication is that the critical concentration of one nutrient may be influenced by the supply of others. In practice, the most important factor for grassland is the effect of stage of maturity of the herbage, though this effect can be largely overcome by sampling the plant material at a specific and uniform stage. Information on the critical concentrations of the various nutrient elements in grasses and legumes is included in later chapters.

Evidence of interactions between nutrient elements has led to the suggestion that ratios between the concentrations of certain elements, as well as their actual concentrations, may be important. This concept has been used in the diagnosis and recommendation integrated system (DRIS), in which various indices are derived by comparing ratios of nutrients in the plant material being assessed with 'norms' based on high-yielding plant material (Walworth and Sumner, 1987). However, although nutrient ratios may provide useful information in some situations, the DRIS assessment has the disadvantage of requiring a number of different analyses. It also depends on the 'norm' of high-yielding plant material not containing nutrients in 'luxury' concentration (Sinclair *et al.*, 1996b), and this condition is often difficult to satisfy.

When a nutrient element is present in excessive amounts in the soil solution, the quantity taken up may result in a herbage concentration that exceeds the 'luxury' range and that may have a toxic effect. The nutrient elements most likely to show toxic effects are Mn, Zn and Cu and, for these elements, it is possible to define an upper critical concentration in the plant material, which is the lowest concentration at which growth is impaired. The critical concentrations for toxic effects from these elements are usually between five and 25 times the critical concentrations for deficiency (see Chapter 9).

Influence of Plant Species and Variety on Herbage Concentrations of Nutrient Elements

Although there are consistent differences between grassland species in the concentrations of some nutrient elements, the differences between species grown under uniform conditions are usually much smaller than the differences within individual species grown under a range of different conditions. The differences between species are generally of most significance in natural and semi-natural grasslands that contain a variety of grasses and legumes, together with forb species, and which are often poorly supplied with at least some of the nutrient elements. Although grass and legume species differ in their root development and other aspects of morphology, they often show little difference in their concentrations of most of the nutrient elements. Only with Ca is the difference clear and consistent (Table 3.7). With many of the other nutrient elements, differences between grasses and legumes are influenced by nutrient supply. For example, grasses tend to contain more K than clovers on soils deficient in K, but less than clovers on soils well supplied with K; and similar effects occur with Cu and Mo (Fleming, 1973). Comparisons between grasses and clovers may also be influenced by the stage of maturity at which the plant material is sampled, since the changes that occur with maturity are often much greater in grasses.

In general, the commonly sown grass species are broadly similar in their concentrations of nutrient elements, though the data in Table 3.8 show

Table 3.7. Typical concentration ranges of 17 nutrient elements in perennial ryegrass and white clover at a leafy stage of growth.

	Perennial ryegrass	White clover
N (%)	2.0–3.5	3.5–5.2
P (%)	0.2–0.6	0.2–0.6
S (%)	0.2–0.5	0.2–0.5
K (%)	1.5–3.5	1.5–3.5
Na (%)	0.05–0.40	0.05–0.40
Ca (%)	0.4–0.8	1.0–2.0
Mg (%)	0.10–0.30	0.15–0.40
Cl (%)	0.2–2.5	0.2–2.5
Fe (mg kg^{-1})	100–200	100–300
Mn (mg kg^{-1})	30–300	30–200
Zn (mg kg^{-1})	15–60	20–40
Cu (mg kg^{-1})	3–15	5–12
Co (mg kg^{-1})	0.03–0.20	0.06–0.40
I (mg kg^{-1})	0.10–0.50	0.10–0.50
B (mg kg^{-1})	2–8	10–40
Mo (mg kg^{-1})	0.1–4.0	0.1–10.5
Se (mg kg^{-1})	0.02–0.4	0.02–1.2

Table 3.8. Concentrations of 11 nutrient elements in the leaf material of four grass species grown in plots: mean values of three replicates (from Fleming, 1963).

	Perennial ryegrass	Cocksfoot	Timothy	Meadow fescue
N (%)	2.1	2.8	2.5	2.6
P (%)	0.32	0.32	0.13	0.30
K (%)	2.3	2.6	1.7	2.1
Ca (%)	0.87	0.57	0.88	0.87
Mg (%)	0.17	0.15	0.27	0.18
Fe (mg kg^{-1})	101	67	42	109
Mn (mg kg^{-1})	41	105	38	29
Zn (mg kg^{-1})	20	23	19	16
Cu (mg kg^{-1})	5.0	7.1	4.6	4.9
B (mg kg^{-1})	9	10	17	10
Mo (mg kg^{-1})	0.47	0.77	0.58	0.60

timothy to be relatively low in P and K and relatively high in Mg. However, in another comparison, three varieties of timothy were no higher in Mg than were ryegrass and cocksfoot, and the concentration of P depended on variety (Patil and Jones, 1970). In a comparison in New Zealand, prairie grass was on average lower in P, Na and Mg and higher in K than perennial ryegrass (Crush *et al.*, 1989). With some elements, there may be appreciable differences between varieties within a species. Thus, in six varieties of perennial ryegrass, the highest concentration of K was approximately twice the lowest, and the highest concentration of Na was approximately five times the lowest (Miles *et al.*, 1964). Similar differences were found amongst varieties of cocksfoot, but varieties of meadow fescue and tall fescue were rather more uniform in composition. In other investigations, clones of cocksfoot in the USA differed significantly in P, K, Ca and Mg (Stratton and Sleper, 1979); and 12 lines of prairie grass in New Zealand differed appreciably in concentrations of Na, Fe, Mn, Zn and I, though less so in other nutrient elements (Rumball *et al.*, 1972). In a study in Ireland, no significant differences in N, K or Mg were found between ryegrass cultivars (Culleton and Fleming, 1983). In another study, although genotypic variation in Mg was found in ryegrass, it appeared to be greater within cultivars than between them (Crush, 1983).

Amongst the legume species, white clover generally has relatively high concentrations of most nutrient elements, in comparison with red clover, lucerne and sainfoin (Table 3.9). The higher concentrations in white clover are probably due, at least partly, to its higher ratio of leaf to stem. Varietal differences within white clover and red clover, though sometimes significant, are often small (Davies *et al.*, 1966).

Despite these rather mixed findings, the evidence of some consistent differences amongst varieties and cultivars of both grasses and legumes has

Table 3.9. Concentrations of 15 nutrient elements in four legume species grown on adjacent plots: mean values for eight successive stages of maturity (from Whitehead and Jones, 1969).

	White clover	Red clover	Lucerne	Sainfoin
N (%)	4.42	3.40	2.94	2.87
P (%)	0.38	0.27	0.26	0.28
S (%)	0.29	0.21	0.27	0.24
K (%)	2.26	2.07	1.65	1.73
Na (%)	0.12	0.04	0.06	0.01
Ca (%)	2.10	1.84	1.82	0.94
Mg (%)	0.18	0.21	0.15	0.17
Cl (%)	0.76	0.53	0.60	0.38
Fe (mg kg^{-1})	117	85	108	93
Mn (mg kg^{-1})	49	44	42	52
Zn (mg kg^{-1})	25	24	24	28
Cu (mg kg^{-1})	7.3	7.4	7.0	7.0
Co (mg kg^{-1})	0.19	0.21	0.16	0.17
B (mg kg^{-1})	31	27	38	41
Mo (mg kg^{-1})	0.64	0.44	0.18	0.18

led to attempts to breed productive cultivars with increased concentrations of specific nutrient elements. Most attention has been given to Mg, due to the importance of hypomagnesaemia in livestock; and cultivars relatively high in Mg have been developed (e.g. Moseley and Baker, 1991). More recent studies have indicated that, while it may well be possible to breed cultivars of perennial ryegrass high in Mg, Na or P, it would probably not be feasible to develop a cultivar low in K (Easton *et al.*, 1997). A major problem is to combine the desired characteristic, such as a high concentration of Mg, with high yield.

Forb species, such as dandelion, chicory and plantain, are often richer than the commonly grown grasses in some nutrient elements, as illustrated by Table 3.10. However, the data on forb species are limited and not entirely consistent. In a sward containing a mixture of species, interactions between the species may affect the concentrations of nutrients in individual species. For example, when brome-grass was grown either with or without associated lucerne, the presence of the lucerne was found to reduce the concentration of S in the brome-grass; it also tended to increase the concentration of P, though it had little effect on other elements (Casler *et al.*, 1987).

Influence of Stage of Maturity on Herbage Concentrations of Nutrient Elements

The concentrations of nutrient elements often change with advancing maturity, but the pattern of change varies with the element and with the plant

Table 3.10. Concentrations of six macronutrient elements (% in DM) in several forb species, in comparison with perennial ryegrass, as reported in four investigations; all species in individual plots in (i), (ii) and (iv), single forb species + ryegrass in individual plots in (iii); (from, (i) Thomas *et al.*, 1952; (ii) Forbes and Gelman, 1981; (iii) Frame *et al.*, 1994; (iv) Wilman and Derrick, 1994).

		N	P	K	Na	Ca	Mg
(i)	Plantain	–	0.33	2.76	0.30	1.66	0.61
	Yarrow	–	0.40	3.68	0.04	1.28	0.59
	Burnet	–	0.27	1.83	0.05	1.31	1.10
	Chicory	–	0.48	4.59	0.27	1.43	0.65
	Ryegrass	–	0.26	1.98	0.14	0.46	0.21
(ii)	Plantain	–	0.32	1.94	0.41	1.03	0.15
	Yarrow	–	0.40	2.97	0.14	0.69	0.23
	Chicory	–	0.30	2.39	0.36	0.86	0.25
	Dandelion	–	0.30	2.18	0.67	0.84	0.30
	Ryegrass	–	0.27	1.54	0.29	0.39	0.11
(iii)	Plantain	2.0	0.35	2.3	–	2.6	0.19
	Yarrow	2.7	0.43	4.3	–	1.2	0.22
	Burnet	2.0	0.32	1.8	–	1.8	0.47
	Chicory	2.3	0.42	5.1	–	1.6	0.27
	Ryegrass	1.5	0.29	2.5	–	0.4	0.14
(iv)	Plantain	1.52	0.36	2.89	0.11	1.09	0.23
	Chickweed	2.44	0.66	6.61	0.30	0.74	0.37
	Dock	3.39	0.48	4.20	0.12	0.52	0.42
	Dandelion	2.72	0.47	4.76	0.14	1.25	0.43
	Ryegrass	2.07	0.40	2.74	0.16	0.65	0.25

species. For most, but not all, of the nutrients, concentrations tend to be higher in young than in older tissues. Nitrogen shows a marked reduction with advancing maturity, particularly in grasses, and there may well be a reduction of about 70% from early spring to midsummer, if the herbage is not harvested during this period (see p. 115). The concentrations of P and S also decline with advancing maturity, often to the extent of about 40–60% of the concentrations in early spring. The concentrations of K, Ca, Mg and Na in grasses are generally higher in young than in mature leaves, but sometimes peak values occur at an intermediate stage of maturity. A factor that may contribute to the decline in concentration with increasing maturity of N, P and K is that fertilizer is normally applied at the beginning of the season, and there is therefore an increasingly long time interval since the application. With many of the micronutrients, the changes with advancing maturity in grasses are inconsistent and often show an erratic pattern.

In legume species, the declines in concentrations of N, P and S with advancing maturity are less than in the grasses. White clover, in particular, shows little change in N, P or S, while red clover and lucerne are intermediate between white clover and the grasses. There is often a decline in K with advancing maturity in legumes, but this may be due, at least partly, to an increasing time interval since the application of fertilizer K. With Ca, Mg and Na, there appears to be no consistent trend and, in some instances, increases in concentration have been reported. The micronutrient cations in ladino clover, red clover and lucerne were found to show a fairly rapid decline during the very early stages of growth, but no consistent trend thereafter (Loper and Smith, 1961). The concentrations of B and Mo remained fairly constant through the year in these same legume species (Baker and Reid, 1977).

As leaves senesce, there is some loss of nutrient elements due to remobilization and/or leaching by rain. However, in terms of nutrient concentrations, these losses are offset to some extent by the simultaneous loss of carbohydrate and protein. Consequently, although the concentrations of the more mobile nutrients may be lower in old leaves than in younger leaves, nutrients that are relatively immobile may show little difference or even an increase in concentration during senescence. Elements which are normally at a low concentration in dead leaf tissue include N, P, S and K, whereas elements which show little change or higher concentrations include Ca and the micronutrient cations (Table 3.11; Waughman and Bellamy, 1981; Greene *et al.*, 1987; Clark and Woodmansee, 1992). The onset of senescence can be accelerated by shading and, under these conditions, both N and Cu are remobilized more rapidly (Marschner, 1995).

Table 3.11. Some comparative data on concentrations of macronutrient elements in living and dead herbage from the same swards (% in DM).

	Type of sward	Living	Dead	Reference
N	Perennial ryegrass	2.2–3.4	1.1–2.5	Whitehead, 1986
N	Perennial ryegrass	2.8	1.7	Wilman *et al.*, 1994
N	Long-term, mixed species	2.0–3.0	0.8–1.3	Clark and Woodmansee, 1992
P	Perennial ryegrass	0.44	0.22	Wilman *et al.*, 1994
P	Mixed grasses	0.12	0.04	Greene *et al.*, 1987
P	Long-term, mixed species	0.22–0.33	0.09–0.14	Clark and Woodmansee, 1992
S	Long-term, mixed species	0.11–0.17	0.05–0.07	Clark and Woodmansee, 1992
K	Perennial ryegrass	2.8	1.2	Wilman *et al.*, 1994
K	Mixed grasses	2.02	0.57	Greene *et al.*, 1987
Ca	Perennial ryegrass	0.3	1.8	Wilman *et al.*, 1994
Ca	Mixed grasses	0.73	0.52	Greene *et al.*, 1987
Mg	Perennial ryegrass	0.15	0.28	Wilman *et al.*, 1994
Mg	Mixed grasses	0.13	0.09	Greene *et al.*, 1987

Influence of Weather and Seasonal Factors on Herbage Concentrations of Nutrient Elements

When the effects of advancing maturity are avoided by regular defoliation at intervals during the growing season, changes in the concentration of many nutrient elements are often small or inconsistent (Reith *et al.*, 1964; Thomas, 1967; Fleming and Murphy, 1968; Metson and Saunders, 1978a; Metson *et al.*, 1979; Crush *et al.*, 1989). However, both Ca and Mg tend to reach peak values in mid- to late summer (Fleming and Murphy, 1968; Metson and Saunders, 1978a; Crush *et al.*, 1989). When young leaf blade material of perennial ryegrass was analysed at intervals during the season, in order to assess whether seasonal changes occurred in leaves that were at a similar stage of growth, it was found that P decreased, Ca increased and there was little change in N, K, Mg, Fe, Mn, Zn and Cu (Reay and Marsh, 1976). In a parallel study with red clover, there was a decrease in N and an increase in Mg, but otherwise there were no significant differences. In a subsequent study, in which more mature leaf blades of ryegrass from a grazed grass–clover sward were analysed, fluctuations in the concentrations of nutrient elements were attributed partly to variation in the supply of plant-available N (Reay and Waugh, 1983).

When seasonal changes in nutrient concentrations do occur in grasses and legumes, they are no doubt related to weather factors, such as temperature, water-supply and light intensity. A few studies have been carried out to examine the effects of these factors individually. Thus, raising the soil temperature from 11 to 28°C greatly increased the concentrations of Ca and Mg in Italian ryegrass, though it had little effect on N, P, K, S or Na (Nielsen and Cunningham, 1964). However, a rather different pattern was found in a study with lucerne, in which increasing the soil temperature over the range from 5 to 27°C tended to increase the concentrations of N and P but to decrease concentrations of Ca and Mg (Nielsen *et al.*, 1960). Increasing the day/night temperatures from 15/10°C to 27/21°C increased the average concentration of K in five legume species from 1.4% to 2.3% and tended to increase Cu, but had little effect on the concentrations of Fe, Mn and Zn (Smith, 1970).

Differences in water-supply appear to affect P more than the other nutrient elements. There are several reports that the concentration of P in both grasses and legumes is reduced by drought (e.g. Kilmer *et al.*, 1960; Rahman *et al.*, 1971), though, in one investigation, drought was reported to increase P in lucerne (Kidambi *et al.*, 1990). However, the effects of water-supply on other elements, including N, K, Ca, Mg, Na, Fe, Mn, Cu and B, are less consistent (Kilmer *et al.*, 1960). Drought has been reported to reduce the concentration of K (Herriott and Wells, 1963) and S (Begg and Freney, 1960); but in another investigation drought increased K, Na, Ca, Mg and Cl in several grasses and legumes (Rahman *et al.*, 1971). In both lucerne and sainfoin, increasing dryness increased concentrations of Ca, Mg and Zn

(Kidambi *et al.*, 1990). In four grass species, concentrations of N, P, K and Mg increased with increasing soil oxygen (i.e. increasing dryness), and the effect on Ca was variable (Rogers and Davis, 1973).

The effects of light intensity have received little attention, but shading, sufficiently severe to reduce the yield of DM by about one-third, increased the concentrations of N, P, Ca and Mg in Coastal Bermudagrass by 25–50% (Burton *et al.*, 1959).

Another factor that may contribute to seasonal changes in nutrient concentration, as well as to changes with maturity, is the timing of fertilizer application. Fertilizer P and K are often applied only once per year, at the beginning of the growing season, and the available supply of these nutrients therefore tends to decline as the season advances, with corresponding effects on the herbage concentrations of P and K.

Changes in the Herbage Concentrations of Nutrient Elements during Conversion to Hay or Silage

The conversion of herbage into silage or hay, if carried out during good weather conditions, results in little change in the concentrations of nutrient elements on a DM basis. However, when the cutting of herbage for silage or hay is followed by wet or showery weather, there is likely to be some loss of nutrients, especially from hay that is present in the field for a longer time. This effect is illustrated by a study of hay made under different weather conditions (Wilman and Mzamane, 1986). When the hay was made during a period with little or no rain, there was no change in the concentrations of N, P, K, Ca, Mg or Na between the time of cutting and collection of the hay (at a moisture content of 33% of DM), whatever the stage of maturity, rate of application of fertilizer N or swath thickness. However, when rain fell during the later stages of drying, there were large reductions in the concentrations of K and Na, though rain at earlier stages had less effect on these elements. The rain had relatively little effect on concentrations of N and P. In another investigation, involving red clover and lucerne, rainfall during hay-making caused little change in concentrations of P, S, K, Ca and Mg, though heavy rain when the hay was dry did reduce K by about 50% (Collins, 1985).

Concentrations of the nutrient elements are often higher in silage than in hay, for a number of reasons. Silage often receives higher rates of fertilizer and is generally cut at an earlier stage of growth than hay, and therefore tends to contain higher concentrations of some nutrient elements. Silage also spends a shorter period in the field after cutting and is subject to a smaller loss of leaf material during harvesting. In a comparison of more than 100 silages and hays, the concentrations of P, Ca, Mg, Mn and Cu were, on average, between 15 and 50% higher in the silages (Hemingway *et al.*, 1968). Nevertheless, some loss of nutrient elements from silage can occur, particularly if the herbage is left to wilt in the field before transfer to the silo and there is rain during this period, and also if the silage releases liquid effluent. The losses of P, K, Ca, Mg and Cl in silage effluent were found to be smaller when the grass had been wilted to reduce its moisture content (Salo and Sormunen, 1976). In general, the losses of these elements were somewhat greater than those of dry matter, resulting in slightly lower concentrations than in the original grass, but, with Mn, Zn and Cu, there was little change in concentration. Similar results were obtained in another study, in which concentrations of P, K, Na, Ca, Mg and Zn, though not of Fe or Cu, were lower in silage than in the grass from which it was produced (Ettala and Kossila, 1979).

Influence of Soil Type on Herbage Concentrations of Nutrient Elements

Nutrient elements supplied via the soil are derived partly from the soil itself, partly from deposition from the atmosphere, partly from decomposing organic materials and, in some instances, partly from fertilizers. With intensively managed grassland, the amounts of plant-available N, P and K provided by fertilizers often exceed the amounts derived from the other sources, and fertilizer application is then a major influence on herbage concentrations of N, P and K. NPK fertilizers may also influence the concentrations of other macronutrient elements through the presence of incidental constituents, or through synergistic or antagonistic effects (see p. 46). Only when little or no fertilizer is applied are herbage concentrations of the macronutrient elements influenced appreciably by soil type. Deficiencies of N are then generally most severe on soils low in OM. Deficiencies of P are generally most severe on soils that are either highly acid or calcareous, since the availability of soil P is least under these conditions. Absolute deficiencies of K, Ca and Mg occur most commonly on acid soils with low cation exchange capacity; nevertheless, herbage concentrations of Mg and K are sometimes higher on acid than on calcareous soils, presumably due to there being less competition from Ca. In various herbage species, sampled from a number of both calcareous and acid soils (pH 4.5–6.0), concentrations of Ca in the herbage were about 80% greater on the calcareous soils, while concentrations of K and Mg were not consistently different between the two soil categories (Tyler and Zohlen, 1998). With some elements, particularly S and Na, the effect of soil type is often masked by variation in the supply from the atmosphere. Proximity to industrial or urban centres or to power stations fuelled by coal or oil is important for S, while proximity to the coast is important for Na.

With the micronutrients, inputs from fertilizers and the atmosphere are often negligible in relation to the amount in the soil, and herbage concentrations are more closely related to supply from the soil itself. However, the available supply depends not only on the total amount present but also on factors such as pH and the degree of aeration, which influence availability.

Increasing acidity increases the availability and herbage concentrations of Fe, Mn and Co and decreases the concentrations of Mo and Se, but has little effect on Zn and Cu. In a study of the influence of soil aeration, herbage from a poorly drained site contained about seven times more Co and about 30% more Mn than herbage from an adjacent well-drained site: Cu was higher from the poorly drained site in red clover but not in ryegrass, and there were no consistent differences in concentrations of Fe, Zn and Mo (Mitchell *et al.*, 1957b).

Differences in soil texture may have little effect on herbage concentrations of nutrient elements, as shown with timothy grown on a range of soils categorized into coarse mineral soils, clay and silt soils and organic soils (Kähäri and Nissinen, 1978).

Influence of Fertilizers, Lime and Sewage Sludge on Herbage Concentrations of Nutrient Elements

When fertilizer N, P or K is applied, there is usually, though not always, an increase in the herbage concentration of that particular nutrient, and there may also be secondary effects on the concentrations of other nutrients. Secondary effects may arise due to one or more of several factors. There may be synergism or antagonism involving fertilizer ions; the increase in growth due to the fertilizer may dilute the concentrations of elements not added in the fertilizer; nutrient elements other than N, P and K may be present as incidental constituents in the fertilizer; and the fertilizer application may change soil pH. The synergistic effect of a fertilizer ion sometimes exceeds the dilution effect, thus actually increasing the concentration of an element not added in the fertilizer. On the other hand, an antagonistic effect may accentuate the dilution effect. However, the magnitude of synergistic and antagonistic effects depends on the particular combination of ions, as well as on soil factors and on the plant species. In a study, with Italian ryegrass, of the effects of three concentrations of each of ten nutrient ions, there were synergistic effects between NO_3^- and each of Na^+, K^+ and Mg^{2+}, and between Cl^- and K^+, while antagonistic effects occurred between K^+ and each of Na^+, Ca^{2+} and Mg^{2+}, between Na^+ and both K^+ and Ca^{2+}, between NH_4^+ and K^+, and between Cl^- and SO_4^{2-} (Cunningham, 1963). In field investigations, the application of fertilizer K has generally caused a large reduction in the herbage concentration of Na and had a smaller effect on Ca and Mg (see p. 204).

In farming practice, another factor that may contribute to the effect of fertilizer on herbage composition is its influence on the date of harvesting. Fertilizer N, in particular, increases the growth rate of the grass and therefore results in the herbage being cut or grazed at an earlier stage and having higher concentrations of some of the nutrient elements than would occur in older herbage. In general, herbage concentrations of nutrient elements are influenced more by fertilizer N than by fertilizer P and K. Fertilizer N usually induces a large increase in herbage yield and, whatever the form applied, is often taken up mainly as nitrate, which promotes the uptake of those cations present in plentiful amounts. Some forms of fertilizer N also have an appreciable effect on soil pH and therefore on the uptake of some micro-nutrients. For example, in a comparison of N applied to a mixed grass–clover sward as either ammonium sulphate or calcium ammonium nitrate, the herbage receiving ammonium sulphate contained more Mn and less Mo, probably due to a decrease in soil pH, as well as containing less Ca and more S than the herbage receiving calcium ammonium nitrate (Havre and Dishington, 1962). If the regular application of fertilizer N, P and K is stopped, herbage concentrations of these elements tend to decline from year to year for a period of several years (Pegtel *et al.*, 1996).

The effects of lime on herbage concentrations of nutrient elements are due partly to the addition of Ca and partly to the increase in soil pH. The addition of Ca naturally tends to increase herbage Ca, as shown in studies involving ryegrass and white clover (Edmeades *et al.*, 1983) and long-term grassland (Stevens and Laughlin, 1996). Lime may also reduce the concentration of Mg, while having little effect on K or Na (see p. 205). When liming results in an appreciable increase in soil pH (e.g. from < 5.5 to 7.0 or above), there is normally a substantial reduction in the concentrations of Fe, Mn and Co, a small reduction or little difference in Zn and Cu, and an increase in Mo (Mitchell, 1963; Stewart and McConaghy, 1963; Price and Moschler, 1965, 1970; John *et al.*, 1972; Edmeades *et al.*, 1983).

Applications of sewage sludge also influence herbage concentrations of nutrient elements, due partly to the uptake of nutrients via the soil and partly to the adhesion of sludge to the herbage. When sewage sludge is applied repeatedly to grassland, some nutrient elements tend to accumulate in the surface soil, leading to a cumulative increase in herbage concentrations. Such an accumulation is of most significance for Zn, Cu and Mo and also for Cd and Ni, all of which are potentially toxic to grazing animals (Davis *et al.*, 1988). The extent to which sewage sludge adheres to leaf surfaces depends on factors such as the proportion of solid material in the sludge, the weather conditions following application and the rate of growth of the grass, and these various factors therefore influence the ingestion of potentially toxic elements by grazing animals.

Influence of Livestock Excreta on Herbage Concentrations of Nutrient Elements

When swards are grazed, the return of nutrients in dung and urine has an influence on the composition of the herbage, at least in localized patches. There may be changes in the relative numbers of different plant species, as well as changes in the chemical composition of individual species. With grass–clover swards, the N in urine promotes the growth of grass and tends to curtail the growth of clover; it may therefore reduce the overall concentrations of those elements which are at higher concentration in the clover. With all-grass swards, the increased growth may have a dilution effect on nutrient elements in short supply. In studies with ryegrass in New Zealand, urine consistently produced an increase in herbage N and K and a decrease in P and Mg; when applied in spring, there was no significant effect on Ca, Na, Fe or Mn (Joblin, 1981), whereas, in autumn, there was a decrease in Ca and Mn (Joblin and Keogh, 1979). Other studies of the effects of urine have confirmed increases in herbage concentrations of N and K and decreases in P and Ca (e.g. Ledgard *et al.*, 1982). The increase in the herbage K concentration is due not only to the return of K, but also to the synergistic effect of a greater uptake of nitrate derived from urea in the urine (Carran, 1988).

When dung, especially cattle dung, is deposited on grassland, the underlying plants may be smothered, but growth is often increased at the edges of the patch. However, the palatability to livestock of the herbage immediately surrounding a dung patch is reduced, so that it becomes more mature. Sheep dung has much less smothering effect (Williams and Haynes, 1995). Once the dung has decomposed, there may be an increase in plant growth in the patch and the area immediately surrounding it. Dung has been found to increase the herbage concentrations of N, P, K and Mg, while having little effect on Ca (Weeda, 1977; Saunders, 1984). However, in another investigation, dung had little or no effect on herbage concentrations of P, S, Ca or Mg, though there was a small increase in K and some increase in growth after a few weeks (Williams and Haynes, 1995). Some inconsistency in the effects of both dung and urine is to be expected, due to the variation that occurs in their composition (see p. 91).

Chemical Forms of Nutrient Elements in Herbage

The elements N and S are present in herbage mainly in the form of proteins, with smaller amounts of free amino acids and amides. However, when supplies exceed plant requirements, there may be an accumulation of the inorganic forms, nitrate and sulphate. Phosphorus is a constituent of a wide range of metabolically important compounds, including phosphate esters, such as ADP and ATP, nucleic acids and phospholipids. Much of the Ca in plant tissues is present as Ca bound to polygalacturonic acids in the form of calcium pectate, a component of the middle lamella of plant cell walls. Magnesium is a constituent of chlorophyll; and Mg and most of the micronutrient elements are constituents of plant enzymes. Potassium, Na and Cl are the only plant nutrient elements that are not incorporated to a large extent into organic molecules, and their presence as free ions reflects their main functions, namely, regulating osmotic pressure and maintaining pH. Although the forms of the nutrient elements outlined above appear to be the most important in terms of their metabolic functions, substantial proportions of many of the nutrients exist in other forms. Thus, controlled ashing followed by scanning electron microscopy has shown the existence of a partially inorganic component distributed over the outer surfaces of the cell walls of forage plants, with the major elements being P, Ca, S and K (McManus *et al.*, 1977). A number of the micronutrient cations, particularly Mn, Zn and Cu, also appear to be associated with the cell walls (Whitehead *et al.*, 1985), though not necessarily having a function there.

The extent to which the nutrient elements in herbage are soluble in water varies with the element and with plant species. In various grasses and legumes, the solubility of K and Cl is generally > 90%, while the solubility of P and Mg is 70–90%, of Ca, 45–70%, and of S, 35–70% (Makela, 1967; Whitehead *et al.*, 1985). The solubility of the micronutrient cations was found to vary widely within the range 2–80%, depending partly on the cation and partly on the plant species (Whitehead *et al.*, 1985). A problem in studying in more detail the chemical forms in which nutrient elements occur in plant tissues is that some procedures, such as extraction with various reagents, are liable to result in the formation of artefacts. However, for most nutrient elements, the chemical forms in which they exist appear to have little effect on their availability to ruminant animals, due to the solubilizing effects of the digestive enzymes.

Chapter 4

Nutrient Elements in Ruminant Animals

Outline of Digestive Physiology in Ruminant Animals

In contrast to most other animals, ruminants have a complex stomach, which consists of four compartments, the rumen, reticulum, omasum and abomasum (Fig. 4.1). However, the rumen and the much smaller reticulum are not clearly separated, and are often referred to jointly as the reticulorumen or sometimes just as the rumen. This has a capacity of about 100–200 l in dairy cattle and about 8–10 l in sheep. The rumen contains a large population of microorganisms, mainly bacteria but with some protozoa and anaerobic fungi, and these various microorganisms secrete enzymes that carry out a partial digestion of the animal's food. Digestion in ruminants thus has two important stages. The first stage is effected by the rumen microorganisms, while the second stage, in the abomasum and intestines, is similar to digestion

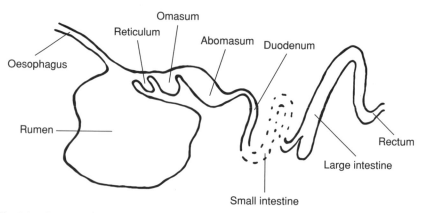

Fig. 4.1. Diagram of the digestive system of a ruminant animal.

in simple-stomached animals, and is effected by enzymes secreted by the animal itself. In the second stage, the substrates for the enzymes consist of dietary material, which has been partially digested, plus microbial biomass which has been synthesized in the rumen. As well as enabling ruminants to utilize diets that consist entirely of rather fibrous material, the two-stage process of digestion has additional advantages. One is that the micro-organisms in the rumen synthesize certain specific nutrients, such as some B vitamins and individual amino acids, which may be inadequate in the diet. A second advantage is that the microorganisms also incorporate non-protein forms of N and S into the proteins of their biomass, enabling these forms of N and S to be utilized in the second stage of digestion (Miller, 1979).

The digestion process begins with the food being chewed and mixed with saliva. Ruminant animals produce large amounts of saliva, about $150 \, l \, day^{-1}$ in cattle and about $10 \, l \, day^{-1}$ in sheep (McDonald $et \, al.$, 1995), which facilitates swallowing and also contains ions such as bicarbonate and phosphate, which buffer the pH of the rumen contents within the range 5.5–7.2. In addition, saliva provides nutrient elements for the rumen microorganisms in forms that are readily available (Miller, 1979). Of the soluble compounds produced in the rumen during the first stage of digestion, some are absorbed directly from the rumen into the bloodstream, while others are incorporated into the microbial biomass. After leaving the rumen, the partially digested food passes to the omasum, where much of the water is removed, and then into the abomasum or true stomach. Gastric juice containing hydrochloric acid is secreted, and this reduces the pH of the digesta to about 2.5, as required for the action of the digestive enzyme, pepsin. In the next portion of the digestive tract, the small intestine, the acidity is neutralized by secretions from the pancreas and liver, and partially digested proteins are hydrolysed to amino acids. Enzymes are secreted into the small intestine from the pancreas, gall-bladder and the walls of the intestine, and many of the soluble products of digestion are absorbed into the bloodstream. There is little further digestion in the large intestine, though water and bicarbonate are absorbed from this portion of the tract. Although, for many of the nutrient elements, absorption occurs mainly from the small intestine, for some elements, the relative importance of different absorption zones is difficult to establish, because substantial amounts are secreted at various points between the mouth and the large intestine (Miller, 1979). Material that remains undigested after passing through the large intestine is voided as faeces and, of this, about 10–20% consists of living and dead microbial cells (Marsh and Campling, 1970).

The extent to which any particular type of plant material is digested by ruminant animals is influenced by the relative amounts of cell-wall fraction and cytoplasm. Plant cell walls, which are composed mainly of cellulose, hemicellulose and lignin, but with small amounts of protein and, in grasses, silica, have a relatively low digestibility. In addition to having a low digestibility themselves, the plant cell walls, especially the lignin and associated phenolic components, curtail the digestibility of the cytoplasm (Buxton et al., 1996). Thus, in general, the digestibility of plant material declines as the content of lignin increases. As a large proportion of each of the nutrient elements occurs in the cytoplasm, their release during digestion depends partly on the physical breakdown of the cell walls, as well as on the enzymic breakdown of components of the cytoplasm, such as proteins and nucleic acids.

Metabolic Requirements for Nutrient Elements

An element is considered to be an essential nutrient for animals if a deficiency results, consistently, in a metabolic function being impaired and if this impairment is prevented or overcome by supplementation with the element (Underwood and Mertz, 1987). The elements known to be essential and which are sometimes deficient in farm livestock are N, P, S, K, Ca, Mg, Na, Cl, Fe, Mn, Zn, Cu, Co, I, Mo and Se. In addition to these 16 elements, several others, including Cr, Ni and Si, have been shown to be essential by the use of highly purified diets, but at such low dietary concentrations that no deficiencies have been identified under normal farming conditions (Mills, 1992; Underwood and Suttle, 1999).

Of the 16 nutrient elements listed above, the first eight are macronutrients and the others are micronutrients. However, within both the macronutrient and micronutrient categories, there are wide differences in concentration in the animal body. For example, the concentration of Ca in ruminants is about 25–35 times that of Mg, and the concentration of Zn about 60–100 times that of Mn (Table 4.1). Differences in concentration reflect, to a large extent, differences between the elements in their functions in the animal body. Some elements (e.g. N, P, S, Ca) are major constituents of body tissues, such as muscle and bone, while others (e.g. Na, K, Mg, Cl) are essential because they are involved in buffering the pH and maintaining the osmotic pressure of the body fluids, and others (e.g. Mn, Zn, Cu) are constituents of enzymes. Many nutrient elements have more than one function. Quantitatively, both N and S are most important as constituents of the proteins in muscles and enzymes, but both are also constituents of certain vitamins. Calcium and P are most important as constituents of bone, and the skeleton contains about 99% of the total Ca and 78% of the total P in the body, mainly in the form of hydroxyapatite (Yano et al., 1991). However, in addition to its role as a constituent of bone, P contributes to the biochemical transfer of energy through the reactions of ADP and ATP, and it is also a constituent of the nucleic acids, which are involved in cell division and the composition of genes. Calcium, together with K and Mg, is also involved in the transmission of nerve impulses, as well as in the activation of enzymes. Iron is essential as a constituent of haemoglobin and therefore for the transport of oxygen, and is also a constituent of a number of enzymes. Most of the other micronutrient elements function either as constituents of enzymes (e.g.

Table 4.1. Typical concentrations reported for the nutrient elements in the body tissue of ruminant animals, excluding digesta (% or mg kg^{-1} in live weight).

	Ruminant animals (McDonald et al., 1995)	Cattle (Miller, 1979; Miller et al., 1991)	Cattle (Agricultural Research Council, 1980)
N (%)	2.8[a]	–	2.7
P (%)	1.0	0.7	0.8
S (%)	0.15	0.15	–
K (%)	0.2	0.17	0.2
Na (%)	0.16	0.14	0.15
Ca (%)	1.5	1.2	1.4
Mg (%)	0.04	0.05	0.045
Cl (%)	0.11	0.10	0.10
Fe (mg kg^{-1})	20–80	50	30
Mn (mg kg^{-1})	0.2–0.5	0.3	–
Zn (mg kg^{-1})	10–50	20–30	24
Cu (mg kg^{-1})	1–5	5	1.2
Co (mg kg^{-1})	0.02–0.1	< 0.06	–
I (mg kg^{-1})	0.3–0.6	0.3	–
Mo (mg kg^{-1})	0.1–4.0[b]	0.03	–
Se (mg kg^{-1})	0.1–2.0[c]	0.2–0.5[d]	–

[a]Assuming 20% fat in live weight.
[b]Taking into account Mills and Davis, 1987; Pais and Benton Jones, 1997.
[c]Taking into account Pais and Benton Jones, 1997.
[d]Confirmed by Ullrey, 1987.

Zn, Cu, Mo, Se) or as cofactors, which are required for the activation of enzymes, though I and Co are exceptions. Iodine is essential as a constituent of the metabolically important hormones triiodothyronine and thyroxine, and Co is essential as a constituent of vitamin B_{12}, and these appear to be their only functions (Underwood and Suttle, 1999).

The dietary requirement for a nutrient element can be regarded as having separate components for maintenance, for live-weight gain, for pregnancy and for lactation. The maintenance component reflects the amount of the nutrient needed to keep the tissues of the animal intact and functioning. It includes the needs for metabolic processes, such as digestion, circulation of the blood, respiration and the muscular movement of the animal. Although nutrient elements, released in soluble forms by digestion and other metabolic processes, are recycled to some extent, there are inevitable losses, through the intestines, kidneys and skin, which it is necessary to replace. For most nutrient elements, the main loss of this type occurs through the intestines, and represents part of the excretion in the faeces (Underwood and Suttle, 1999). The other components of the dietary requirement, for live-weight gain, pregnancy and lactation, apply only to appropriate categories of animal, and reflect the amounts of the nutrient element incorporated into the weight gain, fetus or milk.

All the nutrient elements can have harmful effects on animals if they are present in the diet in excessive concentration, but, in practice, this problem is restricted to N, K and a few of the micronutrients. Although excessive concentrations of N causing metabolic disturbances in animals are unusual, it is possible for them to occur in herbage when high rates of fertilizer N are applied. Excessive concentrations of K in herbage, which occur only when high rates of fertilizer K are applied, are liable to impair the absorption of dietary Mg. With the micronutrients Cu, Mo and Se, excessive concentrations in herbage are usually restricted to localized areas, and are due either to geochemical factors or to industrial pollution.

For optimum nutrition, it is clearly important that the supply of each nutrient element in the diet should be adequate but not excessive. However, it is often difficult to assess whether a particular diet supplies an adequate amount, as, with many of the elements, a slight to moderate deficiency results only in a slower rate of growth or a reduced amount of milk or wool production, but no visible symptoms. Another problem is that for some elements, particularly Ca, it is important for the animal to maintain a reserve, which can be used when the physiological need is greatest, e.g. during lactation. Also, although the requirement for some nutrient elements increases during pregnancy and lactation, these physiological conditions often result in an increase in the extent to which the element is absorbed from the digestive tract, and they do not necessarily require a corresponding increase in the concentration in the diet.

A number of the micronutrient elements (Fe, Mn, Zn, Cu, Co, I, Se) apparently influence processes that are involved in reproduction, and deficiencies or imbalances of these elements may result in a failure of animals to reproduce (Hidiroglou, 1979). There is also evidence that some micronutrient elements, particularly Cu and possibly Co and Se, may be involved in maintaining the animal's resistance to infectious diseases (Suttle and Jones, 1989; Spears, 1991).

Homeostatic Control of Nutrient Elements in Animal Tissues

The concentrations of the nutrient elements in animal tissues and in blood, at least in the forms which are metabolically active, are maintained relatively constant, whatever the composition of the diet. For any particular element, the constancy is maintained by homeostatic control mechanisms, which modify the inputs to and outputs from the body fluids. The various mechanisms enable an animal to adapt to changes in diet and to changes in its requirement for a particular nutrient due to its physiological state. In general, the nutrient elements are conserved when they are in relatively short supply, while their excretion is increased when the dietary concentration is high relative to need. However, at least five mechanisms are involved in homeostatic control:

1. Variation in the extent of absorption from the digestive tract.
2. Variation in the extent of excretion via the faeces.
3. Variation in the extent of excretion via the urine.
4. Deposition of an excess in the tissues in harmless and/or reserve forms.
5. Variation in the extent of secretion into milk.

The importance of these various homeostatic mechanisms varies with the element, as illustrated in Table 4.2. With several elements, including Ca, Fe, Mn and Zn, variation in the extent of absorption is the most important mechanism, and percentage absorption decreases as the concentration in the diet increases. With other elements, especially K, Na and Cl, the percentage absorption is always high, but any excess is rapidly excreted in the urine. Excretion in the urine is also an important mechanism for Mg, I and Se, but not for Zn, Mn or Cu. Deposition in reserve form is important for Ca, which is deposited in bone, and for Fe and Cu, which are stored in the liver, while secretion into milk is important for I, Mo and Se.

The degree of control that can be achieved by these homeostatic mechanisms is influenced by the quantity of the element circulating in the body fluids, i.e. the 'pool size' of the element. With large inputs and outputs but a small pool size, stability is relatively poor. Thus, a high-yielding dairy cow typically has a Ca input, reflecting absorption from the diet, of about 34 g Ca day^{-1}, but has only 3 g Ca in a reserve form that is available rapidly (Table 4.3). Similarly, Mg has an input of 4 g day^{-1}, with a metabolically active reserve of only 0.75 g. Consequently, a small change in the rate of input or output of Ca or Mg can result, quite rapidly, in a serious deficiency of the metabolically active fraction; and this explains why milk fever (due to Ca deficiency) and grass tetany (due to Mg deficiency) are often fatal. In contrast, Na and K have large metabolically active pool sizes relative to the daily input, and so short-term deficiencies are buffered for much longer periods (Payne, 1989).

As well as providing protection against deficiencies and excesses, homeostatic mechanisms are involved in the ability of animals to retain certain elements preferentially in comparison with others. This ability is most apparent with some of the micronutrients. For example, Zn is retained preferentially in comparison with Mn, as shown by their animal : plant concentration ratios (see p. 80). Herbage often contains about four times as much Mn as Zn, whereas the animal body contains about 70 times as much Zn as Mn.

Homeostatic control normally results in the pH of both blood and urine being maintained within narrow limits. However, when there is a serious imbalance in the diet between the amounts of the major cations and anions that remain in free ionic form, the homeostatic mechanisms may be unable to

Table 4.2. Homeostatic mechanisms involved in adaptation to varying intakes of nutrient elements (based on Miller, 1979, with other data).

	Degree of importance			
	Major	Moderate	Minor	Little or none
Change in % absorption	Ca Fe Mn Zn	P Cu	N Mg Se	K Na Cl I Mo
Change in endogenous loss in faeces	P Mn Zn	I		S Co Se
Change in loss in urine	N S K Na Mg Cl I Mo Se	Ca Co	P	Mn Cu Zn
Change in tissue deposition	Ca Fe Cu Mo	I	Na Mg Mn Zn Co Se	N P S
Change in secretion in milk	I Mo Se	Zn	N S Mn Cu Co Se	P K Na Ca Cl Fe

prevent a change in the pH of the body fluids, with possible harmful effects. Variations amongst the ions in the extent of absorption and metabolism make it difficult to calculate accurately the effective dietary cation–anion difference (DCAD), which is usually expressed in terms of mEq 100 g^{-1}, but a number of formulae have been proposed (Roche *et al.*, 2000). These include:

$$DCAD = (K + Na) - Cl$$
$$DCAD = (K + Na) - (Cl + SO_4)$$
$$DCAD = (K + Na + 0.15\ Ca + 0.15\ Mg) - (Cl + 0.5H_2PO_4 + 0.25SO_4).$$

Table 4.3. Typical daily inputs and outputs (g day^{-1}), and reserves, of Ca, Mg, K and Na in a high-yielding dairy cow (from Payne, 1989).

	Input		Reserve		Output	
	Dietary intake (g day^{-1})	Absorbed intake (g day^{-1})	Total reserve (g)	Available reserve (g)	In faeces and urine (g day^{-1})	In milk (g day^{-1})
Ca	100.0	34.0	6000	3.0	8.0	26.0
Mg	20.0	4.0	175	0.75	1.5	2.5
K	50.5	50.5	820	185.0	22.5	28.0
Na	19.5	19.5	700	35.0	6.5	13.0

A recent study with dairy cattle in Australia showed that only when the DCAD was less than +15 mEq 100 g^{-1} for a period of about 3–4 weeks was there a substantial decrease in urine pH (Roche *et al.*, 2000).

Concentrations of Nutrient Elements in Animal Tissues and Milk

In contrast to the usual practice with plant materials, concentrations of the nutrient elements in animal tissues are generally expressed on the basis of live weight. Concentrations reported on this basis as typical for the body tissue of ruminant animals are shown in Table 4.1. The nutrients having the highest concentrations in body tissue and therefore present in the greatest amounts in animal live-weight gain are N, P and Ca. Beef cattle, grazing at a stocking rate of 5.5 animals ha^{-1} for approximately 6 months, might show a total gain of 1000 kg live weight ha^{-1}, which would remove about 27 kg N, 8 kg P and 14 kg Ca from the system.

Typical concentrations of the nutrient elements in cows' milk are shown in Table 4.4. Some elements show a considerable range, due to the concentration changing in response to factors such as the stage of lactation, the nutrition of the animal and disease. Also there may be some contamination of milk by Fe and Cu from dairy equipment, and by I through the use of iodophors for udder washing or for cleaning equipment (Renner *et al.*, 1989). The concentrations of many, but not all, of the nutrient elements in milk are maintained almost constant during periods of deficiency, though yields of milk may be reduced. The elements that show little change in concentration include P, Ca, Na and Fe, while the concentrations of Cu, I and Se in milk decline markedly with deficiency (Underwood and Suttle, 1999). The concentrations of I and Mo can be increased above the normal range by high intakes in the diet. Nitrogen is the nutrient element present in greatest amount in milk, at a concentration four to six times greater than those of P, K, Ca and Cl, which are present in roughly similar amounts (Table 4.4).

Table 4.4. Typical concentrations (g or mg l⁻¹) of the nutrient elements in the milk of dairy cattle.

N (g l⁻¹)	5.3–5.6	Agricultural Research Council, 1980; National Research Council, Subcommittee, 1989
P (g l⁻¹)	0.9–1.0	Underwood, 1981; National Research Council, Subcommittee, 1989; Renner *et al.*, 1989
S (g l⁻¹)	0.3	National Research Council, Subcommittee, 1989
K (g l⁻¹)	1.0–2.0	Underwood, 1981; National Research Council, Subcommittee, 1989; Renner *et al.*, 1989
Na (g l⁻¹)	0.3–0.7	Underwood, 1981; National Research Council, Subcommittee, 1989; Renner *et al.*, 1989
Ca (g l⁻¹)	0.9–1.4	Underwood, 1981; National Research Council, Subcommittee, 1989; Renner *et al.*, 1989
Mg (g l⁻¹)	0.1–0.20	Underwood, 1981; National Research Council, Subcommittee, 1989; Renner *et al.*, 1989
Cl (g l⁻¹)	0.8–1.4	Underwood, 1981; National Research Council, Subcommittee, 1989; Renner *et al.*, 1989
Fe (mg l⁻¹)	0.2–0.6	Morris, 1987; Renner *et al.*, 1989; Grace and Clark, 1991; Anderson, 1992
Mn (mg l⁻¹)	0.01–0.05	Hurley and Keen, 1987; Renner *et al.*, 1989; Grace and Clark, 1991; Anderson, 1992
Zn (mg l⁻¹)	3.0–5.0	Renner *et al.*, 1989; Grace and Clark, 1991; Miller *et al.*, 1991; Anderson, 1992
Cu (mg l⁻¹)	0.03–0.25	Underwood, 1981; Renner *et al.*, 1989; Grace and Clark, 1991; Anderson, 1992
Co (mg l⁻¹)	0.0003–0.001	Smith, 1987; National Research Council, Subcommittee, 1989; Renner *et al.*, 1989; Grace and Clark, 1991
I (mg l⁻¹)	0.005–0.4	Underwood, 1981; National Research Council, Subcommittee, 1989; Renner *et al.*, 1989; Grace and Clark, 1991
B (mg l⁻¹)	0.3–1.0	Nielsen, 1986; Anderson, 1992
Mo (mg l⁻¹)	0.015–0.15	Mills and Davis, 1987; Renner *et al.*, 1989; Miller *et al.*, 1991; Anderson, 1992
Se (mg l⁻¹)	0.01–0.03	National Research Council, Subcommittee, 1989; Renner *et al.*, 1989; Grace and Clark, 1991; Miller *et al.*, 1991

Dairy cows stocked at five animals per hectare for 6 months on a sward receiving moderate to high rates of fertilizer, might well produce 12,000 l of milk ha⁻¹, which would remove about 64 kg N, 12 kg P and 14 kg Ca from the system. The concentrations of some elements are appreciably higher in sheep's milk than in cows' milk, and this is so for P, K, Ca, Fe and Zn (Underwood, 1981; Grace and Clark, 1991).

Where sheep are reared for wool production, the removal of the wool results in an appreciable loss of nutrient elements from the system. Average concentrations of 14 elements in wool are shown in Table 4.5, but concentrations vary to some extent depending on the time period during which the wool was grown and on the season of the year (Grace and Lee, 1992). In

Table 4.5. Concentrations of nutrient elements (% or mg kg^{-1}) reported in sheep wool (washed, and with moisture content of about 5%): (i) 50 samples from USA and Australia (Burns *et al.*, 1964); and (ii) three samples from New Zealand (Grace and Lee, 1992).

	(i)	(ii)
N (%)	15.8	–
P (%)	0.014	0.012
S (%)	3.21	3.05
K (%)	0.005	0.33
Na (%)	0.03	0.006
Ca (%)	0.23	0.032
Mg (%)	0.02	0.006
Fe (mg kg^{-1})	50	42
Mn (mg kg^{-1})	25	3.5
Zn (mg kg^{-1})	115	114
Cu (mg kg^{-1})	25	2.9
Co (mg kg^{-1})	0.003	–
B (mg kg^{-1})	0.53	–
Mo (mg kg^{-1})	0.045	–

comparison with other body tissues, wool is particularly high in S and is also high in Zn, but has relatively low concentrations of P, Ca and Mg.

Animal : Plant Concentration Ratios

Differences between animals and plants in their requirements for nutrient elements are reflected to some extent in the concentrations in their respective biomass, though, in both plant and animals, there is some variation with the type of tissue. For example, leaves differ from stems and muscle differs from liver. However, when whole plant material is compared with the whole-body tissue of animals, all the elements show a much wider variation in plants than in animals. In calculating animal : plant concentration ratios, it is necessary to express the animal concentrations on a DM basis, and the concentrations shown in Table 4.6 were calculated from the data in Table 4.1, assuming an average water content in the animal body of 70%. The content of water is rather more than 70% in young animals, and is less in fully mature animals (Naylor, 1991). Animal : plant concentration ratios, based on typical concentrations in grass herbage material and typical concentrations in the whole-body tissue of ruminants, as in Table 4.6, provide some indication of whether or not a particular nutrient is preferentially retained by the animal. Amongst the nutrient elements, the greatest preferential retention is shown for Se, Ca, I and P, which have animal : plant ratios of > 5 : 1, while the greatest discrimination is shown against Mn and B, which have animal : plant ratios of 0.2 : 1 or less.

Table 4.6. Typical animal : plant concentration ratios, based on grass herbage and ruminant animals.

	Typical concentration in plant DM	Typical concentration in animal body DM[a]	Animal : plant concentration ratio (x : 1)
N (%)	2.8	9.00	3.2
P (%)	0.4	2.66	6.7
S (%)	0.35	0.50	1.4
K (%)	2.5	0.67	0.27
Na (%)	0.25	0.50	2.0
Ca (%)	0.6	4.66	7.8
Mg (%)	0.2	0.15	0.75
Cl (%)	0.35	0.33	0.94
Fe (mg kg^{-1})	150	133	0.89
Mn (mg kg^{-1})	165	1.2	0.007
Zn (mg kg^{-1})	37	83	2.2
Cu (mg kg^{-1})	9	9	1.0
Co (mg kg^{-1})	0.1	0.13	1.3
B (mg kg^{-1})	5	1.0	0.2
Mo (mg kg^{-1})	0.8	0.66	0.83
I (mg kg^{-1})	0.2	1.43	7.2
Se (mg kg^{-1})	0.05	1.2	24.0

[a]Derived from data in Table 4.1 assuming live weight to contain 30% DM.

Absorption of Nutrient Elements from the Diet

The nutrient elements present in herbage may be present in one of three broad categories. They may be present in organic compounds of high molecular weight that are insoluble in water, such as proteins, nucleic acids and pectin; they may occur in soluble compounds of low molecular weight, such as amino acids and metallo-organic complexes; or they may be present as inorganic ions in solution. The first category is not absorbed through the digestive tract, the second is absorbed to a variable extent and, in general, the third is absorbed readily. With most elements, there is some conversion from one category to another during digestion, and the extent of absorption is therefore influenced not only by the form in which the element occurs in the diet, but also by the extent to which it is converted, within the digestive tract, to other forms, which may be either more or less soluble.

The mechanism of actual absorption through the wall of the digestive tract varies among the elements. Those present in the digesta as monovalent ions, e.g. K, Na and Cl, are absorbed readily by simple diffusion, whereas the absorption of others, especially those occurring as trivalent or divalent cations, depends on metabolically active transport, often involving special carrier proteins (Miller, 1979). The extent of absorption of those elements depending on metabolically active transport is influenced by a range of factors, which are linked to the mechanisms of homeostatic control (Bremner and Mills, 1981). They include:

1. Current reserves in body tissues.
2. Metabolic demands as influenced by:
    (a) growth rate;
    (b) age;
    (c) pregnancy;
    (d) lactation.
3. Disease – the stress of infection tends to increase losses.
4. Genetic variables (for example, breeds of sheep differ in their concentrations of Cu in blood and liver).
5. Physical nature of the diet.
6. Chemical composition of the diet, in particular:
    (a) the form and concentration of the nutrient element;
    (b) the presence of complexing compounds that enhance absorption;
    (c) the presence of antagonists that decrease absorption.

Factors Influencing the Availability of Nutrient Elements in the Diet

The concept of the availability of a nutrient element in a diet or component of a diet is more complex than it might appear. This is mainly because the most effective way to express availability is in terms of absorption and, as indicated above, the extent of absorption is not just a property of the diet, but is influenced by the nature of the animal consuming the diet, by the actual intake of the element and by the amounts of other dietary components with which it may interact. Even the extent of absorption for a specific diet and with a specific class of animal is not readily defined, and there is more than one possible method of calculation (Reid, 1980). Thus, if:

I = intake of the element
F = total excretion in the faeces
U = excretion in the urine
Fz = net excretion of the element from body origin into the intestinal tract, i.e. endogenous faecal excretion
Uz = excretion of the element in urine at zero net retention

then:

% apparent absorption = $I - F$ as % of intake
% true absorption = $I - (F - Fz)$ as % of intake
% net retention = $I - (F + U)$ as % of intake
% utilization (in essential metabolism plus live-weight gain or milk or wool production) = $I - (F - Fz) - (U - Uz)$ as % of intake.

The determination of the apparent absorption of an element is straight-forward, requiring only the dietary intake and the amount excreted in the faeces to be measured, though to be accurate, it is necessary for the measurements to be continued for a period that is long enough to allow for fluctuations in body reserves. The determination of true absorption, sometimes regarded as availability (e.g. McDonald *et al.*, 1995), is more complicated. This is because, with most, if not all, elements, there is some endogenous excretion into the digestive tract and, as a result, the faeces contain variable amounts of elements that have been absorbed, metabolized and excreted, as well as the proportion that has not been absorbed. An additional complication is that there may be some reabsorption of an element after its excretion (or secretion) into the digestive tract (Suttle, 1987; Underwood and Suttle, 1999). Distinguishing the endogenous fraction of an element that is absorbed and then excreted in the faeces (*Fz* above) from the fraction that is present in the undigested portion of the diet is difficult. One way of assessing *Fz* is to measure faecal excretion on a diet free of the element, so that endogenous excretion (*Fz*) then equals total excretion (*F*). For some elements, a second means of assessing endogenous excretion is to use a radioisotope technique, which can differentiate between the unabsorbed fraction in the faeces and the fraction that has been absorbed and re-excreted, for example, by comparing two sources of the element labelled with different radioisotopes. More detail is provided in a recent review (Underwood and Suttle, 1999).

Many of the estimates that have been reported for availability are of apparent absorption and are therefore lower than the true absorption. Even when true absorption has been assessed, the values obtained will have been influenced by the various factors described above. For these reasons, it is doubtful whether 'availability', assessed by absorption, can be regarded as an attribute of the particular diet (Suttle, 1987). Certainly, for absorption to approximate to potential availability, it is important that the animals should have a substantial physiological requirement for the nutrient element being studied, due to growth or lactation or to the element being marginally deficient.

Another approach that has been used in assessing the availability of nutrient elements in herbage involves placing samples of the herbage in mesh bags of a non-degradable material, such as nylon, and incubating them either in an animal's digestive system or in conditions that simulate one or more stages of digestion. In one study of this type, samples of several grasses and lucerne were incubated first in the rumen of a cow for 24 h and then in an acid–pepsin solution, simulating the abomasum, for 1 h, after which they were inserted into the duodenum (via a cannula) and subsequently collected in the faeces. The total release of P during this procedure ranged from 84 to 98%, while the release of K was 100%, the release of Ca was (with one exception) 72–91% and the release of Mg was 86–95% (Emanuele *et al.*, 1991). Despite some nutrient elements being associated with the cell walls or other insoluble components of plant material, all the elements except Mn appear to be solubilized to a large extent in the rumen, suggesting that permanent retention by plant components is normally not a major factor limiting absorption. However, after solubilization in the rumen, a nutrient element may form less soluble products with other constituents of the diet and/or with microorganisms, and these interactions may result in some curtailment of absorption (Spears, 1991).

Assessment of the Dietary Requirements of Nutrient Elements

The minimum requirement for a nutrient element is usually defined as the intake needed to maintain a normal physiological concentration of the element in the bloodstream and/or to avoid any impairment of its essential functions in the body (Suttle, 1986). As indicated above, the required concentration in the diet may be influenced by interactions with other constituents of the diet; and one important example of this type of interaction is the increase in the requirement for Cu caused by high concentrations of Mo and S (see p. 247).

In general, two approaches can be used to assess the dietary requirement for a nutrient element: (i) feeding trials in which graded amounts of the element are present in, or added to, the diet, and (ii) calculations, for a particular class of livestock, involving a factorial approach based on the needs for particular physiological functions (Grace, 1984; Underwood and Suttle, 1999). In making an assessment based on feeding trials, it is necessary to relate the dietary supply of the element to measurements of live-weight gain or of some aspect of body composition, or to the appearance (or non-appearance) of characteristic symptoms. However, any feeding trial has the disadvantage that the results apply precisely only to the type of animal and conditions of the trial: to obtain results that apply widely, trials must be repeated for various classes of livestock under a range of conditions.

The factorial approach to assessing the requirement for a nutrient element involves adding together the components for different physiological needs, i.e. for maintenance and for live-weight gain, fetal development and/or lactation and/or wool growth, as appropriate (Price, 1989; Grace and Clark, 1991). The component for maintenance at a constant physiological state is equal to the inevitable endogenous loss in faeces and urine that occurs even when the intake of the nutrient element is extremely low. The components for live-weight gain, fetal development, lactation and wool growth are derived from the concentrations of the element in these products. Summation of these various items then provides the total net requirement. However, as most nutrient elements are not completely absorbed, it is necessary to apply an absorption coefficient in order to convert the net requirement to the dietary requirement. The dietary requirement can then be converted to a dietary concentration on the basis of a known or an assumed intake of DM

(Fig. 4.2). As indicated above, the absorption coefficient differs markedly from one element to another and, for any one element, is influenced by the age and physiological state of the animal and by the composition of the diet. It is generally higher for young animals than for adults. The coefficient is usually > 0.9 for K, Na and Cl, but may be less than 0.1 for Mn, Cu and Co and between 0.1 and 0.5 for Fe, Zn and Se (Grace and Clark, 1991). Deciding on the value of the absorption coefficient represents the greatest uncertainty in the factorial approach, but information on such coefficients is slowly becoming more reliable. The factorial approach has been used to determine the requirements for P, Ca and Mg (e.g. Agricultural Research Council, 1980; National Research Council, Subcommittee, 1989), and also for some of the micronutrients (Grace and Clark, 1991). However, it may not be applicable for Co and I (Grace, 1984). As indicated above, there are uncertainties in the values of absorption coefficients and in the requirements for maintenance, and these have resulted in some differences in recommendations in different countries. Another factor that sometimes contributes to an apparent lack of consistency between recommendations is that the 'recommended' daily intake or dietary concentration may include a safety margin, to allow for factors such as individual animal variation and differences in the composition of the diet.

The dietary concentrations recommended in the USA for dairy cattle at three physiological stages of development are shown in Table 4.7. Broadly similar concentrations are recommended in the UK (Agricultural Research Council, 1980; Agricultural and Food Research Council, 1991), in Australia (Australian Agricultural Council, 1990) and in France (Gueguen *et al.*, 1989), though recommendations for the micronutrients Fe, Mn, Zn and Se are rather lower in the UK than in the USA, and recommendations for P and S are rather higher in France.

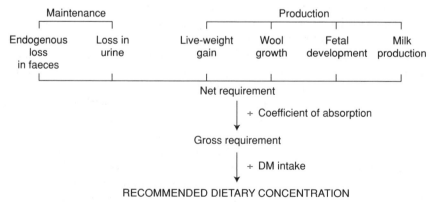

Fig. 4.2.   An outline of the factorial approach for assessing the recommended dietary concentration of a nutrient element for livestock (based on Grace and Clark, 1991).

Table 4.7. Dietary concentrations of nutrient elements for three types of dairy cattle recommended in the USA (from National Research Council, Subcommittee, 1989).

	Heifer, growing > 1 year old	Cow, pregnant, non-lactating	Cow (600 kg) yielding 30 kg milk day^{-1}
N (%)	2.56	1.92	1.92
P (%)	0.37	0.23	0.24
S (%)	0.20	0.16	0.16
K (%)	0.90	0.65	0.65
Na (%)	0.18	0.10	0.10
Ca (%)	0.58	0.29	0.39
Mg (%)	0.20	0.16	0.16
Cl (%)	0.25	0.20	0.20
Fe (mg kg^{-1})	50	50	50
Mn (mg kg^{-1})	40	40	40
Zn (mg kg^{-1})	40	40	40
Cu (mg kg^{-1})	10	10	10
Co (mg kg^{-1})	0.10	0.10	0.10
I (mg kg^{-1})	0.60	0.60	0.60
Se (mg kg^{-1})	0.30	0.30	0.30

Occurrence and Diagnosis of Deficiencies in Ruminant Animals

When grassland receives fertilizer applications that produce a high yield of herbage, the concentrations of N, P and K in the herbage are normally more than sufficient to meet the requirements of animals that are grazing. However, where livestock are fed largely or entirely on hay (which is more mature than the herbage normally used for grazing), N and/or P may be less than adequate, due to the decline in these elements with increasing maturity of the herbage. With grassland that is managed extensively and receives little or no fertilizer, herbage that is grazed is liable to contain less than optimal amounts of N and P. In contrast to N and P, deficiencies of Ca, Mg and S are more likely to occur with intensive than with extensive management. Calcium deficiency usually occurs only in cows producing high yields of milk and only at the beginning of a lactation period, when the physiological demand is greatest and when the animal may not be able to meet its needs from a combination of dietary supply and mobilization from bone reserves. The effect of intensive management on Mg deficiency is due partly to the plant uptake of Mg being reduced by high rates of fertilizer K, and partly to the absorption of Mg by the animal being reduced by high concentrations of K and N in the herbage. In addition, the herbage concentrations of both Ca and Mg are lower in all-grass than in grass–clover swards. A deficiency of Na is most likely to occur in areas where atmospheric inputs are low, but the risk is accentuated by fertilizer K, which tends to depress the plant uptake of Na and also increases the leaching of Na from the soil.

Deficiencies of the micronutrient elements are generally restricted to localized areas, though, in some parts of the world, deficiencies of certain elements, including Cu, Co, I and Se, are widespread. Micronutrient deficiencies occur most frequently in areas where the soils are generally infertile and, if the intensity of management is increased by the application of fertilizers, the increase in herbage yield will inevitably accentuate any deficiency. However, most soils have large reserves of micronutrients and, in general, there is little reduction in the concentrations of micronutrients in herbage as a result of fertilizer use. Despite the generally localized occurrence of micronutrient deficiencies, a study of such deficiencies in New Zealand found that only with Co and Se was there a clear relationship between the total concentration in the soil and the incidence of deficiency in animals; with other elements, various soil, plant and management factors masked the relationship (Grace and Clark, 1991). Even the concentration in the herbage is not always closely related to the incidence of deficiency in grazing animals, as a deficiency may be due to poor absorption rather than low concentration (Spears, 1991).

Mild deficiencies or excesses of nutrient elements in animals are difficult to identify, because their effects are often similar to those due to undernutrition or the presence of internal parasites (Underwood and Suttle, 1999), and, although pathological symptoms may occur, these are often not specific for any one element. For example, anaemia is a symptom of deficiencies of Fe, Cu and Co, and reproductive disorders are liable to occur with deficiencies of Cu, I, Mn, P, Se and Zn (Jones and Thomas, 1987; Underwood and Suttle, 1999). Although some symptoms are specific, they may become apparent only when the deficiency is serious and the effects well advanced. Examples include goitre resulting from I deficiency, white muscle disease from Se deficiency and tetany from Mg deficiency. An important research objective is therefore to develop better means of predicting the incidence of deficiencies and of diagnosing subclinical deficiencies before visible symptoms appear. Although the chemical analysis of herbage or soil for a particular nutrient element may be useful in indicating the likelihood of a deficiency in livestock, it cannot provide definite evidence of a deficiency. Also, if herbage is used, it is important to analyse samples that are as similar as possible to the ingested herbage, since grazing animals tend to select herbage of higher nutritional quality than the average of that available (Little, 1981). A more reliable indication of a subclinical deficiency in livestock is often provided by the analysis of blood, saliva or milk. Such an analysis may be either for the nutrient element itself, or for a metabolite or biologically active form of the nutrient, such as an enzyme, a hormone or a vitamin. Two examples of critical concentrations of actual nutrient elements in the blood serum that have been proposed for cattle and sheep are, for Cu, a concentration of 0.6 mg l^{-1} and, for Se, a concentration of 50 μg l^{-1} (Suttle *et al.*, 1984). Examples of biologically active molecules whose concentration in blood has been used to indicate nutrient status include glutathione peroxidase for Se, the ratio of triiodothyronine : thyroxine for I, and vitamin B_{12} for Co. For each of these assessments, the lower the actual concentration or ratio and the longer the time period during which low values occur, the greater the risk that the deficiency will cause a reduction in growth or milk production. However, although the analysis of blood may provide evidence that an element is deficient, the results are often rather uncertain, because either the analytical methods or the criteria for deficiency are insufficiently precise (Suttle, 1994). When there is a clear restriction in growth or there are visible symptoms, apparently due to a nutrient deficiency, the most effective way to confirm the diagnosis is to carry out a response trial involving supplementation with the nutrient element suspected of being deficient (Jones and Thomas, 1987; Grace and Clark, 1991). There is, however, the disadvantage that such trials are time-consuming and expensive.

Deficiencies of nutrient elements in grazing animals can sometimes be prevented or overcome by the application of an appropriate source of the element to the sward, though this approach is more feasible for intensive than for extensive management. Also, it is usually not appropriate for the rapid treatment of a serious deficiency. Where P deficiency in grazing animals is widespread, the use of fertilizer P is effective in the long term, and, where Mg is deficient, the application of either dolomitic limestone or MgO is generally effective after several days or weeks. However, with some elements, application to the sward is rarely a suitable approach (Underwood and Suttle, 1999). Thus, with Cu, although Cu-containing fertilizers are sometimes effective, it is not feasible to apply the amount necessary to prevent deficiency when there is a high concentration of Mo in the herbage and, in this situation, other measures are necessary. The provision of Cu in mineral licks results in a highly variable intake, and a more effective method for overcoming Cu deficiency is to dose or inject each animal with a Cu compound, such as the glycinate or oxide. When Co is deficient, the application of $CoSO_4$ to the sward generally results in satisfactory uptake by the herbage, but it is expensive and a more cost-effective treatment is to dose each animal with a cobalt 'bullet', which is retained in the reticulum and releases Co slowly (Blaxter, 1980). With Se, application to the sward has the potentially serious disadvantage that it may result in toxic concentrations in the herbage for a period of several weeks (Thompson, 1980).

In order to minimize the risk of a deficiency arising, grazing animals are sometimes provided with free-access mineral licks or blocks containing a number of nutrient elements. The elements most commonly included in these supplements are P, Ca, Mg, Na, Cl, Cu, Co, I and Se. Mineral licks and blocks have the disadvantage that there is wide variation amongst individual animals in the amount consumed (Thompson, 1980), though consumption is generally increased when they are made more palatable by the incorporation of ingredients such as NaCl and/or molasses (McDowell, 1992). In order to minimize the variation in consumption, several other measures have been developed for providing supplementary nutrient elements where deficiencies are likely. These measures include: (i) treating the water supply, which is effective for cattle if a pressurized supply of water is available; (ii) dosing individual animals with soluble glass boluses containing the required element(s) in a slow-release form; (iii) dosing with other slow-release forms of nutrient elements; and (iv) injecting the animals with appropriate compounds, such as copper glycinate for Cu, or sodium selenite for Se (Thompson, 1980; Suttle *et al.*, 1984; Underwood and Suttle, 1999). When the animals receive concentrate feeds in addition to grassland herbage, it is often effective to incorporate supplementary nutrient elements into this component of the diet.

Soil Ingestion by Ruminant Animals

The ingestion of soil is, potentially, an important influence on the amounts of several nutrient elements consumed and absorbed by grazing animals. When herbage is grazed, there is almost inevitably some ingestion of soil, often amounting to between 2% and 10% of the DM intake. In wet and muddy conditions, soil ingestion by cattle may amount to as much as 20% of the DM intake, and soil ingestion by sheep may be even greater (Healy, 1973; Thornton and Abrahams, 1981). The soil that is consumed generally occurs as surface contamination on the herbage, is attached to plant roots pulled from the ground or is in earthworm casts. Factors that influence the amount of soil ingested therefore include: (i) weather and seasonal conditions; (ii) soil type; (iii) the stocking rate of grazing animals; (iv) the earthworm population; and (v) individual animal differences (Healy, 1973). In humid regions, most of the ingested soil is soil that has been splashed on to herbage by rainfall and the treading of livestock. Rainfall is clearly an important factor in the extent of splashing, but it also has an influence, together with temperature, on the rate of herbage growth, and increased herbage growth may dilute the effect of splashed soil. The effect of soil type is due mainly to there being more splashing when the soil has poor aggregation and is poorly drained. In general, soil ingestion is greater when the number of grazing animals per unit area is high. Soil ingestion is also greater when there is a large population of earthworms, and earthworms tend to deposit more casts on the soil surface in wet conditions, especially if the soil is poorly drained (Healy, 1973).

There are often marked seasonal variations in soil ingestion, due some-times to differences in weather conditions and sometimes to differences in the extent to which animals are housed during the winter. Soil ingestion by cattle in New Zealand is generally highest in the winter months, since the animals generally remain outdoors and rainfall is relatively high but grass growth is slow due to low temperature (Healy, 1973). However, in many countries, soil ingestion is highest during spring and autumn, since the ani-mals are housed during the winter and fed silage or hay, which usually have little soil contamination. In semi-arid regions with extensive grazing, soil ingestion results mainly from the deposition of fine soil particles on to the herbage, a process that is at a maximum at the driest times of year, when plant growth is poor, and especially when wind and the trampling of livestock accentuate the movement of dust. The presence of shallow-rooted plants may also be important, as soil is eaten together with roots when these are pulled from the soil (Mayland *et al.*, 1975).

An approximate estimate of the amount of soil ingested can be obtained from the ash content of the faeces, making allowance for the inherent ash content of herbage, typically about 8%. A more accurate estimate of soil ingestion can be obtained from the content of titanium (Ti) in the faeces, as Ti is present in soils in concentrations of several thousand mg kg^{-1}, whereas concentrations in herbage are usually < 10 mg kg^{-1} (Healy, 1973; Thornton and Abrahams, 1981). Using one of these methods, a number of estimates have been made of the amounts of soil ingested by livestock on an annual basis. Thus, in New Zealand, dairy cattle have been estimated to consume between 150 and 650 kg soil $year^{-1}$ and sheep up to 75 kg soil $year^{-1}$ (Healy, 1973). In Ireland, the amount of soil ingested by sheep weighing 60 kg was also estimated to be about 75 kg $year^{-1}$, with rainfall and stocking rate being important factors in seasonal variation (McGrath *et al.*, 1982). Under semi-arid conditions, the amount of soil ingested tends to increase as the amount of herbage diminishes, and therefore as the severity of grazing is increased. In range conditions in Idaho, USA, cattle were found to ingest 35–550 kg soil $year^{-1}$, with a mean of 180 kg, and the daily rate ranged from 100 to 1500 g with a mean of 500 g (Mayland *et al.*, 1975). This consumption represented a range of about 1.2–18.8% of DM intake, with a median value of 6.2%.

The ingestion of soil is sometimes important in the supply of nutrient elements, especially Co, I and Se, whose concentration in soil is much greater than their concentration in herbage. However, there is little information on the availability to animals of the nutrient elements in soil, though it is likely to be less than the availability of the corresponding elements in herbage. Nevertheless, soil ingestion by sheep has been observed to increase the amounts of Ca, Mg and P retained, suggesting that some absorption occurs (Grace and Healy, 1974). There is also evidence that the incidence of I deficiency in animals is reduced when they consume relatively large amounts of soil (Statham and Bray, 1975). With Se, the ingestion by lambs of 100 g soil day^{-1} of two soils resulted in increased concentrations of Se in both blood plasma and liver; and, with one soil, there was also a significant increase in the concentration of vitamin B_{12} in the liver. However, neither soil influenced the concentrations in the liver of Fe, Mn, Zn or Cu (Grace *et al.*, 1996a). In some instances, soil ingestion has been found to have adverse effects on the availability of nutrient elements. For example, it has sometimes, though not always, reduced the availability of dietary Cu (Langlands *et al.*, 1982; Suttle,

1988; Grace *et al.*, 1996a). Clearly, the effects of soil ingestion vary with soil type.

Nutrient Elements Supplied by Drinking-water

In some locations, drinking-water may provide livestock with significant amounts of nutrient elements, particularly Na, Ca, Mg, S and Cl (Table 4.8; Langlands, 1987). Although the presence of nutrient elements in drinking-water may sometimes be beneficial to livestock, in arid and semi-arid areas of the world there is often a high degree of salinity, due to ions such as Na^+, Ca^{2+}, Mg^{2+}, SO_4^{2-}, Cl^- and HCO_3^-, and this can be detrimental. A concentration of total soluble salts of up to 1000 mg l^{-1} is generally acceptable for livestock, but water containing between 1000 and 3000 mg l^{-1} may cause temporary diarrhoea, or may be refused by animals not accustomed to such high concentrations (McDowell, 1992).

Excretion of Nutrient Elements

Only a small proportion of the amount of a nutrient element consumed by an animal is utilized for producing live-weight gain, milk or wool, and the remainder is excreted. The amount excreted includes the proportion that is undigested, the proportion absorbed but not utilized and the proportion utilized for maintaining metabolic processes, which is excreted, after some delay, as a result of the turnover of various tissues and/or secretion into the

Table 4.8. Concentrations of 14 nutrient elements in the water of rivers and lakes in USA (from McDowell, 1992).

	Range	Mean
P (mg l^{-1})	0.001–5.0	0.087
S (as SO_4) (mg l^{-1})	0.0–3,383.0	136
K (mg l^{-1})	0.06–370.0	4.3
Na (mg l^{-1})	0.2–7,500.0	55.1
Ca (mg l^{-1})	11.0–173.0	57.1
Mg (mg l^{-1})	8.5–137.0	14.3
Cl (mg l^{-1})	0.0–19,000.0	478
Fe (µg l^{-1})	0.10–4,600.0	43.9
Mn (µg l^{-1})	0.20–3,230.0	29.4
Zn (µg l^{-1})	1.0–1,183.0	51.8
Cu (µg l^{-1})	0.8–280.0	13.8
Co (µg l^{-1})	0.0–5.0	1.0
I (µg l^{-1})	4.0–336.0	46.1
Se (µg l^{-1})	0.01–1.0	0.016

digestive system. Some elements, including P, Fe and the divalent cations, are excreted mainly in the faeces, whereas other elements, such as N, S, the monovalent cations and B, are excreted mainly in the urine (Table 4.9). In addition, Na and Cl are excreted, to some extent, in sweat; and with Se there may be some loss via the breath (Miller, 1979). Reported average concentrations of a number of elements in cattle faeces are shown in Table 4.10 and concentrations in cattle urine in Table 4.11. However, these concentrations may vary considerably even between individual animals consuming the same herbage, and, for any individual animal, they may vary on different days and at different times of the same day. The concentrations of the nutrient elements in the faeces and urine of sheep are similar to those of cattle if similar diets are consumed (Haynes and Williams, 1993). In faeces, the high concentrations of Ca and Mg result in an excess of cations over the anions phosphate, sulphate and chloride, and the difference is balanced mainly by carbonate.

In the context of assessing the role of excreta in the recycling of nutrient elements in grassland systems, it is important to have information not only on composition but also on the frequency of defecation and urination, on the amounts of dung and urine produced per day and on the areas affected by single depositions of dung and urine. Typical data for these characteristics for cattle and sheep (Table 4.12) make it possible to calculate the average rates at which nutrient elements are returned to grazed swards, and the proportion of the grazed area affected, as discussed on p. 27. In the context of assessing the

Table 4.9. Typical partitioning of excreted nutrient elements between faeces and urine.

	Proportion (%) in faeces	Proportion (%) in urine	Reference
N	20–55	45–80	See p. 122
P	95+	< 5	Safley *et al.*, 1984
S	10–94	6–90	Barrow, 1987
K	10–30	70–90	Wilkinson and Lowrey, 1973; Safley *et al.*, 1984; Barrow, 1987
Na	*c.* 40	*c.* 60	Safley *et al.*, 1984
Ca	96–99	1–4	Wilkinson and Lowrey, 1973; Safley *et al.*, 1984
Mg	70–90	10–30	Wilkinson and Lowrey, 1973; Safley *et al.*, 1984
Cl	*c.* 45	*c.* 55	Safley *et al.*, 1984
Fe	95+	< 5	Wilkinson and Lowrey, 1973; Safley *et al.*, 1984
Mn	95+	< 5	Wilkinson and Lowrey, 1973; Safley *et al.*, 1984
Zn	95+	< 5	Wilkinson and Lowrey, 1973; Safley *et al.*, 1984
Cu	95+	< 5	Wilkinson and Lowrey, 1973; Safley *et al.*, 1984
Co	< 50	> 50	Smith, 1987
I	40–45	55–60	Miller *et al.*, 1975
B	30–50	50–70	Nielsen, 1986
Mo	30–70	30–70	Barrow, 1987
Se	60–80	20–40	Levander, 1986; Langlands *et al.*, 1986

Table 4.10. Average concentrations of some nutrient elements in the faeces of dairy cattle (% or mg kg^{-1} in DM): (i) from cattle on seven farms in North Carolina, USA (Safley *et al.*, 1984); (ii) from cattle grazing grass–clover in Germany (Kirchmann and Witter, 1992); and (iii) from cattle grazing grass–clover in New Zealand (Marsh and Campling, 1970).

	(i)	(ii)	(iii)
N (%)	2.92	2.35	–
P (%)	1.20	0.90	0.70
S (%)	–	0.36	–
K (%)	0.84	0.73	0.8
Na (%)	0.22	0.00	0.36
Ca (%)	1.28	1.96	2.4
Mg (%)	0.63	0.6	0.8
Cl (%)	0.61	–	–
Fe (mg kg^{-1})	1600	2030	–
Mn (mg kg^{-1})	200	150	–
Zn (mg kg^{-1})	200	130	–
Cu (mg kg^{-1})	50	40	–
B (mg kg^{-1})	–	15	–

Table 4.11. Average concentrations (g or mg l^{-1}) of some nutrient elements in the urine of cattle: (i) from dairy cattle on seven farms in North Carolina, USA (Safley *et al.*, 1984); and (ii) from beef cattle grazing grass–clover in New Zealand (Ledgard *et al.*, 1982).

	(i)	(ii)
Total solids (g l^{-1})	60.7	–
N (g l^{-1})	10.8	7.7
P (g l^{-1})	0.19	0.34
S (g l^{-1})	–	0.73
K (g l^{-1})	7.47	9.5
Na (g l^{-1})	1.11	0.22
Ca (g l^{-1})	0.16	0.73
Mg (g l^{-1})	0.52	0.12
Cl (g l^{-1})	2.38	–
Fe (mg l^{-1})	5.49	–
Mn (mg l^{-1})	0.18	–
Zn (mg l^{-1})	1.83	–
Cu (mg l^{-1})	0.61	–

extent of uptake for new plant growth and the potential for loss by leaching, it is important to assess the rates at which the nutrient elements are returned to the actual urine and dung patches in grassland swards, and estimates for the rates, equivalent to kg ha^{-1}, of the macronutrient elements typically provided by grazing cattle are shown in Table 4.13.

Partitioning of Ingested Nutrients to Animal Products and Excreta

The retention of nutrients in live-weight gain and milk is greatest in lactating and actively growing animals, and least in non-lactating mature animals. With sheep, there is also some retention in wool. The amounts can be calculated, for given yields of the various products, from the data in Tables 4.1, 4.4 and 4.5. An example of the partitioning of some ingested macronutrients between milk production and excretion in urine and faeces for dairy cattle (Table 4.14) shows that, for each element, the amount excreted is at least twice the amount secreted in milk. For other classes of livestock, the proportions retained in live-weight gain, milk or wool are likely to be generally lower and the proportions excreted higher than in this example.

Table 4.12. Typical values for some characteristics of the excreta of cattle and sheep that influence their impact on the sward.

	Cattle	Sheep
Urinations per day	8–12[a,b]	15–20[a,b]
Volume per urination (l)	1.6–2.8[a,b]	0.10–0.18[a,b]
Volume of urine per day (l)	12–34	1.7–3.6
Area covered by urine patch (m²)	0.16–0.50[a,b]	0.03–0.05[a,b]
Area affected by urine patch (m²)	0.5–2.5[b]	0.06–0.15
Defecations per day	11–16[a,b,c,d]	7–26[a,b]
Weight of DM per defecation (g)	250–380[a,b,d]	6–34[a,b]
Weight of faecal DM per day (kg)	2.7–5.9	0.2–0.6[b]
Area covered by single defecation (m²)	0.05–0.09[a,b]	0.007–0.025[a,b]
Area affected by single defecation (m²)	0.25–0.54[b]	0.015–0.075[a,b]

[a]Haynes and Williams, 1993.
[b]Nguyen and Goh, 1994a.
[c]MacDiarmid and Watkin, 1972.
[d]Marsh and Campling, 1970.

Table 4.13. Approximate amounts of macronutrient elements supplied to the urine and dung patches produced by grazing cattle, in kg ha⁻¹ to actual area of patch, assuming patch areas of 0.35 m² for urine (with 2 l per urination) and 0.07 m² for dung (with 310 g DM per dung patch), and concentrations of nutrients in urine and dung typical of productive grassland.

	N	P	S	K	Na	Ca	Mg
Urine	510	15	31	490	34	26	18
Dung	1040	350	150	380	80	970	310

Table 4.14. Approximate distribution of seven macronutrient elements between secretion in milk and excretion in faeces and urine, by dairy cattle consuming fertilized grass herbage (from Kemp and Geurink, 1978).

	Concentration in herbage (%)	Dietary intake (g day^{-1})	Secretion in 25 kg milk (g day^{-1})	Excretion in faeces (g day^{-1})	Excretion in urine (g day^{-1})
P	0.41	66	24	48	0.2
S	0.42	67	7	18	42
K	3.02	483	41	53	389
Na	0.37	59	10	9	40
Ca	0.61	98	30	68	0.5
Mg	0.23	37	3	31	3
Cl	1.08	173	29	21	123

Chapter 5

Nitrogen

Natural Sources and Amounts of N in Soils

The N in soils has been derived, over the course of soil development, almost entirely from the N_2 in the atmosphere, with only negligible amounts from the weathering of rocks. Until recent decades, almost all fixation of atmospheric N_2 occurred through the activity of soil microorganisms, but additional fixation now occurs through human activities. In some of these activities, such as the manufacture of fertilizers, the fixation of N_2 is deliberate, while in others, such as the combustion of fossil fuels, N_2 fixation is incidental. The microorganisms involved in the fixation of N_2 can be divided into two main categories: (i) symbiotic organisms, of which the most important are the rhizobial bacteria associated with the roots of leguminous plants; and (ii) free-living soil organisms, such as *Azotobacter*. In favourable conditions, the rhizobial bacteria associated with clovers are able to fix 200–400 kg N ha^{-1} year^{-1}, though, in natural and semi-natural grasslands, the amounts are usually much less, possibly less than 10 kg ha^{-1} year^{-1}. The free-living soil bacteria rarely fix more than about 15 kg N ha^{-1} year^{-1}, and often only 1–2 kg N ha^{-1} year^{-1} (Clark and Woodmansee, 1992). Nevertheless, in natural and semi-natural grasslands, the small amounts of N fixed annually by symbiotic and free-living bacteria have provided much of the N currently present in the soils.

Usually, more than 95% of the total soil N occurs in the soil OM, and less than 5% occurs in the inorganic forms, ammonium and nitrate. The concentration of N in soil OM is fairly constant at about 5% of dry weight, and the OM content of soil is therefore reflected in the content of total soil N. Differences in the content of OM are due mainly to the influence, over many years, of factors such as vegetation, soil texture and climate, and the extent to which the soil has been tilled. In general, the content of OM increases with increasing clay content, with low average temperature and with uninterrupted growth of vegetation, i.e. an absence of tillage. Soil N content shows a similar pattern. Thus, in soils of temperate regions, the total concentration of N to a depth of 15 cm ranges from about 0.08% N (1.5% OM) in regularly cultivated sandy soils to about 0.5% N (10% OM) in clay soils under long-term grassland (Table 5.1). The larger amounts of OM and N in clay soils reflect a slower rate of decomposition, due partly to poor aeration, and partly to the formation of complexes between clay and OM. Total soil N normally declines with increasing depth (except in some peat soils) and is often negligible below 30–40 cm. Typically, the total amount of soil N to a depth of 40 cm is between 5000 and 15,000 kg ha^{-1} in long-term grassland soils, and between 2000 and 4000 kg ha^{-1} in soils cultivated for arable crops.

When grass is sown on land that has been cultivated for many years, the contents of OM and N generally increase from year to year, so long as the sward remains unploughed. The time taken to reach an equilibrium content of OM varies with soil type, climate and sward management, but, in temperate climatic conditions, is likely to be between 50 and 200 years. However, when grassland is managed for intensive agricultural production, intermittent changes in the management often prevent an equilibrium content of OM being attained. Factors such as occasional ploughing and reseeding, the improvement of drainage and the application of lime increase the rate at which OM is decomposed and therefore tend to reduce the rate of accumulation.

Agricultural and Atmospheric Inputs of N

The amount of N fixed by the rhizobia associated with the legumes of temperate grassland varies widely: in favourable conditions for white clover or lucerne it may amount to as much as 400 kg N ha^{-1} year^{-1}, and occasionally more than 650 kg N ha^{-1} year^{-1} (Ledgard and Giller, 1995). For conditions to be favourable, there should be plentiful supplies of all nutrients other than N, adequate but not excessive moisture, and grazing or cutting management that prevents the dominance of grasses (Peoples *et al.*, 1995). In much of Europe and in New Zealand, productive grass–clover swards (not given fertilizer N) often fix between 100 and 300 kg N ha^{-1} year^{-1}, depending on the vigour of

Table 5.1. Typical concentration of N in soils under arable cultivation, ley/arable and long-term grass (% in dry weight, 0–15 cm) as influenced by soil texture (derived from Archer, 1988).

Soil texture	Arable	Ley/arable	Long-term grass
Sand	0.05–0.08	0.07–0.09	0.15–0.25
Light loam	0.07–0.12	0.09–0.15	0.20–0.30
Light silt	0.10–0.13	0.12–0.15	0.25–0.35
Medium loam/medium silt	0.12–0.15	0.14–0.16	0.25–0.40
Clay	0.16–0.20	0.18–0.21	0.40–0.50

clover growth. Some of the N fixed by the clover is transferred to the grass, mainly by the decomposition of plant tissue or via the consumption of plant material by livestock and the subsequent excretion of N. In ungrazed grassland, up to 100 kg N ha^{-1} year^{-1} may be transferred from clover to grass, usually accounting for between 2% and 30% of the N fixed; and, in grassland that is grazed intensively, an approximately equal amount may be transferred via excreta (Ledgard and Giller, 1995). However, as noted above, in lightly grazed natural and semi-natural grasslands not supplied with fertilizer P or K, the amount of N fixed by legumes is often < 25 kg N ha^{-1} year^{-1}.

Intensively managed all-grass swards often receive fertilizer N at rates up to 400 kg N ha^{-1} year^{-1}; and in warm climatic conditions, such as those in southern USA, up to 800 kg N ha^{-1} year^{-1} may be applied. Ammonium nitrate (35% N) and urea (46% N) are widely used as sources of fertilizer N. The fertilizer greatly increases herbage growth, though it has less effect on the amount of roots. There is, therefore, an increase in the amount of plant material decomposing *in situ* and, if the sward is grazed, an increase in the amount of excreta returned to the soil by livestock. However, the application of fertilizer N often produces no increase in soil N, because the fertilizer enhances the concentration of N in plant material, and therefore promotes mineralization rather than accumulation of soil OM (see p. 98). Also, when fertilizer N is applied to grass–clover swards, there is a marked reduction in the amount of N$_2$ fixed by the clover, due partly to some substitution of fixation by fertilizer N and partly to the fertilizer increasing the competitive ability of the grass (Ledgard *et al.*, 1996). Results from several field investigations suggest that fertilizer N, even when applied to grassland regularly for a number of years, has no significant effect on soil N, unless the soil is extremely low in OM (Whitehead, 1995).

When sewage sludge is applied to grassland, there is often an increase in the soil contents of N and OM. Sewage sludge usually contains between 2% and 5% N (see Table 2.8, p. 32), most of which is in organic forms and not readily lost from the soil.

All soils receive some input of N from the atmosphere, in addition to that derived from the microbial fixation of N$_2$. This atmospheric input is partly in the form of ammonia and nitrogen oxides, which can be absorbed by vegetation or soil, and partly in the form of aerosol particles containing ammonium and nitrate. Most of the ammoniacal N in the atmosphere is present as a result of volatilization from the urea in livestock urine or from the urea and ammonium salts applied as fertilizers (see p. 101), whereas most of the nitrogen oxides and nitrates are derived from the combustion of fuels, particularly in motor vehicles. The total deposition of N (wet plus dry) is thought to amount to about 30–50 kg N ha^{-1} year^{-1} over much of England (Goulding, 1990; Goulding *et al.*, 1998) and Germany (Fangmeier *et al.*, 1994), and to 40–80 kg N ha^{-1} year^{-1} in areas of The Netherlands with intensive livestock production (van Breemen and van Dijk, 1988). However, in areas remote from both intensive agriculture and urban centres, areas such as northern and western parts of the UK, and parts of the USA, inputs from the atmosphere are smaller, sometimes less than 15 kg N ha^{-1} year^{-1} (National Research Council, 1978; Campbell *et al.*, 1990; Brady and Weil, 1999). The various forms of N deposited on grassland from the atmosphere are largely taken up and assimilated by either plants or soil microorganisms, and are thus converted into organic soil N. In areas of natural or semi-natural grassland, the amount of N deposited from the atmosphere is often greater than the amount fixed by microorganisms, but this situation reflects the contribution of the ammonia and nitrogen oxides generated by human activities.

Recycling of N through the Decomposition of Organic Residues

The amounts of OM and N reaching the soil in dead plant material and animal excreta are usually greater in grassland than in soils under arable crops. Dead herbage and root material are major sources of recycled N for grassland soils, the actual amounts being influenced by factors such as sward management, climatic conditions and the plant species present. For a productive grass sward receiving a moderate rate of fertilizer N (about 250 kg N ha^{-1} year^{-1}), the input of dead herbage material to the soil is likely to be about 5000–8000 kg DM ha^{-1} year^{-1} (see p. 26) which, with an average concentration of N of 1.7% (see Table 3.11, p. 62), would return 85–140 kg N ha^{-1}. The input of dead root material from such a sward is likely to be about 5000 kg DM ha^{-1} year^{-1}, which, with a concentration of N in grass roots of about 1%, would provide about 50 kg N ha^{-1}. When swards are grazed, large amounts of OM and N are transferred to the soil in excreta. Assuming an annual consumption by grazing animals of 10,000 kg herbage DM, containing 250 kg N ha^{-1}, and assuming 80% of the N to be excreted (see p. 121), then the return of N in excreta would be 200 kg N ha^{-1} (though some may be deposited in milking parlours, yards, etc.). The actual amount of excretal N returned in any situation will be influenced by factors such as stocking rate, the proportion of time spent by the animals on the grazed area, the composition of the herbage and whether additional feed is provided. With a productive grass sward receiving about 250 kg fertilizer N ha^{-1} year^{-1}, the total amount of N recycled each year through dead herbage and roots and livestock excreta is likely to be between 300 and 400 kg N ha^{-1}, whereas, with an extensively managed sward receiving no fertilizer N, the total amount of N recycled may be less than 50 kg N ha^{-1} year^{-1}.

Some of the N in dead plant materials and animal excreta quickly becomes available for renewed plant uptake through the process of mineralization, while some tends to accumulate in the soil. The balance between mineralization and accumulation is influenced by the C : N ratio of the decomposing material and, to some extent, by soil type. The N in materials with a C : N ratio of more than about 30 : 1 is generally slow to mineralize, whereas, with materials having a C : N ratio of less than about 20 : 1, mineralization begins rapidly. When the C : N ratio is high, there may actually be some immobilization of inorganic N already present in the soil, due to its assimilation by soil microorganisms. Some typical C : N ratios of the plant materials and animal excreta decomposing in grassland soils, together with the C : N ratios of soil microorganisms and earthworms, are shown in Table 5.2.

When organic N is mineralized, the basic process is one of ammonification, i.e. the conversion of organic N to ammonium. Some of the resulting ammonium may then be taken up by plants or microorganisms, and a small amount may be volatilized, but usually most is nitrified to nitrate by soil bacteria, provided that conditions are aerobic:

$$NH_4 \rightarrow NH_2OH \rightarrow (NOH) \rightarrow NO \rightarrow NO_2 \rightarrow NO_3$$

Nitrosomonas spp. are largely responsible for the oxidation of ammonium to nitrite and *Nitrobacter* spp. for the oxidation of nitrite to nitrate.

Urine has a particularly low C : N ratio (Table 5.2), and most of its N is present as urea, which is largely converted ammonium within a few days. Nitrification of the ammonium occurs more slowly, often requiring a period of a few weeks. However, both ammonification and nitrification occur more slowly when conditions are cold. Although there is usually some uptake and assimilation of the ammonium and nitrate by plants and soil microorganisms, there is often some loss of ammonia through volatilization and some loss of nitrate through leaching and/or denitrification. In contrast to the N in urine, most of the N in dung is mineralized slowly, reflecting its higher C : N ratio, particularly once the soluble fraction has been leached away by rain.

Table 5.2. Typical C : N ratios of plant residues, animal excreta and the biomass of soil microorganisms and earthworms decomposing in grassland soils.

	C (% in DM)	N (% in DM)	C : N ratio	Reference
Dead grass herbage, little or no fertilizer N	48.4	1.1	44 : 1	Whitehead, 1995
Dead grass herbage, high rate of fertilizer N	48.4	2.5	19 : 1	Whitehead, 1995
Dead clover herbage	48.4	2.7	18 : 1	Whitehead, 1995
Grass roots, little or no fertilizer N	49.4	1.07	46 : 1	Whitehead, 1970
Grass roots, high rate of fertilizer N	48.5	1.64	30 : 1	Whitehead, 1970
White clover roots	50.2	3.77	13 : 1	Whitehead, 1970
Faeces of cattle or sheep	48	2.4	20 : 1	See p. 122
Urine of cattle or sheep	43	11.0	3.9 : 1	See p. 122
Farmyard manure	37	2.8	13 : 1	Jenkinson, 1981
Bacteria	50	15.0	3.3 : 1	Jenkinson, 1981
Actinomycetes	50	11.0	4.5 : 1	Jenkinson, 1981
Fungi	44	3.4	13 : 1	Jenkinson, 1981
Earthworms	46	10.0	4.6 : 1	Jenkinson, 1981

Nevertheless, dung may induce a slow increase in the concentration of nitrate in the underlying soil, and an increase in the rate of denitrification (Ryden, 1986).

Through their consumption of herbage, grazing animals tend to increase the rate at which N is recycled, since the N consumed is mainly in the form of insoluble organic compounds, whereas much of the excreted N is in soluble compounds, such as urea, which are mineralized rapidly (Clark and Woodmansee, 1992). However, the extent to which the excreted N is taken up again for new grass growth depends on a number of soil and weather factors and on whether or not fertilizer N is applied. When little or no fertilizer N is applied, there is a substantial uptake of excreted N (Day and Detling, 1990), but, when high rates of fertilizer N are applied, the net uptake of excreted N may be almost zero (Deenen and Middelkoop, 1992). In all instances, the effects of dung and urine are confined to relatively small patches within the sward, and these receive extremely high rates of N. In urine patches, the input of N may be equivalent to > 500 kg N ha^{-1} (see Table 4.13, p. 93) and, exceptionally, up to 1000 kg N ha^{-1} (Clough et al., 1996). When not masked by the application of fertilizer, the effect of urine N on grass growth is often visible for about 3 months (Richards and Wolton, 1976). With dung patches, the rate of N received is also typically > 500 kg N ha^{-1} (see Table 4.13), but much of the N is not readily available, and the dung is often dispersed over a wider area before decomposition and mineralization are complete. The proportion of the N in dung that is soluble in water, and therefore readily available, appears to be greater with sheep than cattle, 35% compared with 16% in a comparison of animals grazing grass–clover swards (Williams and Haynes, 1995).

When livestock excreta are applied to grassland in the form of slurry, there is usually a substantial loss of N through the volatilization of ammonia (see p. 101), but there may, nevertheless, be some increase in soil N. Cattle slurry applied to grassland at various rates during a 4-year period was found to increase soil N by between 90 and 570 kg N ha^{-1} year^{-1}, representing between 18 and 30% of the N applied in the slurry (Gostick, 1981). The conversion of livestock excreta to farmyard manure also entails losses of N by ammonia volatilization, denitrification and possibly leaching (Petersen et al., 1998a). Nevertheless, the C : N ratio of the manure is typically about 13 : 1 (Table 5.2).

Forms and Availability of N in Soils

The N present in the soil OM is not immediately available for plant uptake: it becomes available only after decomposition of the OM and mineralization of the organic N to ammonium and nitrate. Usually, at least 95% of the soil N is present in the OM, and almost all of this occurs in complex organic polymers. About 30–40% of this N can be identified, after hydrolysis, as amino acids and about 5–6% as amino sugars. The relative amounts of the individual amino acids and amino sugars indicate that a substantial proportion of the soil N is derived from microbial material. In addition to the amino acids and amino sugars, about 35% of the organic soil N is present in heterocyclic N compounds and about 20% is ammoniacal. It is likely that much of the heterocyclic N occurs in the complex humus polymers, which result partly from the biochemical modification of plant residues and partly from the reactions of amino compounds and ammonia with phenolic compounds derived from lignin (Stevenson, 1994). Although all forms of organic N in the soil are subject to mineralization, the speed of the process varies widely and generally occurs most slowly with the humified OM containing heterocyclic N.

The rate of mineralization is important in relation to the supply of plant-available N and, in general, is enhanced by a low C : N ratio (i.e. a high concentration of N), by the presence of earthworms, by warmth and by soils having a good capacity to retain water without becoming waterlogged. The addition of fertilizer N to a previously unfertilized sward also tends to promote the mineralization of the existing organic N (Gill *et al.*, 1995). On an annual basis, net mineralization is greater when the OM content of the soil is decreasing or at equilibrium, rather than increasing.

The inorganic forms of N, ammonium and nitrate, are both readily available for uptake by plants but they differ in their reactions in the soil. Ammonium is retained by cation exchange on clay and OM, thus restricting its mobility and accessibility to roots, but ensuring that it is not susceptible to loss by leaching. Ammonium that is present in solution, or is readily exchangeable, is subject to nitrification; and when it is present at a relatively high concentration due to the application of fertilizer or the hydrolysis of urea in urine, nitrification is often complete within 3–4 weeks. Nitrate, an anion, is not retained by adsorption and is therefore mobile in the soil solution, readily accessible to plant roots, but also susceptible to leaching and to denitrification.

Losses of N from Soils

There are three main pathways by which N can be lost from soils: the volatilization of ammonia, the leaching of nitrate and the loss of N_2 and nitrogen oxides through denitrification and nitrification.

The volatilization of ammonia occurs mainly from the high concentrations of ammoniacal N that occur temporarily following the deposition of excreta or the application of slurry or fertilizer. In general, volatilization amounts to 5–25% of the urine N from grazing livestock, with a lower proportion from dung (Whitehead, 1995; Petersen *et al.*, 1998b). Ammonia volatilization from grazed swards tends to increase with increasing rate of fertilizer application (Table 5.3; Bussink, 1994), due to the higher concentrations of N in herbage and thus in urine. When slurry is applied to

Table 5.3.   Ammonia volatilization from three grazed swards: (i) ryegrass–white clover receiving no fertilizer N; (ii) ryegrass receiving 210 kg N ha^{-1} year^{-1}; and (iii) ryegrass receiving 420 kg N ha^{-1} year^{-1}; mean values for two grazing seasons, assessed by a micrometeorological method (from Jarvis *et al.*, 1989).

	(i)	(ii)	(iii)
Ammonia volatilized (kg ha^{-1} year^{-1})	6.7	9.6	25.1
Proportion (%) of fertilizer N	–	4.6	6.0
Proportion (%) of estimated urinary N	9.8	11.2	12.1

the surface of grassland, the volatilization of ammonia may amount to as much as 20–80% of the total N, depending on sward and weather conditions. Volatilization from slurry is relatively high, because much of the excreted N is converted to ammonium during storage and because the slurry is applied fairly uniformly over the area, thus minimizing its penetration into the soil and maximizing its exposure to the atmosphere. Volatilization also occurs following the application of some forms of N fertilizer, particularly urea and, on calcareous soils, ammonium salts. Volatilization from fertilizers occurs more from grassland than from arable soils, because the fertilizer is applied to the surface and not incorporated into the soil. In general, such losses probably account for 10–25% of the N from urea, 1–3% of the N from ammonium nitrate and calcium ammonium nitrate and variable proportions, up to about 25%, from diammonium phosphate (Sommer and Ersboll, 1996; Van der Weerden and Jarvis, 1997).

The volatilization of ammonia from whatever source increases with increasing temperature, and is greatest from soils of high pH and low cation exchange capacity. On average, about 30% of the ammonia that is volatilized into the atmosphere is likely to be deposited on to soils or vegetation within 5 km of the source, but some is transported over much greater distances. Much of the ammonia not deposited close to the source reacts in the atmosphere with SO_2 and NO_x, and the resulting aerosols may travel up to 1000 km before deposition is complete (Van der Meer and Van der Putten, 1995).

In some situations, the leaching of nitrate is the major pathway for N loss from grassland soils. This is especially so with sandy soils and in areas where rainfall is high. Leaching is also enhanced when high rates of fertilizer N are applied and when sward management involves grazing, rather than cutting. Urine patches are probably the most important source of leached nitrate in grazed grassland (Clough *et al.*, 1996), while leaching from dung appears to be negligible (Hogg, 1981). However, in any grassland soil, the total amount of nitrate susceptible to leaching is influenced by the balance between the various inputs and the other outputs of inorganic N. Inputs include fertilizer application, the mineralization of organic sources, including urine, and atmospheric deposition, while the outputs include plant uptake, ammonia volatilization, denitrification and immobilization. In practice, for freely drained soils, a major influence is the rate of application of fertilizer N, as illustrated in Table 5.4. However, leaching tends to be rather less in young swards, where there is more immobilization than in old swards. In most temperate regions, leaching occurs mainly during the winter, and the greatest losses of N tend to occur after dry summers, presumably because dry weather restricts plant uptake and denitrification (Tyson *et al.*, 1997). Prolonged winter conditions may also enhance leaching by restricting the uptake of nitrate, and this has been suggested as a contributory factor in the relatively high leaching loss of about 25% of the N returned in urine in an area of north-eastern USA (Stout *et al.*, 1998).

Denitrification is the process by which nitrate is converted to the gaseous compounds NO, N_2O and N_2, which then diffuse into the atmosphere. The process is due mainly to bacterial enzymes, which catalyse the sequence of reactions:

$$NO_3 \rightarrow NO_2 \rightarrow NO \rightarrow N_2O \rightarrow N_2$$

The extent of denitrification depends on the amount of nitrate in the soil solution and on the extent to which the supply of oxygen to the soil microorganisms is restricted by anaerobic soil conditions. Factors that increase denitrification include wet and warm soil conditions, a plentiful supply of decomposable OM and near-neutral pH. Little denitrification occurs in soils in which the concentration of nitrate is consistently low, as in grassland soils receiving little or no fertilizer N and with few grazing animals. In more intensively managed grassland, high concentrations of nitrate inevitably occur from time to time, due to the application of fertilizer and/or the return of urine N, but denitrification occurs only when these coincide with anaerobic conditions. Most denitrification occurs during short periods when conditions are particularly favourable, usually beginning a few hours after

Table 5.4. Influence of drainage and of reseeding of old grassland grazed by cattle on the estimated amounts of nitrate N leached and denitrified (kg N ha^{-1} year^{-1}) at two rates of fertilizer N: mean values, 4 years (from Scholefield *et al.*, 1988).

	Nitrate N leached (kg ha^{-1})	Denitrification (kg ha^{-1})	Proportion (%) of leaching + denitrification as leaching
Old grassland, undrained, 200 kg N ha^{-1}	20	86	19
Old grassland, drained, 200 kg N ha^{-1}	56	56	50
Old grassland, undrained, 400 kg N ha^{-1}	48	110	30
Old grassland, drained, 400 kg N ha^{-1}	187	80	70
Reseeded grass, undrained, 400 kg N ha^{-1}	24	113	13
Reseeded grass, drained, 400 kg N ha^{-1}	74	78	49

the onset of rainfall, though some denitrification may occur over much longer periods in localized zones of the soil. Comparisons of the rate of denitrification from cut and grazed swards have shown the process to be greatly increased by the return of excreta. On poorly drained soils, nitrate accumulated during a summer grazing season is particularly vulnerable to denitrification in autumn. Often, the amounts of N lost by denitrification and leaching during the autumn/winter period are inversely related, with the ratio between the two loss pathways being determined by soil type and rainfall. Although the N_2 lost from soil is necessarily due to denitrification, NO and N_2O can be produced by both denitrification and nitrification, and these two processes can occur simultaneously in coexisting anaerobic and aerobic microsites. In well-drained soils, the volatilization of NO is usually ten to 100 times larger than that of N_2O, suggesting that NO results mainly from nitrification.

In grazed swards, each of the three main routes for the loss of N (ammonia volatilization, nitrate leaching and denitrification/nitrification) occurs mainly from the high concentrations of soluble N present in urine patches, and is enhanced when these overlap or when they receive fertilizer N. Management practices that promote as uniform a distribution as possible of urine patches, e.g. by strip grazing and the use of movable troughs for water and supplementary feeding, therefore tend to reduce losses. However, as ammonia volatilization, nitrate leaching and denitrification/nitrification are all interrelated, measures to curtail one loss pathway, for example by changing the form of fertilizer or changing the time of application of slurry, may well increase losses from another pathway (Jarvis, 1997).

Assessment of Plant-available N in Soils

If no fertilizer N is applied, if there is no appreciable biological fixation of N_2 and if atmospheric deposition is ignored, the amount of N becoming available during a growing season reflects the amount mineralized from the soil OM. With soils under all-grass swards, this can be assessed retrospectively from the amount of N contained in the herbage harvested at intervals during the growing season from unfertilized plots. When the supply is low, grass takes up virtually all the inorganic N available through its extensive root system and, in temperate regions with a substantial winter rainfall, there is usually little carry-over of inorganic N from one season to the next. Any ammonium present in the autumn is nitrified to nitrate and, in general, this is either denitrified or leached during the winter. However, in dry winters, some inorganic N may remain in the root zone and contribute to uptake in the following spring. In assessing plant-available N from the amount taken up, the amount of N present in roots is usually ignored as, with an established sward, the amount of root material shows little change from year to year and N released by decomposition is balanced by uptake for new root growth.

Assessments made on the basis of measuring N in grass herbage in several multicentre experiments carried out in the UK showed that plant-available N varied widely from 6 to about 250 kg N ha^{-1} year^{-1}, though in some instances the amount may have been augmented slightly by the presence of clover (Whitehead, 1995). Amounts of plant-available N greater than 100 kg ha^{-1} year^{-1} are generally associated with long-term grassland or with swards that have been ploughed and reseeded. While the retrospective assessment of plant-available N for a particular field is clearly of no predictive use for that season, it does enable a prediction to be made for the future, assuming that weather conditions are similar.

Although numerous attempts have been made to develop an effective method of soil analysis for assessing plant-available N, such attempts have had only limited success. This is partly because it is difficult to relate the amount of N dissolved by an extractant to the amount that is mineralizable over a period of time by the soil microorganisms. Another problem is that the amount of N that is mineralized and taken up is influenced by weather, particularly rainfall and temperature during the growing season.

The capacity of soils to release N by mineralization can also be assessed, though much more slowly, by measuring the amount of NH_4-N + NO_3-N produced during a period of incubation, either under uniform laboratory conditions or in the field. In either situation, a potential problem is that some of the nitrate may be lost through denitrification and, in some procedures in the field, by leaching. However, such losses can be prevented by carrying out the incubation either in the presence of acetylene, to inhibit nitrification, or in an atmosphere of 20% oxygen in helium, which has a similar effect (Jarvis *et al.*, 1995). Recent assessments of mineralization involving the incubation of soil cores under field conditions, with nitrification being inhibited, have shown that, with long-term grassland swards, there was much more mineralization when fertilizer N had been applied regularly in past years and when the soil was well drained rather than poorly drained (Jarvis *et al.*, 1995). Incubation procedures normally provide an assessment of mineralization that is net of immobilization, but values for gross mineralization and immobilization can be obtained by incubations incorporating the use of ^{15}N (Barraclough, 1995; Jarvis *et al.*, 1996a).

Uptake of N by Herbage Plants

Nitrogen is required by plants in larger amounts than any of the other nutrient elements. With the exception of the N acquired by legumes by symbiotic fixation, plants absorb almost all their N through the roots as nitrate and ammonium ions. Both ions are absorbed readily, but plants grown in soil usually take up more of their N as nitrate than as ammonium, partly as a result of nitrification and partly because nitrate is more mobile in the soil solution. However, the rate of nitrification is curtailed by soil acidity and by low temperatures and, under these conditions, much of the total N may be taken up in the form of ammonium. If the microbial population in the soil is increasing, there may be short-term competition for ammonium and nitrate between plants and the soil microorganisms, but much of the N immobilized by the microorganisms will be mineralized later when the microbial population declines. In addition to ammonium and nitrate, urea and amino acids can also be absorbed by plant roots, though much less readily, and, because they are converted rapidly to ammonium by soil microorganisms, there is little uptake of urea and amino acids as intact molecules.

In addition to uptake through the roots, plants can absorb some gaseous forms of N through the stomata in their leaves. Both ammonia and nitrogen dioxide can be absorbed, but, in most situations, absorption by the leaves contributes only a small proportion (< 5%) of the total N uptake.

When nitrate is taken up by plant roots, some is translocated to the shoots as nitrate and some is converted in the roots to amino acids and amides before translocation. When ammonium is absorbed, it is normally converted in the roots to amino acids and amides. The conversion of nitrate to ammonium in plant tissues occurs in two stages and requires an input of energy from photosynthesis.

$$NO_3 + 2e^- \rightarrow NO_2$$

$$NO_2 + 6e^- \rightarrow NH_4$$

As the first stage, the conversion of nitrate to nitrite, is the slower of the two, there is no accumulation of nitrite in the plant. Ammonium is also metabolized rapidly and is present in only small amounts. The main reaction of ammonium is with glutamate to form glutamine, which, in turn, reacts with oxoglutarate to form two molecules of glutamate via the glutamate synthase cycle (Fig. 5.1). One of the molecules of glutamate is then utilized for the synthesis of other amino acids and amides. If the total supply of N is less than that required to match the plant's potential rate of photosynthesis, the amino acids are converted rapidly into proteins and the formation of amides is negligible. However, if the supply of N exceeds the amount needed to match photosynthesis, some will accumulate as nitrate (if this form is taken up) and some will accumulate in the form of amides.

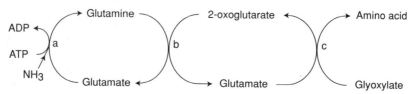

Fig. 5.1.   The glutamate synthase cycle (enzymes: a, glutamine synthetase; b, glutamate synthase; c, transaminase).

During the growth of grass leaves, the basal zone of the leaf, in which new cells are produced, may contain as much as 7.5% N, but the concentration decreases as the cells grow, and an increasing proportion of the N becomes incorporated into the proteins of photosynthetic enzymes (Gastal and Nelson, 1994). Young grass leaves as a whole may contain about 4.5% N. Some of the N absorbed by plants may be re-utilized several times during a growing season, partly as a result of photorespiration which releases ammonia from glycine, and partly due to the remobilization of proteins from senescent leaves (Haynes, 1986).

In legume species, the fixation of N_2 involves a symbiotic relationship between the legume and the rhizobial bacteria, in which the legume provides the rhizobia with a supply of energy through photosynthesis, and the rhizobia provide the legume with ammonia or a related compound synthesized by the nitrogenase enzymes. Under field conditions, all legume species show a wide variation in the amount of N_2 fixed (see p. 96). However, in situations favourable for fixation, the amount fixed may exceed the capacity of the plant to assimilate it fully; in white clover, for example, the concentration of N in the leaf material is often more than 3.5% and may be as high as 5.8% (Whitehead, 1995).

Functions of N in Plants

Most plant tissues contain between 1% and 5% N on a dry weight basis. This concentration is higher than that of other nutrient elements, with the possible exception of K, and reflects the need for N as a constituent of proteins, nucleic acids, chlorophyll and various minor plant constituents. A large proportion of the protein in plant tissues is enzyme material, of which about half is the photosynthetic enzyme, ribulose biphosphate carboxylase-oxygenase. Each naturally occurring protein involves a specific sequence of amino acids, and this sequence determines the physical and chemical properties of the protein. As many as 22 amino acids may contribute to each protein, and different proteins differ in their proportions of the individual amino acids (Fig. 5.2). However, in general, the proteins of grasses and legumes are similar to one another in amino acid composition (Lyttleton, 1973). Glutamic acid, aspartic acid and arginine are major amino acids in the proteins of both grasses and legumes, with lysine, alanine and glycine also being present in substantial amounts. The properties of the various proteins are determined, to a large extent, by the side chains of the amino acids, particularly those containing SH, NH_2, OH, COOH and/or aromatic groups.

The nucleic acids are polymers composed of a series of nucleotides, each consisting of a nucleoside linked to a phosphate group. Nitrogen is present in nucleosides in the form of either pyrimidine or purine, attached to ribose, which, in the nucleotides, is linked to phosphate (see p. 135).

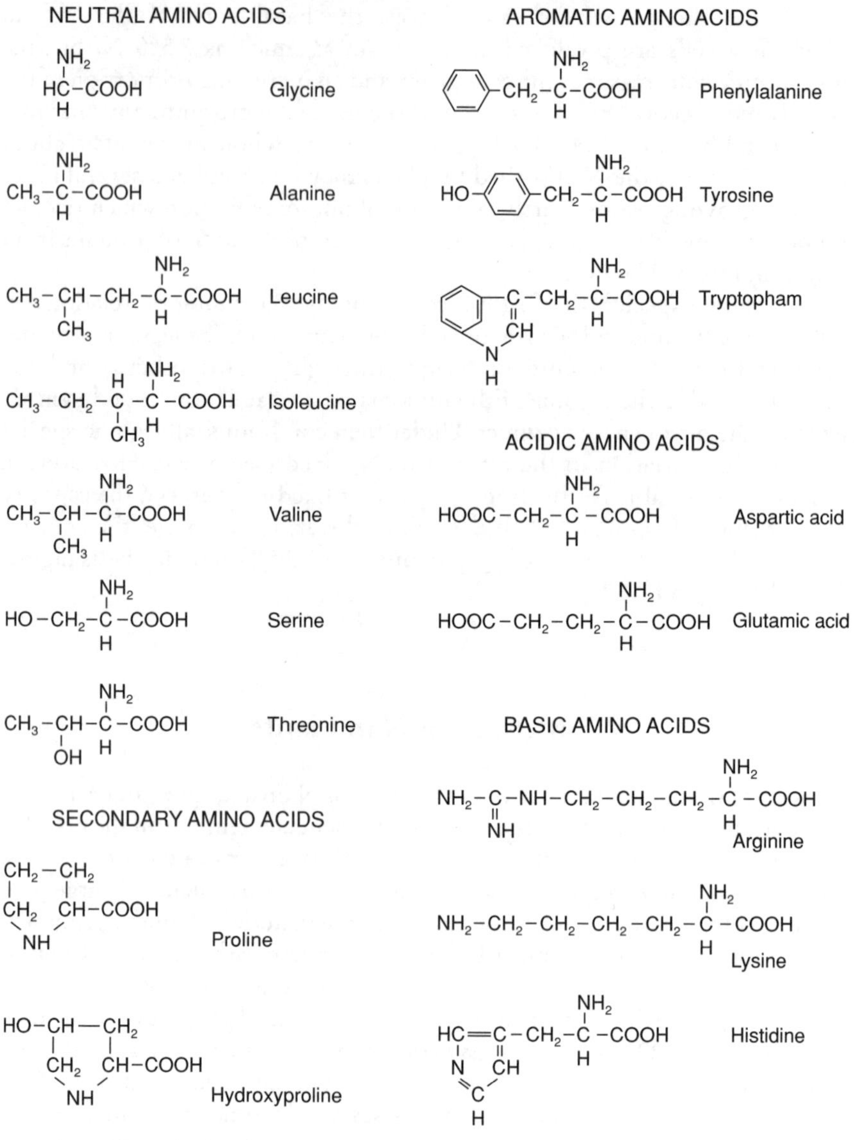

NEUTRAL AMINO ACIDS   AROMATIC AMINO ACIDS

$$\underset{H}{\overset{NH_2}{H\overset{|}{C}-COOH}}$$ Glycine

$$\underset{H}{\overset{NH_2}{CH_3-\overset{|}{C}-COOH}}$$ Alanine

$$CH_3-CH-CH_2-\underset{H}{\overset{NH_2}{\overset{|}{C}}-COOH}$$ Leucine
$$\qquad\overset{|}{CH_3}$$

$$CH_3-CH_2-\overset{H}{\underset{\overset{|}{CH_3}}{C}}-\underset{H}{\overset{NH_2}{\overset{|}{C}}-COOH}$$ Isoleucine

$$CH_3-CH-\underset{H}{\overset{NH_2}{\overset{|}{C}-COOH}}$$ Valine
$$\quad\overset{|}{CH_3}$$

$$HO-CH_2-\underset{H}{\overset{NH_2}{\overset{|}{C}}-COOH}$$ Serine

$$CH_3-CH-\underset{H}{\overset{NH_2}{\overset{|}{C}}-COOH}$$ Threonine
$$\quad\overset{|}{OH}$$

SECONDARY AMINO ACIDS

$$\begin{array}{l} CH_2-CH_2 \\ | \qquad | \\ CH_2 \quad CH-COOH \\ \;\;\diagdown NH \diagup \end{array}$$ Proline

$$\begin{array}{l} HO-CH-CH_2 \\ \quad\;\; | \qquad\;\; | \\ \quad\;\; CH_2 \quad CH-COOH \\ \quad\;\;\diagdown NH \diagup \end{array}$$ Hydroxyproline

Phenylalanine
$$\text{⬡}-CH_2-\underset{H}{\overset{NH_2}{\overset{|}{C}-COOH}}$$ Phenylalanine

$$HO-\text{⬡}-CH_2-\underset{H}{\overset{NH_2}{\overset{|}{C}-COOH}}$$ Tyrosine

Tryptopham

ACIDIC AMINO ACIDS

$$HOOC-CH_2-\underset{H}{\overset{NH_2}{\overset{|}{C}}-COOH}$$ Aspartic acid

$$HOOC-CH_2-CH_2-\underset{H}{\overset{NH_2}{\overset{|}{C}}-COOH}$$ Glutamic acid

BASIC AMINO ACIDS

$$NH_2-\underset{NH}{\overset{||}{C}}-NH-CH_2-CH_2-CH_2-\underset{H}{\overset{NH_2}{\overset{|}{C}}-COOH}$$ Arginine

$$NH_2-CH_2-CH_2-CH_2-CH_2-\underset{H}{\overset{NH_2}{\overset{|}{C}}-COOH}$$ Lysine

Histidine

Fig. 5.2. Chemical structure of some amino acids (from Stevenson, 1994). (For amino acids containing S, see Fig. 7.1, p. 165.)

Influence of Fertilizer N on Herbage Yield and on Milk Production

The response of grass swards to fertilizer N, applied at a range of different rates, has been examined in numerous field trials, discussed in more detail elsewhere (Whitehead, 1995). In general, grass swards fertilized and cut at intervals during the growing season show large increases in the yield of herbage with increasing rate of fertilizer N, as illustrated in Fig. 5.3. When no fertilizer is applied (and with little or no clover in the sward), the amount of harvestable herbage in temperate lowland regions is usually between 1 and 5 tonnes DM ha^{-1} year^{-1}, though yields are less in upland areas. The actual yield at a particular location depends mainly on the supply of N from the soil and on weather conditions. As the rate of fertilizer N is increased, there is often an almost linear increase in yield, amounting to between 20 and 30 kg DM kg^{-1} N, until the rate of application is at some point between 250 and 400 kg N ha^{-1} year^{-1}, after which the response declines. The actual response during the linear phase depends on soil and weather factors, with water-supply from soil plus rainfall during the growing season being particularly important. The response will also be restricted when another nutrient, especially P, S or K, is deficient (Whitehead, 1995; Brown *et al.*, 1999). At even higher rates of fertilizer N, the response per kilogram of additional fertilizer N decreases until the maximum yield is attained, at which point the response to additional fertilizer N is zero. Eventually, at excessively high

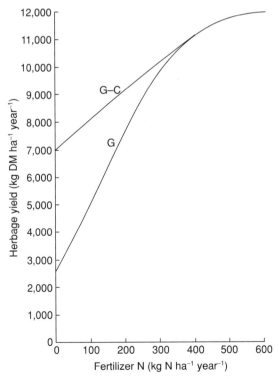

Fig. 5.3. Generalized response to fertilizer N of an all-grass sward (G) and a grass–clover sward (G–C) in conditions similar to those of lowland UK (based on data in Whitehead, 1995).

rates of application, there is an actual decrease in yield, i.e. the response becomes negative.

When fertilizer N is applied to a grass–clover sward, N_2 fixation by the clover is reduced and the yield response per unit of fertilizer N is therefore less than with a comparable all-grass sward. Grass–clover swards are more variable than all-grass swards in their response, because the growth of clover varies more widely than that of grass from site to site and from year to year. When the proportion of clover is small or conditions are unfavourable for clover growth, the herbage yield and response to fertilizer N are similar to those of an all-grass sward. On the other hand, when there is a large proportion of clover in the sward and the herbage yield without fertilizer N is relatively high, the response to fertilizer N is inevitably much smaller than that of an all-grass sward (Fig. 5.3).

When swards are grazed, a large proportion of the N consumed in the herbage is returned in excreta, enabling some of the N derived from fertilizer to be utilized more than once during a growing season. Taking this into account, grazed swards might be expected to show a greater response than cut swards to fertilizer N. However, the return of excreta often entails large losses of N, through the volatilization of ammonia, denitrification and leaching, and grazing animals also have adverse effects on the sward through treading, selective grazing and the contamination of herbage by dung. These adverse effects tend to offset the positive effect of the return of nutrients, though their impact is influenced by soil and weather conditions, and the net effect of grazing animals on the response to fertilizer N is often small.

Intensively managed grassland, except where clover grows vigorously, generally receives routine applications of moderate to high rates of fertilizer N in order to ensure that yield is not seriously limited by the supply of N. However, such routine applications almost inevitably involve substantial gaseous or leaching losses of N, either directly from grazed swards and/or from slurry. As these losses represent an inefficient use of fertilizer N and also have detrimental effects on the wider environment (see p. 11), various methods have been examined for adjusting the rate of fertilizer application more closely to the actual need in a particular situation. One method is to assess the optimum rate of application on the basis of soil texture, soil depth and summer rainfall, and to take into account the effects of previous cropping history and of N recycled during grazing (Baker et al., 1991). Another method is to adjust the rate of fertilizer application to the concentration of soil inorganic N, determined at appropriate intervals throughout the growing season. In a comparison involving six farms in the UK, in which fields were divided into two halves, with one half receiving fertilizer on the basis of soil analysis and the other half receiving fertilizer according to the farm's normal procedure, the yield of herbage did not differ between the two approaches but, on average, the 'tactical' approach showed a reduction in fertilizer input from 300 to 211 kg N ha^{-1} (Wilkins, 1995). A similar approach, in which the rate of fertilizer application is decided on the basis of the amount of N made available in the soil from non-fertilizer sources, has been developed in The Netherlands (Hassink, 1995). This entails estimating the initial content of inorganic N in the soil, plus the amounts likely to be provided by mineralization, deposition from the atmosphere and biological fixation.

The increase in herbage yield due to fertilizer N is normally reflected in an increased production by animals of either milk or live-weight gain. With intensive management, the response in terms of milk production may well be about 16 kg milk kg^{-1} fertilizer N, and this arises not from any significant increase in yield per cow but from an increase in stocking rate, i.e. the number of animals grazed per hectare (Jarvis, 1998).

Distribution of N within Herbage Plants

In grasses, the concentration of N in the roots is often about half that in the herbage. When little or no fertilizer N is applied, the concentration in grass roots is usually about 1.0% or less (Power, 1968, 1981; Whitehead, 1970), though a range of 0.5% to 1.8% N was reported for five grass species grown without fertilizer N in Minnesota, USA (Wedin and Tilman, 1990). In another study in the USA, the concentration of N in the roots of several grass species was highest, at about 1.0–1.2%, in roots near to the soil surface, decreased to a minimum value of about 0.6% at a depth between 20 and 50 cm, and then increased below 50 cm: this pattern was thought to reflect differences in the ratio of young to old roots (Power, 1986). The concentration of N in grass roots increases with increasing rate of fertilizer N, to a maximum of about 2.2%, and tends to be greater in grazed than in cut swards (Whitehead, 1995). Legumes that are actively fixing N$_2$ have a high concentration of N in their root nodules (4.5–9.0% N), and the overall root concentration is therefore higher in legumes than in grasses. The concentration of N in the roots of white clover is generally between 2 and 4%, while concentrations in lucerne and red clover are rather lower. With tap roots, such as those of lucerne and, to a lesser extent, red clover, there is a seasonal variation in N concentration, with N tending to accumulate during late autumn and winter and being subsequently depleted when herbage growth begins in spring (Volenec and Nelson, 1995). Defoliation also causes a remobilization of N from roots to new leaf growth in both grasses and legumes.

Within the shoots of grasses, the concentration of N is much higher in the leaf laminae than in the leaf sheath and stem, and is higher in young leaves than in older leaves (Simpson and Stobbs, 1981; Parsons *et al.*, 1991). The mean concentration of N in three progressively older leaf laminae of individual tillers of ryegrass was 4.1%, 3.8% and 3.4%, while the mean concentration in the leaf sheath was 1.4% and that in the stem was 1.5% N (Parsons *et al.*, 1991). As leaves senesce, N is remobilized to young leaves, and this process may account for 70–75% of the maximum amount of N contained in the older leaves (Lemaire and Chapman, 1996).

Within the shoots of legumes, the concentration of N is also highest in the leaf laminae. Mean values for white clover from a grazed grass–clover sward were 5.6% N in the laminae, 2.9% N in the petioles and 2.7% N in the stolons (Parsons *et al.*, 1991).

Ranges of Concentration and Critical Concentrations of N in Herbage Plants

Concentrations of N in the harvested herbage of grasses and legumes are usually between 1.0 and 5.0% (see Table 3.4, p. 54). The actual concentration is influenced by three main factors: (i) the supply of available N from the soil, which is often governed by the rate of fertilizer application; (ii) differences between grasses and legumes; and (iii) the stage of maturity of the herbage. The effects of these factors are discussed in more detail below.

The range in the concentration of N in herbage from a multicentre trial in the UK, involving ryegrass-only swards, receiving fertilizer N at rates between nil and 450 kg ha^{-1} year^{-1} and harvested six times per year over 4 years, was 1.1–4.6% (Table 5.5). The concentration of N in white clover is usually > 2.5% and occasionally exceeds 5.0% (Whitehead and Jones, 1969; Wilman and Hollington, 1985).

The critical concentration of N in herbage material is very dependent on the stage of maturity, and any proposed critical concentration must be related to a specific portion of the plant and its age at the time of sampling. For grasses, it is usual to select either the total herbage cut at a specified height after a certain period of regrowth or individual leaves at a certain stage of growth. The critical concentration in grass, assessed with 4-week old regrowth, cut at a height of 2.5 cm, is about 3.2–3.5% (De Wit *et al.*, 1963; Smith *et al.*, 1985); with 6-week old regrowth, it is about 2.5%. At the critical concentration, almost all the N is in organic forms and little nitrate is present. However, when nitrate is the main source of N, the concentration of nitrate N in the herbage, though low, can be used to indicate whether the supply is adequate. Estimates of the critical concentration of nitrate N in young grass leaves have ranged from about 0.05% to 0.15% NO_3-N (Van Burg, 1966; Reid and Strachan, 1974; Smith *et al.*, 1985).

Table 5.5. Concentrations of N in the herbage of ryegrass (% in DM) as influenced by the rate of fertilizer N at 20 sites in the UK: range and mean values for six successive cuts (Morrison *et al.*, 1980).

Rate of fertilizer N	Herbage N concentration (range)	Herbage N concentration (mean)
0	1.08–3.34	2.2
150	1.53–3.70	2.5
300	1.80–4.19	3.0
450	2.14–4.57	3.4

When legumes are dependent on N_2 fixation, their growth may be limited, either because fixation is insufficient to match potential photosynthesis or because photosynthesis is insufficient to match potential fixation. When N_2 fixation itself is restricted, the concentration of N in the plant tissue is likely to be less than a critical value, whereas, when photosynthesis is restricted (e.g. by light or nutrient supply), the concentration of N is likely to be higher than the critical value. With white clover, the critical concentration of N in the leaf + petiole material after 4 weeks' regrowth is thought to be about 3.5% (Whitehead, 1982) and in 15-week-old whole plants (including roots) about 3.2% (Haydock and Norris, 1966).

Influence of N Supply from Soil and Fertilizer on Concentrations of N in Herbage

In the absence of fertilizer N, the supply of N from the soil (plus deposition from the atmosphere) might be expected to exert a major influence on the concentration of N in grass herbage, but often this is not so. Variations in the supply of N from the soil, though influencing yield, often have little effect on herbage N concentration. Similarly, the application of low rates of fertilizer N, while usually increasing growth, may also have little effect on the concentration of herbage N. However, as the rate of fertilizer N is increased, both herbage yield and the concentration of N increase, and the concentration continues to increase even after the maximum yield is attained. An example of the effect of increasing rate of fertilizer N on the concentration of N in ryegrass herbage is shown in Table 5.5. Fertilizer N usually has little, if any, effect on the concentration of N in clovers (Whitehead *et al.*, 1983; Simpson *et al.*, 1988).

Influence of Plant Species and Variety on Concentrations of N in Herbage

Unless they receive a large input of fertilizer N, grasses are normally lower than legumes in N concentration. The concentration in young grass herbage ranges from about 1% without fertilizer N to about 4.5% when large amounts of fertilizer N are applied, whereas the concentration in white clover is usually between 2.5% and 5.5% N. Other legume species are usually lower than white clover in N concentration, but higher than grasses grown without fertilizer N but at a similar stage of maturity. When grass and white clover are grown together in mixed swards without fertilizer N, the concentration of N in the clover is consistently higher than that in the grass (Cowling and Lockyer, 1967; Metson and Saunders, 1978b; Whitehead *et al.*, 1983; Wilman and Hollington, 1985). However, grass alone that receives a high rate of fertilizer N may have a higher concentration than clover not receiving fertilizer N (Cowling and Lockyer, 1967; Wilman and Hollington, 1985).

Differences in herbage N concentration between grass species are usually small, at least when comparisons are made in regularly cut swards (Table 5.6). However, when comparisons are made at a particular date during primary growth in spring, there may be differences between species, and even varieties within a species, due to differences in the rate at which the plant matures. For example, the concentration of N at any particular date during primary growth is higher in the late-flowering S23 ryegrass than in the earlier-flowering S24 ryegrass or S37 cocksfoot (Minson, 1990). However, comparisons at the same growth stage are usually influenced by differences in the time interval since the application of fertilizer or the onset of growth in spring. Comparisons made on this basis showed little difference between perennial ryegrass, cocksfoot, timothy and smooth-stalked meadow-grass (Fleming and Coulter, 1963). In a comparison of seven perennial grass species, herbage concentrations of N were generally lowest in the higher-yielding species: when no fertilizer N was applied, concentrations were in the range 0.9–1.4% N and, when fertilizer was applied at 225 kg N ha^{-1}, the range was 1.5–2.2% N (Power, 1986). There is little information on varietal differences in herbage N concentration, but the difference between two varieties of each of three species (brome-grass, cocksfoot and timothy) was often more than 10% of the lower value, when comparisons were made at the same stage of maturity, though the difference was smaller when comparisons were made on the same dates (Fulkerson et al., 1967). There were only small differences in average N concentration amongst eight varieties of perennial ryegrass harvested on eight successive occasions during the growing season though, in general, the late-maturing varieties had higher concentrations than did the varieties that matured early (Wilkins et al., 1999).

In the absence of fertilizer N, the presence of a legume in a grass–legume mixture may well increase the concentration of N in the grass component, in comparison with grass grown alone. For example, with timothy–lucerne mixtures under various different managements, there was a 15–40% increase in the concentration of N in the timothy (Ta and Faris, 1987).

Table 5.6. Mean concentrations of N (% in DM) in the herbage of five grass species as reported in three investigations ((i) Fleming, 1963; (ii) Waite, 1965; and (iii) J.W. Dent, 1967 (personal communication – data from the National Institute of Agricultural Botany, Cambridge)).

	(i)	(ii)	(iii)
Perennial ryegrass	2.1	2.07	3.05
Italian ryegrass	–	3.08	2.82
Timothy	2.5	2.52	3.39
Cocksfoot	2.8	3.24	3.26
Meadow fescue	2.6	3.04	3.33

Table 5.7. Mean concentrations of N (% in DM) in the herbage of four legume species as reported in four investigations ((i) Van Riper and Smith, 1959; (ii) Davies *et al.*, 1966; (iii) Whitehead and Jones, 1969; (iv) Baker and Reid, 1977).

	(i)	(ii)	(iii)	(iv)
White clover	3.94	3.50	4.42	3.9
Red clover	3.34	2.93	3.40	3.2
Lucerne	3.47	2.82	2.94	4.0
Sainfoin	–	–	2.87	–

Comparative data for the herbage N concentration of white clover, red clover, lucerne and sainfoin (Table 5.7) show the concentration of N to be considerably higher in white clover than in the other species. This probably reflects a higher proportion of leaf material and a lower proportion of stem in the harvested herbage of white clover.

Influence of Stage of Maturity on Concentrations of N in Herbage

In grasses, the concentration of N in the herbage declines markedly with increasing maturity, due mainly to the relative increase in cell wall material and the relative decrease in cytoplasm. The decline is most apparent in primary growth sampled at intervals from early spring to the post-flowering stage, as illustrated in Fig. 5.4. Dead herbage may contain as little as 0.8% N (see Table 3.11).

In legumes, the decline is less marked than in grasses and shows more variation between species. It occurs to a greater extent in lucerne, red clover and sainfoin than in white clover which may show a decline only from about 5% to about 3.5% N (Whitehead and Jones, 1969).

Influence of Season of the Year and Weather Factors on Concentrations of N in Herbage

When grass is defoliated at intervals during the growing season, changes in N concentration are less marked than those during prolonged primary growth. However, with grass that is cut or grazed regularly, the concentration in herbage is usually lower in summer than in spring and autumn. In a multi-centre trial at 20 sites in the UK, in which ryegrass swards were cut six times during the season and fertilizer N was applied at various rates at the beginning of the season, and then after each cut, the concentration of N in the herbage usually decreased from the first harvest in May to the second harvest in June, and then increased steadily to reach a maximum at the sixth

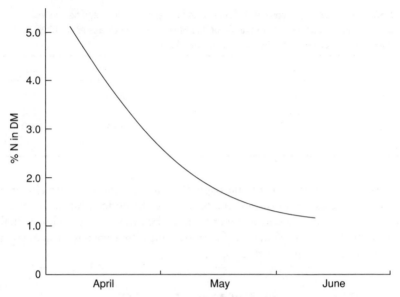

Fig. 5.4. Typical change in the concentration of N with advancing maturity in the herbage of perennial ryegrass (based on data in Whitehead, 1995).

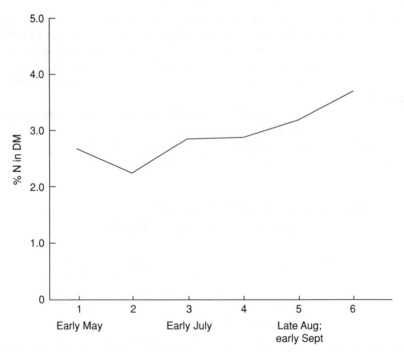

Fig. 5.5. Concentration of N in successive harvests of perennial ryegrass receiving fertilizer N in early spring, and subsequently after each harvest, at a rate equivalent to 300 kg N ha^{-1} year^{-1}: mean values from 21 sites in the UK (from Morrison *et al.*, 1980).

harvest in October (Fig. 5.5). A marked reduction in N concentration during the summer was also observed in New Zealand, with both grass and clover components of mixed swards, and this was followed by a relatively high concentration of N during late autumn to early spring (Metson and Saunders, 1978b). Concentrations of N in the grass component were generally in the range 2.5–5.0% and in the clover component 4.0–5.8%.

The ambient temperature sometimes has an effect on the concentration of N in grass herbage. Thus, increasing day/night temperature from 15/10°C to 25/20°C tended to reduce the concentration of N, in experiments with fertilized ryegrass grown in pots (Deinum, 1966). However, in other experiments, increasing temperature has resulted in increases in herbage N (Minson, 1990), and sometimes the effect of temperature varies with the supply of N (Nowakowski et al., 1965). The effect of soil water status is also inconsistent, but long-term drought appears to reduce the concentration of N in herbage (Minson, 1990). Light intensity may also have an effect, with greater light intensity tending to reduce the concentration of N in herbage (Deinum, 1966; Nowakowski and Cunningham, 1966).

White clover shows little seasonal change with regular defoliation. Thus, when sampled at monthly intervals over a 6-month period in New Zealand, white clover contained 4.2–5.2% N, with no particular trend (Bailey, 1964).

Influence of Fertilizer P, K and S and Lime on Concentrations of N in Herbage

The application of fertilizer P or K usually has little or no effect on the concentration of N in grass herbage, though, when fertilizer N is also applied, the application of K may reduce herbage N (Reith et al., 1964; Laughlin et al., 1973; Banwart and Pierre, 1975). With lucerne, high rates of fertilizer K tended to reduce herbage N concentration (Smith, 1975). In a study in New Zealand, the application of fertilizer P had no appreciable effect on the concentration of N in white clover, while the application of fertilizer S caused a small increase, despite the initial concentration being > 4% N (Sinclair et al., 1996b). Superphosphate increased the concentration of N in both ryegrass and subterranean clover, but whether this was due to the P, S or Ca component was uncertain (Saul et al., 1999). Liming an acid soil (pH 5.4) increased the average concentration of N in white clover from 2.6% to 3.2% (Bailey and Laidlaw, 1999).

Influence of Livestock Excreta on Concentrations of N in Herbage

The addition of urine substantially increases the concentration of N in grass herbage (Joblin and Keogh, 1979; Joblin, 1981; Ledgard et al., 1982; Williams and Haynes, 1994), though not in the clover component of grass–clover swards with active N_2 fixation (Ledgard *et al.*, 1982). Urine also increases the concentration of N in grass roots (Day and Detling, 1990). However, the uptake of N is reduced if the grass is scorched by the urine addition (see p. 28).

The effect of dung occurs more slowly, but the herbage growing close to or through the patch several weeks after the deposition may have an increased concentration of N (Weeda, 1977; Williams and Haynes, 1995).

Chemical Forms and Availability of Herbage N to Ruminant Animals

In the context of the nutritional value of herbage for livestock, the concentration of total N is often expressed as 'crude protein', which is the concentration of N multiplied by 6.25. This factor is derived from the average concentration of N in plant proteins, namely 16%. However, because it is based on total N, the 'crude protein' includes both genuine protein and other nitrogenous constituents. Usually, between 70% and 90% of the total N in grass and legume herbage is present as genuine protein and between 10% and 30% is present in other forms, which include amino acids, peptides, amides and sometimes nitrate (Van Straalen and Tamminga, 1990). Neither the stage of growth nor the application of fertilizer N has any appreciable effect on the amino acid composition of the protein (Lyttleton, 1973; Reid and Strachan, 1974). Even severe deficiencies of N or S had no effect on the amino acid composition of the protein in perennial ryegrass (Bolton *et al.*, 1976).

When non-protein N is present in appreciable amounts, it represents an accumulation of N that is surplus to immediate requirements. The main forms of non-protein N are amides, particularly glutamine and asparagine, and nitrate. The amount of ammoniacal N in herbage is usually negligible. An accumulation of non-protein N occurs most commonly when growth is restricted, for example by a deficiency of K (Griffith *et al.*, 1964) or S (Adams and Sheard, 1966; Millard *et al.*, 1985) or by low light intensity (Deinum and Sibma, 1980). Defoliation, especially when followed by the application of fertilizer N, may also result in temporarily high concentrations of nitrate in the regrowth (Deinum and Sibma, 1980), depending on the rate of fertilizer applied and on the time interval between application and sampling. High rates of fertilizer N may induce an accumulation of non-protein N, even without defoliation (Goswami and Willcox, 1969), and nitrate may account for more than 5% of the total N, especially within 3–4 weeks of fertilizer application. Occasionally, in young plants fertilized with N, nitrate may account for more than half the total N (Nowakowski *et al.*, 1965). However, the concentration of nitrate in herbage depends on the time interval between application and sampling. With single applications of 80–140 kg N ha^{-1} as calcium ammonium nitrate in spring, the maximum concentration of nitrate in the herbage occurred about 2 weeks later (Wilman, 1965). The concentration of nitrate is also influenced by the form of fertilizer N, with nitrate producing higher concentrations than ammonium (Nowakowski, 1961). There are also some differences between species in the tendency to accumulate nitrate, with cocksfoot and tall fescue accumulating more than timothy and brome-grass (Murphy and Smith, 1967).

Concentrations of nitrate in legumes are normally lower than in grasses, and are often negligible. However, ladino clover was found to contain more than 0.3% NO_3-N when receiving fertilizer N at a rate of > 300 kg N ha^{-1} in spring (Murphy and Smith, 1967).

Metabolic Functions of N in Ruminant Animals

Most of the N in animal tissues is present in the proteins, which provide the basic structures of muscles and enzymes. Proteins are also important components of milk and wool. In addition, N is a constituent of nucleic acids and of various metabolically important compounds, including some vitamins and hormones. However, in quantitative terms, N is needed mainly for the synthesis of proteins, and the dietary requirement is therefore greatest during periods of rapid growth and during pregnancy and lactation. Milk contains about 0.6 g N l^{-1}, and a dairy cow producing 1000 l milk year^{-1} will therefore secrete about 600 g N year^{-1} in this form. Muscle, milk and wool differ in their proportions of the various amino acids, with muscle protein containing almost twice as much arginine as does milk protein, and wool protein containing relatively large proportions of the S amino acids, particularly cystine (Minson, 1990).

Absorption of Dietary N by Ruminant Animals

Some of the protein in the diet of ruminant animals is hydrolysed in the first stage of digestion by the microbial population of the rumen (see p. 70), and some remains intact until it reaches the small intestine. The hydrolysis of protein by protease and peptidase enzymes in the rumen results in peptides and amino acids, which are then subject to deamination and decarboxylation, with the release of ammonia (Van Straalen and Tamminga, 1990). Normally, more than half the protein is hydrolysed in the rumen, and most of the resulting peptides and amino acids are deaminated (Ørskov, 1992). Much of the ammonia, together with some free amino acid, is then assimilated into microbial protein in the rumen. Ammonia that is not assimilated diffuses through the rumen wall and is transported in the bloodstream to the liver, where it is converted to urea. Some of the urea is then returned to the rumen in saliva, or via the bloodstream, but most is removed from the blood by the kidneys and is excreted in the urine. In the second stage of digestion, the residual dietary protein and the microbial protein synthesized in the rumen are both subject to hydrolysis in the small intestine, with the release of small peptides and amino acids, which are absorbed into the bloodstream (Webb and Bergman, 1991). The amino acids and peptides are then utilized for the synthesis of proteins in body tissues or milk. Undigested protein passes into the large intestine, where there is a small amount of further digestion, but most is excreted in the faeces.

The maximum utilization of the dietary N occurs when the ratio between available nitrogen and available energy is close to the optimum for the rate of live-weight gain and/or the amount of milk or wool being produced. With young grass herbage, the ratio between available nitrogen and energy is often higher than the optimum and, as a result, excess protein is used as a source of energy and a relatively high proportion of the ammonia released in the rumen is excreted as urea. On the other hand, with mature grass herbage, especially that receiving little or no fertilizer N, the ratio of available N to energy may be too low for the available energy to be utilized completely and, in this situation, live-weight gain and the production of milk or wool may be restricted. Animal productivity may then benefit from the provision of a protein-rich supplement, such as soybean meal, to the diet.

Even with an optimal ratio between N and energy, the proportion of the dietary N actually converted into protein by the animal is generally no more than 25%. The conversion is greater for milk production than for live-weight gain or wool, and is usually in the range 16–23% for dairy cows (Van Vuuren and Meijs, 1987), 5–10% for beef cattle and 3–15% for sheep and lambs (Henzell and Ross, 1973). Increases in the utilization of dietary N can sometimes be achieved by measures that involve slowing down the rate of hydrolysis of N compounds in the rumen and/or increasing the incorporation of N into microbial biomass in the rumen by providing an additional source of metabolizable energy (Wilkins, 1995).

Of the non-protein forms of N in herbage, amide N can be utilized effectively by the rumen microorganisms, and hence by the host animal, but nitrate is poorly utilized. High concentrations of nitrate N in herbage may actually be harmful to livestock. Although the nitrate ion itself is relatively non-toxic, it is converted in the digestive tract to nitrite, which is much more toxic. When nitrite is absorbed into the blood, it converts haemoglobin to methaemoglobin and, as this lacks the ability to transport oxygen to the tissues, the animal suffers from a lack of oxygen. However, nitrite is absorbed only when it is formed more quickly than it is converted to ammonia, and differences in the rates of these two processes may well account for the conflicting results that have been reported for the effects of moderate concentrations of nitrate in the diet. Although forages containing more than 0.34% nitrate N are often regarded as potentially toxic (Miller, 1979), several studies have shown herbage containing about 0.7% nitrate N to have no harmful effects on sheep or dairy cows (e.g. Phipps, 1975; Dickson and Macpherson,

1976). It is possible that animals adapt to some extent to high-nitrate diets (Miller, 1979).

Nutritional Requirements for N in Ruminant Animals

The requirement for N is greatly influenced by the physiological state of the animal, being highest in lactating animals and in those that are young and actively growing. The requirement is somewhat less in pregnant animals and those growing at a moderate rate, and is least in animals that are neither lactating, nor pregnant, nor growing.

The assessment of the N requirement of ruminant animals is often based on a factorial calculation, involving the current body weight and the needs for the anticipated live-weight gain, milk production, etc. (see p. 83). The requirement is then expressed in terms of 'metabolizable protein', which takes into account, for any type of feed, the degradability of the protein in the rumen and the extent to which it provides for microbial protein synthesis, together with the digestibility, in the second stage of digestion, of the dietary protein that is undigested in the rumen (Agricultural and Food Research Council, 1992; Webster, 1996). This type of assessment is particularly useful when livestock diets comprise a number of different components, such as silage, barley, beet pulp, rapeseed meal, etc., as it enables the components to be combined in the optimum proportions. However, when the diet consists entirely of grassland herbage, adequate amounts of protein are normally provided when the concentration of N is at least 1.6% (10% protein) for non-productive animals and is in the range 1.9–3.0% N (12–19% protein), depending on milk yield, for lactating animals (Table 5.8).

Excretion of N by Ruminant Animals

Dietary N that is not actually utilized by the animal for the synthesis of body tissue or milk is excreted, and this proportion is normally 75–80% for dairy cattle, 90–95% for beef cattle and 85–95% for sheep. A 250 kg steer consuming, per day, 6 kg herbage containing 3% N, and gaining 0.8 kg, will ingest

Table 5.8. Recommended dietary concentrations of N for dairy cattle, as influenced by physiological state (from National Research Council, Subcommittee, 1989).

	% N in diet	% protein in diet
Maintenance only	1.6	10
During pregnancy	1.9	12
During lactation	1.9–3.0	12–19
During active growth of young animals	1.9–2.6	12–16

180 g N day^{-1}, retain about 20 g in the live-weight gain and excrete the remainder (Follett and Wilkinson, 1995). The distribution of the excreted N between dung and urine depends mainly on the concentration of N in the diet. Excretion in the dung of both cattle and sheep is approximately 8 g N kg^{-1} DM consumed, though it may increase slightly with increasing N concentration in the diet (Blaxter et al., 1971). Because the excretion of N in the dung is fairly constant per unit of DM consumed, changes in the concentration of dietary N are reflected mainly in the amount of N excreted in the urine (Barrow and Lambourne, 1962; Betteridge et al., 1986; Lantinga et al., 1987). In general, the proportion of the excreted N that is present in the urine increases from about 45% when the diet contains 1.5% N to about 80% when the diet contains 4.0% N.

The concentration of N in the faeces of cattle and sheep fed on fertilized grassland herbage is usually in the range 1.2–4.0% of DM, and typically about 2.4% (Van Faassen and van Dijk, 1987). The range on a fresh weight basis is about 0.2–0.5% N. Typically, the C : N ratio of faeces is about 20 : 1. Most of the N in faeces consists of non-absorbed dietary N, including microbial material synthesized in the rumen, but some is endogenous and results from enzymes and mucus secreted into the digestive tract, and from the release of epithelial cells. Most of the N in faeces is organic and most is insoluble in water, with about 45–65% being in proteins and related compounds, about 5% in nucleic acids, 3% in ammonia and the remainder consisting of partially degraded nucleic acids, bacterial cell walls and N bound to fibre (National Research Council, Subcommittee, 1985). With sheep, about 25% of the faecal N is generally present as water-soluble products of animal and microbial metabolism, about 10–20% is undigested dietary N and the remainder, about 60%, is thought to be present in bacterial cells and their residues (Mason et al., 1981).

Urine normally contains 4–12% of dissolved solid material, much of which consists of nitrogenous compounds. The concentration of N in urine varies with the N content of the diet and with the amount of water consumed, but reported ranges of concentration agree quite closely. For both cattle and sheep, the concentration in urine is usually between 2 and 20 g N l^{-1}, with an average of 8–10 g l^{-1}. Urea generally accounts for between 60 and 90% of the total urine N, though the proportion may be as low as 25% with sheep fed on extremely low-protein diets. Other nitrogenous compounds, which usually comprise 10–40% of the N in urine, include hippuric acid, allantoin, uric acid, xanthine, hypoxanthine, creatine and creatinine (Whitehead, 1995). The C : N ratio of urine will depend on its composition, but is likely to be within the range 2–5 : 1.

The total amounts of N excreted, on a per hectare basis, by grazing dairy cows and beef cattle, under relatively intensive management, are shown in Table 5.9.

When the excreta of housed livestock are mixed and stored as slurry, the various nitrogenous constituents are subject to hydrolysis by microbial

Table 5.9. Return of N to grazed swards in faeces and urine (kg N ha^{-1} year^{-1}) of grazing dairy cows and beef steers (from Jarvis *et al.*, 1995).

	Fertilizer N (kg ha^{-1} year^{-1})	Herbage N (% in DM)	N in dung (kg ha^{-1})	N in urine (kg ha^{-1})	Urine N as % of N excreted
Cows	250	3.3	86	214	71
	540	4.1	104	354	77
Steers	0 (grass–clover)	2.8	58	74	56
	210	3.1	62	93	60
	420	3.7	84	237	74

enzymes. The urea from urine is hydrolysed rapidly in slurry, with the release of ammonia, and the other nitrogenous constituents also decompose, though more slowly. Usually, between 60 and 75% of the N incorporated into slurry is converted to ammonia, but often between 25 and 40% of this is lost by volatilization during storage. When the slurry is applied in the field, there is a further substantial loss of N through the volatilization of ammonia, unless it is treated to reduce volatilization or is injected below the surface of the soil (Whitehead, 1995). There may also be an appreciable loss by denitrification, and this loss is likely to be increased by injection of the slurry (Stevens and Laughlin, 1997).

Transformations of N in Grassland Systems

The main chemical forms of N involved in the cycling of N through grassland systems are shown in Fig. 5.6. During the sequence from soil to plant to animal and back to soil, the transformations that involve the digestion of plant material by ruminants, the incorporation of plant residues and excreta into the soil OM and the mineralization of N from the soil OM all depend on the activities of microorganisms. Their main role is to break down large insoluble molecules, such as proteins, to smaller soluble molecules. At the same time, they assimilate some of the small molecular compounds into proteins and other components of their biomass, and the biomass is then, in turn, subject to decomposition in the digestive system of ruminant animals or in the soil.

In extensively managed grassland, the supply of N for new plant growth depends partly on remobilization within the plant and partly on the decomposition of dead plant tissues; and these processes continue with little, if any, loss of N from the system. With more intensive management, involving moderate to high rates of fertilizer N and large numbers of ruminant animals, the amounts of N in circulation are much greater, and there are substantial losses from the system through leaching and the volatilization of gaseous compounds. Although the presence of large numbers of animals accelerates

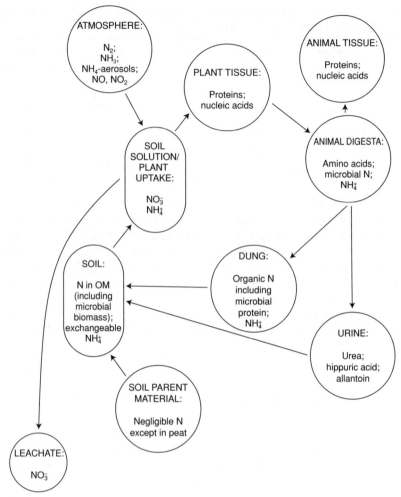

Fig. 5.6.   Outline of the main forms of N involved in the cycling of N in grassland.

the recycling of N, the excreted N is distributed unevenly and this increases the extent of loss. The utilization of recycled N therefore becomes less efficient, in percentage terms, as the intensity of management of grazed swards is increased. When swards are cut and the herbage fed to housed animals, there is a wide variation in the extent to which recycled N is utilized, depending largely on the extent to which the excreta are collected in slurry or manure and on how effectively losses of N are minimized.

Quantitative Balances of N in Grassland Systems

The N balances (Table 5.10) estimated for intensively managed all-grass swards, for moderately intensive grass–clover and for extensively managed

Table 5.10.   Estimated N balances (kg ha^{-1} year^{-1}) for three systems of grassland management: intensively managed grass, moderately intensive grass–clover and extensive grass, all grazed by cattle in UK conditions (based on data from Dampney and Unwin, 1993; Whitehead, 1995).

	Intensive grass (dairy cows)	Grass–clover	Extensive grass
Inputs			
Fixation of atmospheric N$_2$	0	120	8
Deposition from atmosphere	30	30	15
Fertilizer	330	0	0
Supplementary feeds	100	0	0
Aspects of recycling			
Uptake into herbage	400	240	60
Consumption of herbage by animals	280	160	30
Dead herbage to soil	120	80	30
Dead roots to soil	60	50	30
Excreta to soil of grazed area	285	120	25
Outputs			
Milk/live-weight gain	70	25	3
Leaching/runoff	180	40	5
Volatilization of ammonia	80	15	3
Denitrification/nitrification	100	20	2
Loss through excretion off sward	28	10	0
Gain to soil	30	50	10

long-term grassland, each of them subject to grazing, show that there are marked differences between the types of sward in the quantity of N in circulation and in the amount of N lost from the system. The quantity of N in circulation, based on annual uptake by the herbage, may vary from < 60 to > 400 kg N ha^{-1}, while the total loss may vary from < 10 to > 350 kg ha^{-1} year^{-1}. In other N balances prepared for intensively managed grass (Aarts *et al.*, 1992; Dampney and Unwin, 1993; Jarvis *et al.*, 1996b), for intensively managed grass–clover in New Zealand (Ledgard *et al.*, 1996) and for extensive grass in the UK (Batey, 1982), the total amounts of N in circulation and the relationship between leaching, volatilization of ammonia and denitrification/nitrification differ to some extent from those shown in Table 5.10. For any particular intensity of management, the relationship between the three major loss pathways can vary greatly, depending on soil and weather conditions. In intensive systems, it is difficult to achieve a substantial reduction in the total loss of N while maintaining high productivity, though some improvement can be achieved by the tactical adjustment of fertilizer N, and sometimes by the supplementation of grass herbage or silage with maize silage, beet pulp or cereals.

Chapter 6

Phosphorus

Natural Sources and Concentrations of P in Soils

Much of the P in rocks is in the form of apatite, which has the empirical formula $3[Ca_3(PO_4)_2].CaX_2$, where X may be carbonate, hydroxide, fluoride or chloride. Other phosphate minerals that are sometimes present include the Fe and Al phosphates, such as strengite, $FePO_4.2H_2O$, and variscite, $AlPO_4.2H_2O$. The concentration of P is usually higher in basaltic than in granitic rocks (see Table 2.2, p. 20). During the process of weathering, phosphate is released as the soluble ions $H_2PO_4^-$ and HPO_4^{2-}, but a large proportion becomes insoluble again through precipitation or adsorption. Phosphate is often reprecipitated in sedimentary rocks as apatite, though generally in a microcrystalline form, which has a larger surface area per unit weight than the apatite in igneous and metamorphic rocks and is therefore more susceptible to renewed weathering (Sanyal and De Datta, 1991). Of the sedimentary rocks, shales are generally much richer in P than are sandstones (see Table 2.2).

During the development of soils, the phosphate released by weathering is again largely reprecipitated or adsorbed by other soil constituents. In neutral and calcareous soils, the reprecipitation occurs mainly with Ca and tends to result eventually in apatite. In acid soils, the reprecipitation and adsorption occur mainly with compounds of Fe and Al, rather than Ca, and there may be a small amount of adsorption by clay minerals. The phosphate adsorbed by the hydrous oxides of Fe and Al tends to become increasingly insoluble, due to its becoming slowly occluded within the oxide particles. Although derived originally from inorganic sources, P is also a constituent of soil OM, and the C : organic P ratio in soils of humid temperate regions is usually within the range 50 : 1–100 : 1 (Stevenson, 1994). As a result of the various reactions that it undergoes in soils, P is generally retained strongly, and there is little loss by leaching.

The concentration of P in soils of temperate regions is usually within the range of 400–4000 mg kg^{-1} dry weight (see Table 2.3, p. 20; Gressel and

McColl, 1997; Mathews *et al.*, 1998), and tends to reflect the concentration in the soil parent material. In many grassland soils, the concentration of P is highest in the surface horizon, due to the continued deposition of dead leaf litter and animal excreta on the surface, and the retention of P in insoluble forms. The tendency for the concentration to be relatively high at the surface is accentuated when fertilizer P is applied repeatedly.

Agricultural and Atmospheric Inputs of P to Soils

Intensively managed grassland soils often receive inputs of P in fertilizers. Grassland soils also receive recycled P in the excreta of grazing animals or in slurry, sometimes with a net addition of P if the animals are provided with supplementary feed. Of the main fertilizers and fertilizer constituents that contain P (Table 6.1), superphosphate and triple superphosphate are usually applied as individual fertilizers, while the ammonium phosphates are usually included in compound fertilizers together with other sources of N and/or K. The phosphate in each of these types of fertilizer is soluble initially, but is progressively adsorbed or precipitated, or immobilized by the microbial biomass, thus curtailing its uptake by plants and enhancing its retention in the soil. Rock phosphate, in the form of apatite, which is insoluble, is sometimes applied as a fertilizer to acid soils and, in such conditions, slowly becomes available. With intensively managed grassland, the repeated application of fertilizer P for many years may account for a substantial proportion of the total P in the surface soil, as shown by the data in Table 6.2. Similarly, in an investigation in which superphosphate was applied to grassland at a rate of 30 kg P ha^{-1} year^{-1} for more than 100 years, the concentration of P in the top 23 cm of soil increased from 575 mg kg^{-1} to 1425 mg kg^{-1}, with smaller increases at greater depth (Powlson, 1994).

In general, sewage sludge contains between 1% and 5% P on a DM basis and, where this is applied to grassland, it tends to increase the concentration of both total P and plant-available P in the surface soil (Sommers and Sutton, 1980). Of the P in sewage sludge, 70–90% is usually inorganic and 10–30% organic (Sommers, 1977), and the proportion that is organic is higher when the sludge has been digested aerobically rather than anaerobically (Hinedi

Table 6.1. The main types of phosphate fertilizer.

Fertilizer	Chemical formula	% P	Solubility in water
Superphosphate	$Ca(H_2PO_4)_2 + CaSO_4.2H_2O$	8–9	High
Triple superphosphate	$Ca(H_2PO_4)_2$	20	High
Monoammonium phosphate	$NH_4H_2PO_4$	26	High
Diammonium phosphate	$(NH_4)_2HPO_4$	23	High
Rock phosphate	$Ca_3(PO_4)_2$/apatite	12–16	Low

Table 6.2. Effect of applying superphosphate annually for 35 years on the concentration of total P (mg P kg^{-1} soil) in four horizons of a grazed grassland soil in New Zealand (from Nguyen and Goh, 1992a).

Soil horizon, depth (mm)	Initial soil P (mg kg^{-1})	Rate of application (kg P ha^{-1} year^{-1})		
		0	16	32
0–75	710	750	930	1150
75–150	670	715	885	1010
150–225	600	625	680	740
225–300	505	500	500	510

et al., 1989). Some of the inorganic P in sewage sludge occurs as phosphates of Ca, Fe and/or Al, and some as phosphate adsorbed on to hydrous oxides of Fe and Al. The organic P occurs mainly in microbial cells and their degradation products, including inositol phosphates (Sommers, 1977).

The deposition of P from the atmosphere is usually negligible. Rainfall, together with some dry deposition, usually provides between 0.2 and 1.5 kg P ha^{-1} year^{-1} (Wadsworth and Webber, 1980; Newman, 1995; Owens *et al.*, 1998), though 1.7 kg ha^{-1} was reported in a high-rainfall area of northern UK (Harrison, 1978). The P that is deposited from the atmosphere is thought to be derived mainly from dust transferred to the atmosphere by the wind erosion of soil (Sharpley *et al.*, 1995), and there may be a small contribution from the burning of plant material and fossil fuels (Newman, 1995).

Recycling of P through the Decomposition of Organic Residues

Phosphorus is recycled to grassland soils through the death of herbage and roots and in the excreta of grazing animals. Although most of the P in actively growing plants is organic, the proportion in inorganic form tends to increase as plants senesce, and dead plant residues may contain about 60% of their P in inorganic form (Jones and Bromfield, 1969). Although some P may be leached from senescent or dead plant material by rain, leaching is usually curtailed by the growth of microorganisms within the dead material (Mays *et al.*, 1980). Once in contact with the soil, the organic P of dead plant material is mineralized by soil microorganisms, with the release of inorganic phosphate. However, since the microorganisms have a substantial requirement for P, with an average concentration of about 1–2% (Clark and Woodmansee, 1992), net mineralization occurs only when their need has been met. If there is insufficient P in the dead plant material, there may be a net immobilization of P from the soil, with competition between plants and microorganisms for the small amount of P in solution (Tate, 1984). The critical ratio of organic C : organic P for mineralization to occur appears to vary from < 100 : 1, for soils well supplied with available P, to > 200 : 1 for deficient soils. However, when the ratio is > 300 : 1, net immobilization is almost certain (Tate, 1985). In general, the mineralization of organic P is more rapid at warm temperatures, as shown in a study involving white clover herbage, labelled with ^{32}P, incorporated into soil. The rate of mineralization was about 20% greater at a day/night temperature regime of 28/22°C than at 15/10°C (Till and Blair, 1978). The mineralization of P from plant residues also occurs more rapidly when the residues are first consumed by earthworms. This effect was illustrated when dead leaf litter, labelled with ^{32}P, was used to provide a source of P for newly sown ryegrass: uptake was two to three times greater when the litter had been ingested by earthworms and excreted in worm-casts (Mansell et al., 1981).

The amount of P recycled through dead herbage material is likely to be between 1 and 20 kg ha^{-1} year^{-1}, depending on the type of grassland. The lower amount would occur from 1000 kg of dead herbage DM containing 0.1% P (C : P ratio of about 480 : 1), typical of an extensively managed natural or semi-natural grassland, while the higher amount would result from 10,000 kg of dead herbage containing 0.2% P (C : P ratio of about 240 : 1). Similar amounts of P may be recycled through the death of roots. Thus, an annual root turnover of 5000 kg DM ha^{-1}, with a concentration of 0.15% P, would provide 7.5 kg ha^{-1} year^{-1}. In addition, there may be some recycling of P from living roots, as these tend to release inorganic P when desiccated during dry conditions (Tate, 1985).

When grassland is grazed, most of the P consumed by the animals is returned in excreta, since less than 40%, and often only 10–25%, of the P in the diet is converted into live-weight gain or milk. Almost all the excreted P is in the dung and, of this, some is inorganic and some organic. Much of the inorganic P occurs as dicalcium phosphate, which has a low solubility in the conditions of relatively high pH in dung. Although some of the organic P is soluble, most is mineralized only slowly. There is, therefore, little transfer of P into the soil while the dung remains on the surface, but the rate of transfer is increased if the dung is buried by earthworms or beetles (Barrow, 1987). It is also greater when conditions remain moist, rather than alternating between wet and dry (Mathews et al., 1998). In field studies, there was little change in the concentration of P in dung that was placed on the surface of grassland swards, indicating that the release of P occurred at approximately the same rate as the loss of dry matter (Rowarth et al., 1985; Dickinson and Craig, 1990). In another investigation, sheep dung retained 40% of its initial P content after 2 years, during which rainfall amounted to more than 100 cm, and 90% of the residual P was organic (Bromfield and Jones, 1970). Due to the low solubility of the inorganic P, together with the slow mineralization of the organic P and the uneven distribution of dung, excreted P is poorly utilized for new plant growth during any one season, though it provides a cumulative effect over a large proportion of the area when a sward is grazed for many years. Nevertheless, the uneven distribution of dung results in a large spatial variation in the amounts of both total and plant-available P in the soil under grazed grassland (Fisher *et al.*, 1998). When large amounts of supplementary feed are provided – for example, as grain to grazing beef cattle – there is likely to be an overall increase in the concentration of P, and especially of the more labile P, in the soil (Benacchio *et al.*, 1970).

The recycling of excreted P in cattle slurry, which generally contains between 0.5 and 1.0% P on a DM basis (see Table 2.7, p. 31), results in a more uniform distribution than with grazing. Also, the proportion of P in inorganic forms is higher in slurry than in excreta returned directly to the sward. However, a potential disadvantage of slurry is its greater susceptibility to loss by runoff during heavy rainfall.

Forms and Availability of P in Soils

The proportions of inorganic and organic forms of P in soils vary widely. In topsoils, the proportion in either form can vary from about 10 to 90% (Withers and Sharpley, 1995; Paul and Clark, 1996), though, in most agricultural soils, including prairie grassland (Aguilar and Heil, 1988), more than 50% is inorganic. Usually, the proportion in organic form decreases sharply down the profile, though in peats there may be some increase with depth (Sanyal and De Datta, 1991).

The inorganic P in soils includes: (i) phosphate present in the soil solution, mainly as $H_2PO_4^-$ and HPO_4^{2-} ions; (ii) insoluble Ca phosphates, including apatite; (iii) phosphate adsorbed, and possibly occluded, by hydrous oxides of Fe and Al; (iv) phosphate adsorbed by 1 : 1 lattice clays, such as kaolinite; and (v) phosphate in various unweathered minerals. These categories are in dynamic equilibrium with one another, though most transformations are extremely slow. The phosphate in the soil solution also undergoes exchange with the P in the soil OM, through the metabolic activity of the microbial biomass.

The ratio of $H_2PO_4^-$ to HPO_4^{2-} in the soil solution depends on the ambient pH: at pH 6, about 94% of the phosphate occurs as $H_2PO_4^-$ and 6% as HPO_4^{2-}, but, at pH 7, about 60% occurs as $H_2PO_4^-$ and 40% as HPO_4^{2-} (Stevenson, 1986). With both ions, actual concentrations depend on the extent of adsorption and precipitation by other soil constituents. In acid soils, soluble phosphate is adsorbed by hydrous oxides of Fe and Al and by 1 : 1 lattice clays, while, in neutral and calcareous soils, the main reaction is the precipitation of di- and tricalcium phosphates, which change gradually to the carbonate form of apatite. In general, the maximum solubility of phosphate occurs in the slightly acid to neutral soil pH range, 6.0–7.0. However, in soils with a pH of around 7 and above, the localized production of organic acids by plant roots and microorganisms reduces the pH in the rhizosphere, and this tends to increase the solubility of the P in apatite and Ca phosphates.

The solubility of P associated with hydrous oxides of Fe and Al is not increased by the production of acid *per se*, but it is increased by those organic acids that are able to form stable but soluble complexes with the Fe or Al (Kucey *et al.*, 1989). Citric and malic acids are two examples. Liming an acid soil may either increase, decrease or have no effect on the solubility of P. An increase in solubility may result if Fe and Al phosphates become more soluble, but a decrease is possible if exchangeable Al is precipitated as hydrous Al oxide, thus producing new surfaces for the adsorption of P (Frossard *et al.*, 1995).

Some of the organic P in soils is in the form of small molecular compounds, such as inositol phosphates, and some is present in complex humus polymers. There is a general relationship between the amount of soil organic P and the amounts of C, N and organic S, but the C : organic P ratio is more variable than the ratios of C : N and C : organic S. The greater variation is due mainly to the fact that small molecular compounds, such as inositol hexaphosphates (phytic acid and its salts) and inositol pentaphosphates, may account for up to 60% of the organic P in soils, whereas the N and organic S are present mainly in humus polymers. The inositol phosphates in soils are largely bacterial in origin, and occur mainly in the form of Ca, Fe and Al salts that are insoluble. Nucleic acids and nucleotides usually account for only about 1–2% of the soil organic P, and phospholipids account for a similar proportion (Stevenson, 1994; Paul and Clark, 1996). Some of the remaining soil organic P occurs in the form of teichoic acids, which are components of microbial cell walls (Gressel and McColl, 1997). Phosphorus in the living microbial biomass was estimated to account for 5–24% of the organic P in eight grassland soils, with an average annual flux through the biomass of 23 kg P ha^{-1} year^{-1}, compared with an average uptake by the grass herbage of 12 kg P ha^{-1} year^{-1} (Brooks *et al.*, 1984). In another study, microbial P in the top 75 mm of 21 grassland soils in New Zealand ranged from 11 to 57 kg P ha^{-1}, representing between 0.5% and 12.0% of the total P (Perrott and Sarathchandra, 1989). The importance of the microbial contribution to the soil organic P reflects the fact that the concentration of P is much higher in soil bacteria than in plant material. Concentrations are about 1.0–2.0% P in bacteria (Clark and Woodmansee, 1992), compared with 0.1–0.5% P in plant material.

The availability of the soil organic P for plant uptake depends partly on the rate at which it is mineralized and partly on the balance between mineralization and immobilization. The rate of mineralization is influenced by factors that affect microbial activity (e.g. temperature, moisture, aeration and pH), while its relationship to immobilization depends mainly on the C : P ratio of the residues undergoing decomposition. Mineralization is particularly slow in acid soils in cool upland areas, and a mat of partially decomposed organic material tends to accumulate unless the soil is limed or subject to cultivation. Such a mat of OM may contain up to 60 kg P ha^{-1}. On the other hand, OM may increase the availability of the soil inorganic P, through the release of organic acids during decomposition. In calcareous soils, organic acids tend to increase the solubility of calcium phosphates, but they may also have an effect in acid soils by forming complexes with Fe and Al or by displacing phosphate from the hydrous oxides of Fe and Al (Hué, 1995). Phytase enzymes, which are secreted by soil microorganisms and by plant roots, also increase the availability of organic P by hydrolysing inositol hexaphosphate (Li et al., 1997). One effect of liming acid soils is to increase the mineralization of organic P (Perrott and Mansell, 1989; Wheeler, 1998), but, if the soil pH is raised to > 7.0, the availability of the phosphate may be curtailed through the formation of insoluble Ca phosphates (Withers and Sharpley, 1995).

Losses of P from Soils

Most soils show little or no loss of soil P through leaching, as phosphate ions are rapidly adsorbed or precipitated by other soil constituents. Nevertheless, some leaching may occur when fertilizer P is applied to sandy soils or to loam or silty soils whose retention capacity is already saturated. And there may be some flow of soluble inorganic P and particulate P through macropores (Stamm et al., 1998; Sharpley et al., 2000). In addition, small amounts of P are leached in the form of organic P compounds, some of which are more mobile than inorganic P (Schoenau and Bettany, 1987). The P in organic manures has been shown to be more susceptible to leaching than that in inorganic fertilizers (Eghball et al., 1996), and the leaching of P from grazed grassland tends to be increased when the stocking density is high, possibly due to a greater amount of soluble organic P being released from dung (Beauchemin et al., 1996). The injection of slurry greatly increases the leaching of P (Tunney et al., 1997), and even surface application to grassland may result in an appreciable leaching of P, probably mainly through macropores (Hooda et al., 1999).

Surface runoff is often more important than leaching in terms of the loss of P from grassland, and both suspended particulate material and P in soluble forms may contribute (Withers and Sharpley, 1995; Sharpley et al., 2000). The loss is most serious when the runoff, due to heavy rainfall or melting snow, occurs soon after the application of fertilizer or slurry (Gillingham, 1987). Other factors that influence the amount of loss include the degree of slope of the field, the rate and method of application of the fertilizer or slurry, the form of fertilizer P and the height and density of the sward. Losses of about 1% of added fertilizer in runoff have been reported for New Zealand pastures (Parfitt, 1980), but, when no slurry or fertilizer is applied, the loss of P through runoff from grassland is normally negligible, perhaps about 0.1 kg ha^{-1} year^{-1} (Timmons and Holt, 1977).

On sloping fields of fertilized and grazed grassland in Ohio, the combined loss of P through leaching and runoff was estimated to be about

1.5 kg P ha^{-1} year^{-1} (Owens *et al.*, 1998). Similarly, in two catchment areas in South Australia, both supporting annual grasses and legumes subject to grazing, the loss by leaching plus runoff amounted to 1.0–1.1 kg P ha^{-1} year^{-1}, and the proportion of the P in particulate form and present in runoff was much greater in the area with more steeply sloping fields (Nelson *et al.*, 1996). In many situations, the combined loss of P due to leaching and runoff is < 1 kg ha^{-1} year^{-1}, as estimated for several watersheds of mixed land use in the UK (Withers and Sharpley, 1995), but, when slurry is injected into grassland soils, leaching losses may amount to > 15 kg P ha^{-1} year^{-1} (Tunney *et al.*, 1997). Although the amount of P lost by leaching and runoff is generally negligible in terms of soil fertility (except where slurry is injected), it may be sufficient to cause a significant increase in the eutrophication of watercourses. Even small losses of P, of 1–2 kg ha^{-1}, are of concern in this context (Gillingham, 1987; Tunney *et al.*, 1997; Haygarth *et al.*, 1998).

The possibility of P being lost from soils in gaseous form exists only with phosphine (PH_3) and, if this compound is formed in soils, its volatilization appears to be slight (Burford and Bremner, 1972).

Assessment of Plant-available P in Soils

The amount of soil P becoming available during a growing season comprises that present in the soil solution at the beginning of the season, plus the fraction that becomes soluble during the season. In many soils, the concentration of P in the soil solution at any one time is only about 0.05 mg l^{-1}, and the assessment of plant-available P therefore depends mainly on the measurement of the potentially soluble fraction. The methods that have been proposed are based either on dissolving a proportion of the soil P with an extractant solution, or on equilibrating the soil with a source of soluble P labelled with ^{32}P in order to determine the amount of labile phosphate. Of the various extractants, 0.5 M sodium bicarbonate at pH 8.5 is widely used. It is effective when the soils to be compared are of similar type, but is less satisfactory when the soils differ in their major forms of P (Schoenau and Karamanos, 1993). Although it does not extract all the potentially available P, 0.5 M sodium bicarbonate provides a means of categorizing soils in terms of plant-available P, and is used in this way as a basis for fertilizer recommendations (Table 6.4). A number of other extractants for available soil P have been proposed, and some may be more effective than sodium bicarbonate for specific types of soil. For example, a mixture of 0.05 M HCl and 0.0125 M H_2SO_4 has been recommended for acid soils (Fixen and Grove, 1990).

The amount of soil P that equilibrates readily with the P in the soil solution, as estimated by exchange with ^{32}P, provides another means of assessing plant-available P. However, the accuracy of this type of assessment is limited by the extent to which a true equilibrium is reached between ^{32}P and the labile ^{31}P and by whether or not there is appreciable mineralization of organic P (Gillingham, 1987). In response to the importance attached to the movement of P from soil to water, renewed attention is now being given to the development of better procedures for assessing the availability of soil P and its cumulative release in soluble forms throughout the year (Frossard *et al.*, 2000).

Uptake of P by Herbage Plants

Plants can take up P as either the $H_2PO_4^-$ or the HPO_4^{2-} ion, and the relative amounts of the two ions depend on pH. The monovalent $H_2PO_4^-$ is normally the major form, but the proportion of HPO_4^{2-} increases with increasing pH. The rate of uptake is determined partly by the concentration in the soil solution near the root surface and partly by the rate of movement of the ions towards the root surface. Diffusion is more important than mass flow, but movement by both means is slow, due to the precipitation and adsorption of P by soil constituents (Barber, 1980; Frossard *et al.*, 1995). Because movement is slow, the uptake of P occurs mainly from a zone of 2–3 mm diameter around active roots: the rate of root growth and the number and length of root hairs per unit length of root are therefore important factors in uptake (Barber, 1980). In grass–clover swards, the grasses, which have a denser root system, compete for P more effectively than do the clovers (Gillingham, 1987). Amongst the grasses, *Agrostis tenuis* appears to be particularly effective in competing for P, at least in comparison with other species in New Zealand, due to its ability to increase root growth relative to shoot growth when P is deficient (Mouat, 1983).

Some plants secrete organic acids through their roots, and these acids may enhance the solubility of soil phosphates. There is evidence that the secretion of acids, such as citric and malic, increases when the plant is deficient in P (Lipton *et al.*, 1987; Jones, 1998). White lupin (*Lupinus albus*) appears to be an exceptionally effective species in this respect, as its secretion of citric acid was found to amount to as much as 23% of its dry weight at 13 weeks, when it was grown on a calcareous soil (Dinkelaker *et al.*, 1989). However, oxalic acid, which is produced by some species, may be more effective than citric and malic acids in releasing P from Ca minerals, such as apatite, as the resultant Ca oxalate is insoluble (Jones, 1998). Plant roots also secrete the enzyme phytase, which solubilizes a major form of the organic P in soils, and this secretion also appears to increase when P is deficient (Li *et al.*, 1997). Arbuscular mycorrhizae increase the effective size of the root system, and mycorrhizal associations have been shown to increase uptake from soils low in available P (Gillingham, 1987; Paul and Clark, 1996).

The uptake of P by roots occurs against a marked electrochemical gradient, with the concentration of P in root cells and xylem sap usually being 100–1000 times greater than that in the soil solution. It is an active process dependent on a supply of energy (Frossard *et al.*, 1995).

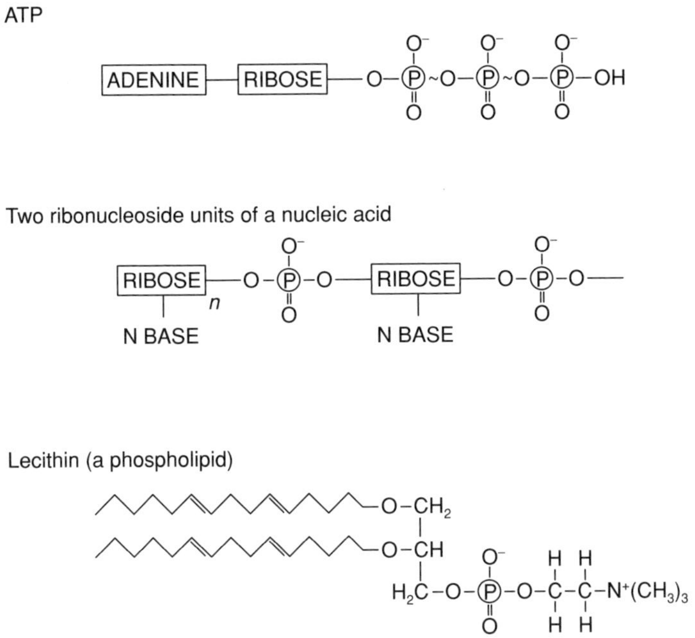

ATP

Two ribonucleoside units of a nucleic acid

Lecithin (a phospholipid)

Fig. 6.1. Chemical structure of some plant and animal constituents containing P.

Functions of P in Herbage Plants

Of the numerous biochemical functions of P in plants, the transfer of energy through ADP and ATP is particularly important. ADP has one, and ATP two, high-energy phosphate bonds, which provide a mechanism for the storage of energy from photosynthesis and aerobic respiration. The stored energy is released subsequently to provide for processes such as the active uptake and transport of ions and the synthesis of organic molecules. Another major function of P is as a constituent of nucleic acids and, in both DNA and RNA, it forms a bridge between ribonucleoside units (Fig. 6.1). As a result of its role in the structure of nucleic acids, P is essential in cell division, and its concentration is relatively high in meristem tissue. Phosphorus is also a constituent of phospholipids (Fig. 6.1), which contribute to the structure of the cytoplasmic membranes. Seeds require relatively large amounts of P during germination, and this requirement is met by the storage of P in the form of inositol hexaphosphate or its Ca or Mg salts (phytic acid and phytates).

Influence of Fertilizer P on Yields of Grassland Herbage

The extent to which grassland shows a yield response to fertilizer P is influenced by soil type, by past fertilizer application and by whether the sward is regularly cut or grazed. With soils in the pH range of about 5.0–7.5 that are deficient in P, the application of fertilizer P will usually produce a cost-effective yield response. In general, the regular application of fertilizer P for a number of years increases the soil supply of plant-available P and, partly for this reason, many areas of grassland in intensively farmed regions now show little response to current applications of fertilizer P. Thus, in a series of trials on 21 lowland farms in England and Wales, phosphate applied at 35 kg P ha^{-1} increased herbage yield at only seven of the 21 sites, and then at only one of the two cuts per year (Paynter and Dampney, 1991). However, in regions that have been farmed less intensively in the past, responses to fertilizer P are likely to be more widespread, as was found in trials in Ireland carried out during 1967–1970 (Ryan and Finn, 1976). Responses are likely to be greater where the herbage has been regularly cut and removed, rather than grazed. In contrast to the situation with slightly acid or neutral soils, highly acid and highly calcareous soils may show little response to fertilizer P, due to its becoming rapidly insoluble. With acid soils, the problem may be overcome by liming, while, in calcareous soils, OM tends to increase the availability of P.

The extent to which grassland responds to fertilizer P is also influenced by supplies of other nutrients, particularly N and S. The application of a high rate of fertilizer N may increase the need for fertilizer P (Wolton et al., 1968), and, in some regions, there is a strong positive interaction between P and S. Such an interaction occurred in a trial involving a grass–clover sward in New Zealand (Table 6.3). The clover responded more than the grass to P, and increasing the rate of fertilizer P from nil to 80 kg P ha^{-1} increased the average proportion of clover in the sward from 35% to 44% (Sinclair et al., 1996a). Different grass species may also differ in their response to fertilizer P. For example, in a comparison of six grass species in the USA, tall fescue and ryegrass showed the greatest response (Mays et al., 1980).

In general, the decision on whether to apply fertilizer P, and if so at what rate, is based on soil analysis. In the UK, the recommended rate of

Table 6.3. Interaction of fertilizer P and S on the yield of herbage (kg DM ha^{-1} year^{-1}) from five cuts of a sward of ryegrass and white clover in its second year in New Zealand (from Sinclair et al., 1996a).

P (kg ha^{-1})	S (kg ha^{-1})		
	0	15	30
0	6,210	9,130	10,300
10	7,260	10,790	11,020
20	8,180	11,640	12,080
40	7,210	11,920	13,610
80	8,350	12,960	14,130

application is based on the category of soil P status, as assessed by extraction with 0.05 M sodium bicarbonate (Table 6.4). Several other countries use the same basis of assessment, though they differ in the amounts of fertilizer P recommended. The differences appear to be due to a number of factors: differences in soil type and climate, differences in past use of fertilizer P and differences in the safety margin incorporated into the recommendations (Tunney *et al.*, 1997). On some soils, the triennial application of fertilizer P may be almost, if not quite, as effective as applications repeated annually (Morton *et al.*, 1995).

Distribution of P within Herbage Plants

In grasses, the concentration of P is usually greater in the shoots than in the roots (Wilkinson and Lowrey, 1973; Mika, 1975; Nguyen and Goh, 1992a). However, this may not be so when the supply of P exceeds that needed for maximum yield (Hylton *et al.*, 1965). Also, when the herbage is mature, its concentration of P may be no greater than that in the roots – about 0.13–0.15% in upland pasture (Macklon *et al.*, 1994). Other factors that may influence the concentration in the roots and the ratio of concentrations between roots and shoots include the time of year, the rate of application of fertilizer P and the time interval since any fertilizer P was applied. Thus, the concentration of P in the roots of South African grasses decreased in spring and early summer and increased during autumn to a maximum in midwinter (Weinmann, 1940). The effect of fertilizer P was illustrated in a study in which superphosphate, applied at 35 kg P ha^{-1} year^{-1} for more than 30 years to a ryegrass–white clover sward, increased root P concentration from 0.12% to 0.22% (Nguyen and Goh, 1992a).

In white clover grown in soil, concentrations of P were higher in roots than shoots, especially when the plants were supplied with a high rate of

Table 6.4. Amounts of fertilizer P (kg P ha^{-1}) recommended for grassland in the UK, based on soil analysis (MAFF, 1994).

	Extractable soil P (mg P l^{-1})[a]				
	0–9	10–15	16–25	26–45	> 45
Grazed grass and grass–clover	26	17	9	0	0
Cut grass and grass–clover					
1st cut	44	26	13	13	0
2nd cut	22	13	13	0	0
3rd cut	0	0	0	0	0
4th cut	0	0	0	0	0

[a]Soil P extracted by 0.05 M NaHCO$_3$ at pH 8.5, mg P l^{-1} of soil.

fertilizer P (Caradus, 1992), but, in studies with lucerne, concentrations of P were similar in roots and shoots (Nielsen *et al.*, 1960) or were lower in the roots than the shoots (Rominger *et al.*, 1975). One factor contributing to the variation in the root : shoot ratio of P concentration in legumes may be the number of root nodules, as these have been shown, at least in white clover grown in solution culture, to be particularly enriched in P (Hart, 1989).

In the shoot material of both grasses and clovers, the concentration of P is highest in the inflorescence, lower in the leaf blade and generally lowest in the leaf sheath and stem (Table 6.5; Fleming, 1963; Wilkinson and Gross, 1967; Davey and Mitchell, 1968). The concentration of P in herbage tends to decrease with age, due partly to the increase in the proportion of cell-wall material and partly to the translocation of P to tissues where metabolism is more active (Gillingham, 1987). Phosphorus is a relatively mobile element in plants and is translocated from old to young leaves when P is deficient, though less so when there is a plentiful supply of P (Ozanne, 1980). Although most translocation probably occurs within the plant, there may be some additional recycling of P from old leaves to new growth through the leaching of P by rainfall and its uptake by roots (Till, 1981).

Range of Concentrations and Critical Concentrations of P in Herbage Plants

The usual range for the concentration of P in grass or legume herbage is 0.1–0.6% (Table 6.6). However, herbage concentrations as low as 0.05% have been reported from regions, such as South Africa, which combine P-deficient soils with a pronounced dry season (Jones and Thomas, 1987). At the upper end of the range, young grass herbage grown with a plentiful supply of phosphate may contain > 0.8% P (Table 6.6). Since the concentration of P declines with increasing maturity of the herbage, the concentration is normally higher in silage than in hay. In a survey in Scotland, the mean concentration in silage was 0.25%, compared with 0.20% in hay (Hemingway *et al.*, 1968).

Table 6.5. Distribution of P (% in DM) in the shoots of three grasses at early anthesis (from Smith, 1973, as cited by Reid, 1980).

	Cocksfoot	Timothy	Tall fescue
Whole shoot	0.25	0.21	0.22
Leaf blade	0.27	0.18	0.21
Leaf sheath	0.19	0.13	0.16
Inflorescence	0.39	0.39	0.34
Internode	0.21	0.20	0.21
Stubble	0.20	0.20	0.15

Critical concentrations of P in a number of grass and legume species, as reported from various investigations, are shown in Table 6.7. The critical concentration of P declines with increasing maturity, and, if the stage of maturity is not well defined, this effect can be taken into account by relating the concentration of P to that of N or fibre. The effect of the herbage concentration of N on the critical concentration of P, as proposed in two investigations, is shown in Fig. 6.2 (Knauer, 1966; Prins *et al.*, 1986). Studies with ryegrass grown in sand culture suggested that the critical concentration of P in the shoots was greatly influenced by the way in which it was defined: the concentration associated with 90% of maximum yield was 0.21% P,

Table 6.6. Concentrations of P in grassland herbage analysed for survey or advisory purposes in several regions: range and mean values (% in DM).

Region	No. of samples	P concentration Range	Mean	Reference
England and Wales	> 2000	0.25–0.49[a]	0.35	MAFF, 1975
The Netherlands	> 700[b]	0.27–0.55[a]	0.41	Kemp and Geurink, 1978
Belgium	> 3000[c]	0.05–0.50	0.29	Lambert and Toussaint, 1978
Finland	> 2000[d]	0.13–0.98	0.29	Kahari and Nissinen, 1978
Pennsylvania, USA	> 9000	0.05–0.81	0.26	Adams, 1975
North Island, New Zealand	> 5000	0.11–0.99	0.41	Smith and Cornforth, 1982

[a]Based on 2 × SD.
[b]All farms intensively managed.
[c]Mainly hay.
[d]All timothy.

Table 6.7. Some reported values for the critical concentration of P in some grasses and legumes (% in DM).

Species and portion of plant	% P	Reference
Perennial ryegrass	0.30 (95% max yield)	Smith *et al.*, 1985
Perennial ryegrass	0.28	Rangeley, 1989
Several grasses, 1-week old regrowth	0.30–0.35	Lunt *et al.*, 1966
Kentucky bluegrass	0.25	Walker and Pesek, 1967
White clover	0.38	Evans *et al.*, 1986
White clover, young shoots	0.20–0.30	Rangeley and Newbould, 1985
White clover, mature leaf + petiole	0.23	Andrew, 1960
White clover, young herbage	0.3–0.4	McNaught, 1970
Red clover	0.28	Davis, 1991
Subterranean clover, whole shoots	0.30–0.35	Lewis, 1992
Subterranean clover	0.28–0.31	Pinkerton and Randall, 1994
Lucerne	0.35	Melsted *et al.*, 1969
Lucerne	0.24	Andrew and Robins, 1969

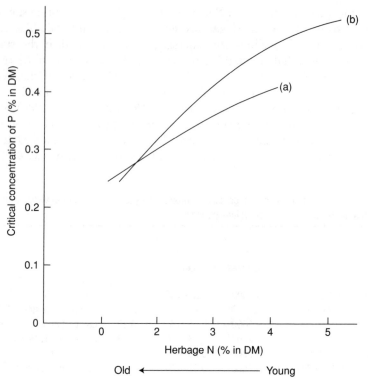

Fig. 6.2. The effect of stage of maturity of grass herbage, as indicated by herbage N concentration, on the critical concentration of P; based on: (a) data of Knauer, 1966; and (b) formula cited by Prins *et al.*, 1986.

while the concentration associated with 99% of maximum yield was as high as 0.44% P (Smith *et al.*, 1985).

Influence of Plant Species and Variety on Concentrations of P in Herbage

Grasses and legumes often contain similar concentrations of P (Baker and Reid, 1977; Lambert and Toussaint, 1978; Metson and Saunders, 1978a; Mackay *et al.*, 1995; Mayland and Wilkinson, 1996). However, when ryegrass and white clover are grown together in a mixed sward, the concentration of P is generally higher in the grass than in the clover (Whitehead *et al.*, 1983; Wilman and Hollington, 1985; Wheeler, 1998). On the other hand, in some comparisons of grasses and legumes grown separately, legumes have shown higher concentrations than grasses (Thomas *et al.*, 1952; Adams, 1975).

The commonly cultivated grass species are generally similar to one another in P concentration, with the exception of timothy, which tends to have a relatively low concentration (Table 6.8). Those grass species typical of unproductive grassland, e.g. *Deschampsia cespitosa, Nardus stricta* and *Molinia caerulea*, are also generally low in P, and have been reported to contain less than other species, such as ryegrass, even when grown in a mixture with them (Lambert and Toussaint, 1978). Amongst the legume species, white clover generally has a higher concentration of P than do red clover, lucerne and sainfoin (Table 6.9), probably reflecting its higher ratio of leaf to stem. However, in a comparison in the USA, the concentration of P was higher in lucerne, at 0.41%, and birdsfoot trefoil, at 0.40%, than in white and red clover, at about 0.30% (Gross and Jung, 1981). There appear to be only small differences in P concentration between different varieties within individual species of both grasses (Forbes and Gelman, 1981) and clovers (Hunt *et al.*, 1976).

Several forb species, particularly chickweed, chicory and yarrow, have been reported to be relatively rich in P, in comparison with ryegrass (see Table 3.10, p. 61).

Influence of Stage of Maturity on Concentrations of P in Herbage

In grasses, the concentration of P declines markedly with advancing maturity (Fig. 6.3; Knauer, 1966; Wilson and McCarrick, 1967; Fleming and Murphy, 1968; Fleming, 1973; Baker and Reid, 1977; Reid, 1980). Typically, the extent of the decline is slightly less than that of N (e.g. Fleming and Murphy, 1968).

Table 6.8. Herbage P concentration in four grass species: mean values from four investigations (% in DM).

	Thomas *et al.*, 1952	Coppenet, 1964	Whitehead, 1966	Lambert and Toussaint, 1978
Perennial ryegrass	0.26	0.42	0.26	0.33
Timothy	0.23	0.37	0.22	–
Cocksfoot	0.26	0.48	0.29	0.31
Meadow fescue	0.25	0.51	0.25	0.30

Table 6.9. Herbage P concentration in four legume species: mean values from four investigations (% in DM).

	Van Riper and Smith, 1959	Davies *et al.*, 1966	Whitehead and Jones, 1969	Gross and Jung, 1981
White clover	0.34	0.25	0.38	0.30
Red clover	0.29	0.19	0.27	0.30
Lucerne	0.32	0.16	0.26	0.43
Sainfoin	–	–	0.28	–

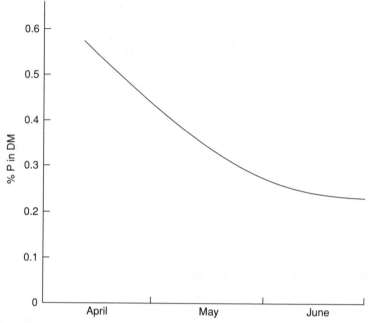

Fig. 6.3. Typical change in the concentration of P with advancing maturity in the herbage of perennial ryegrass (based on data of Fleming and Murphy, 1968; Reid, 1980).

The decline occurs in individual leaf blades, as well as in the whole herbage, as shown in a study with ryegrass (Wilman *et al.*, 1994). Dead grass herbage may contain < 0.1% P when the supply of P is low, but the concentration may be > 0.2% P for plants grown with a plentiful supply (see Table 3.11, p. 62). In the dead leaf litter of long-term grassland in the UK, the concentration of P was about 60% of the concentration in living herbage (Dickinson, 1984).

Amongst the legumes, there is generally a fairly steady decline in herbage P concentration with advancing maturity in red clover, lucerne and sainfoin, but the decline is less marked in white clover (Thomas *et al.*, 1952; Van Riper and Smith, 1959; Fleming and Coulter, 1963; Davies *et al.*, 1966, 1968; Whitehead and Jones, 1969; Baker and Reid, 1977). As an example, in red clover and sainfoin, there was a decline from about 0.38% to about 0.20% P during the period from late April to mid-July, while the corresponding decline in white clover was from 0.47% to 0.32% (Whitehead and Jones, 1969). The decline occurs in regrowth, as well as in primary growth, as shown with swards of red clover and lucerne (in Wisconsin, USA): following a cut in late August, the regrowth, sampled at intervals from early October to mid-November, decreased in herbage P concentration from about 0.30% to 0.22% (Collins, 1983). Although the decline with maturity may be influenced by a change in the leaf : stem ratio, there was a marked difference in P concentration between the young leaves of lucerne, with 0.43% P, and the old leaves, with 0.18% P (Rominger *et al.*, 1975).

Influence of Season of the Year and Weather Factors on Concentrations of P in Herbage

When a sward is cut or grazed at intervals during the growing season, changes in herbage P concentration are inconsistent. Concentrations are often at their lowest in midsummer (Reith *et al.*, 1964; Fleming and Murphy, 1968; Saunders and Metson, 1971; Reid, 1980; Lightner *et al.*, 1981; Umoh *et al.*, 1982; Whitehead *et al.*, 1983; Rowarth *et al.*, 1988; Wheeler, 1998), though sometimes they are lower in early summer than in mid- and late summer (Thompson and Warren, 1979).

Seasonal variation has been found not only in the herbage sampled as a whole but also in leaves sampled separately from stems. Thus, the concentration of P in the leaf blades of both ryegrass and white clover was generally lower in summer than at other times (Reay and Waugh, 1983), and a marked seasonal variation was found in the leaves of red clover, with the lowest levels just after midsummer (Reay and Marsh, 1976).

Differences in soil temperature may contribute to the seasonal variation, though reported effects have been variable and sometimes negligible. In a pot experiment with ryegrass, herbage P concentration was considerably greater at a soil temperature of 20°C than at either 10 or 30°C – an average of 0.21% compared with 0.13% and 0.16% (Parks and Fisher, 1958). Increasing temperature over the range 17–26°C generally decreased the concentration of P in grasses, while having variable and smaller effects in legume species (Gross and Jung, 1981). However, Italian ryegrass showed little effect of temperature on herbage P concentration over the range 11–28°C (Nielsen and Cunningham, 1964). With lucerne, increasing soil temperature over the range 5–19°C was found to increase P concentration, though there was no further increase at 27°C (Nielsen *et al.*, 1960).

Differences in the supply of water appear to have a greater and more consistent effect than differences in temperature. There are several reports of herbage P concentration being decreased by drought (e.g. Saunders and Metson, 1971; Greene *et al.*, 1987) or increased by irrigation (Kilmer *et al.*, 1960). However, in contrast to these results, one investigation showed the concentration of P in lucerne to be much lower in a wet year than in a dry year, an effect attributed to the dilution effect of increased growth (Markus and Battle, 1965). Shading may also have a small effect: it slightly increased the P concentration of Coastal Bermudagrass (Burton *et al.*, 1959).

Influence of Fertilizer N on Concentrations of P in Herbage

The application of fertilizer N often has only a small effect on the concentration of P in herbage (e.g. Stewart and Holmes, 1953; Reith *et al.*, 1964; Lambert and Toussaint, 1978; Whitehead *et al.*, 1978; Rodger, 1982). However, decreases appear to be more common than increases (MacLeod, 1965;

Heddle and Crooks, 1967; Wilman and Mzamane, 1982). The concentration of P is most likely to be increased by fertilizer N when there is a plentiful supply of plant-available P, and is most likely to be reduced when the supply of P is limited. Thus, there was an increase in P concentration in both grass and clover following the application of N as urea, in a study in which the P concentration in both species was initially > 0.4% (Ledgard and Saunders, 1982). Also, in a multicentre trial in the UK, in which fertilizer P was applied only in spring and fertilizer N was applied for each harvest, the fertilizer N generally increased herbage P at the first harvest in May but reduced it at the fourth harvest in August (Hopkins *et al.*, 1994). On the other hand, in a pot experiment involving three soils, increasing the application of N as ammonium nitrate decreased the concentration of P in ryegrass when fertilizer P was also applied, but had little effect in the absence of fertilizer P, when the herbage concentration of P was only about 0.1% (Mouat and Nes, 1983). As the uptake of P is influenced by soil pH, changes in herbage P resulting from the repeated application of fertilizer N, may sometimes be due partly to changes in soil pH.

Influence of Fertilizer P and K and of Lime on Concentrations of P in Herbage

The application of fertilizer P tends to increase herbage P concentration, though there may be little effect if the soil is already well supplied with P (e.g. Rodger, 1982; Sweeney *et al.*, 1996; Morton *et al.*, 1999; Morton *et al.*, 1999). Where there is an effect, it often increases from year to year if fertilizer P is applied annually. On a P-deficient site, the application of about 100 kg P ha^{-1} as superphosphate to grazed long-term grassland raised the average concentration of P in the year of application from 0.23 to 0.43%; smaller applications, although increasing herbage yield, had much less effect on P concentration, though three annual applications of 33 kg P ha^{-1} produced an average of 0.35% P in the herbage in the third year (Norman, 1956). In New Zealand, the annual application of 100 kg P ha^{-1} for 3 years resulted in an increase in herbage concentration from 0.30% to about 0.45% P (Rowarth *et al.*, 1988). Also, in a series of field trials in England and Wales, an application of 35 kg P ha^{-1} as triple superphosphate increased the average concentration of herbage P from 0.29 to 0.32% at the first of two silage cuts and from 0.27% to 0.29% at the second cut (Paynter and Dampney, 1991). Similarly, in Scotland, the application of superphosphate, at about 50 kg P ha^{-1}, usually resulted in small increases in the P concentration of mixed swards (Reith *et al.*, 1964). In Germany, an annual application of 35 kg P ha^{-1} year^{-1} for 50 years, to a sward cut for hay, increased herbage P concentration from 0.16% to 0.32% at the end of this period (Schellberg *et al.*, 1999). Species may vary in the extent to which herbage P concentration responds to differences in P supply via the soil. For example, cocksfoot was found to respond much more than did red fescue (Scott, 1999).

Where legumes are grown alone, fertilizer P often has a marked effect on herbage P concentration, as found with white clover (Evans *et al.*, 1986; Bailey and Laidlaw, 1999), subterranean clover (Ozanne and Howes, 1971a) and lucerne (Markus and Battle, 1965). With white clover, the application of 39 kg P ha^{-1} year^{-1} for 7 years resulted in a herbage P concentration of 0.45%, in comparison with 0.30% P where no fertilizer P was applied (Evans *et al.*, 1986).

Fertilizer K usually has little if any effect on herbage P concentration (e.g. Gardner *et al.*, 1960; Rahman *et al.*, 1960; Banwart and Pierre, 1975; Rodger, 1982; Sweeney *et al.*, 1996), though, in some instances, it has caused a small reduction (Reith *et al.*, 1964; Heddle and Crooks, 1967; Evans *et al.*, 1986; Bailey and Laidlaw, 1999). However, the addition of Na at high rates may increase herbage P (Chiy and Phillips, 1996a).

Liming often has little effect on herbage P concentration, though sometimes it causes small increases and sometimes small decreases (Evans *et al.*, 1986; Holford and Crocker, 1994; Wheeler, 1998; Bailey and Laidlaw, 1999).

Influence of Livestock Excreta on Concentrations of P in Herbage

Urine contains very little P, and is therefore liable to reduce herbage P concentration, due to the dilution effect of increased growth resulting from N and K. The average concentration of P in herbage from urine patches, on which urine had been deposited at least 6 weeks previously, was 0.36% compared with 0.48% P in herbage not affected by urine (Joblin and Keogh, 1979). Urine, applied at a rate supplying 740 kg N ha^{-1}, also reduced the concentration of P in the grass component of a grass–clover sward from 0.48% to 0.41%, though it increased the concentration in the clover from 0.39% to 0.44% P (Ledgard *et al.*, 1982).

Most excreted P is in the dung; and dung was found to increase herbage P by about 10% for 2–3 years in an area about five times greater than the area covered (During and Weeda, 1973). This suggests that, with a stocking rate of three to four cattle per hectare, about half the grazed area might well benefit, at any one time, from the P returned in dung. Increases in extractable P commonly occur in the surface soil below dung patches (Haynes and Williams, 1993), but, in short-term field trials, the uptake of P from both the organic and inorganic fractions of the P in dung appeared to be much less than from monocalcium phosphate (Rowarth *et al.*, 1990).

The application of slurry generally increases the concentration of P in herbage (e.g. Adams, 1973) though the extent of the increase may be influenced by soil type (Smith and van Dijk, 1987).

Chemical Forms of Herbage P

Most of the P in herbage is present in organic compounds, though some occurs as inorganic phosphate ions. The organic compounds include various phosphate esters (e.g. ADP and ATP), nucleic acids and phospholipids. Usually, 70–90% of the P in the herbage of grasses and legumes is soluble in water (Makela, 1967; Mansell *et al.*, 1981; Whitehead *et al.*, 1985), and the proportion tends to increase within this range as the total P concentration increases (Gillingham *et al.*, 1980). A small proportion, about 3–7%, of the herbage P appears to be bound to the cell-wall material (Whitehead *et al.*, 1985). In seeds, but not in vegetative herbage, an appreciable proportion of the total P is present as phytic acid or its Ca or Mg salts.

Occurrence and Metabolic Functions of P in Ruminant Animals

Ruminant animals contain about 0.7–1.0% P on a live weight basis (see Table 4.1, p. 73). About 86% of the total amount of P occurs as a structural component in the skeleton and teeth (National Research Council, Subcommittee, 1989), and the remainder is widely distributed in the body, especially in red blood cells, muscle and nerve tissues. The P contained in the skeleton is subject to a continual turnover, and this enables it to act as a reserve, which can be utilized for other metabolic functions when the diet is low in P. Replenishment of the skeleton occurs when the concentration of P in the diet exceeds the immediate metabolic need. Even with a high concentration of P in the diet, some mobilization from the skeleton is inevitable during early lactation, when the production of milk, which contains about 0.9–1.0 g P l^{-1}, is greatest (National Research Council, Subcommittee, 1989; Follett and Wilkinson, 1995). However, the rate at which P can be mobilized from the skeleton decreases as animals become older.

In the soft tissues, P occurs in the form of phosphate esters, nucleic acids, phospholipids, phosphoproteins and inorganic phosphates. As in plants, P has an important function in energy metabolism through the high-energy phosphate of ATP and, as a constituent of nucleic acids, it is essential for cell growth and division. The phospholipids are involved in forming the lipoprotein component of membranes and in the transport of fatty acids throughout the body (McDonald *et al.*, 1995). Inorganic phosphates are essential in buffering the pH of blood and other body fluids and in maintaining osmotic balance (Underwood and Suttle, 1999).

Whole blood contains about 350–450 mg P l^{-1}, much of which is present in the red cells. The plasma normally contains inorganic P at a concentration in the range 40–80 mg l^{-1}, though this is not closely controlled (National Research Council, Subcommittee, 1989; Underwood and Suttle, 1999), and there is also a small amount of organically bound P associated with proteins, lipids and carbohydrates (Fontenot and Church, 1979).

In ruminant animals, P is required not only for the metabolism of the animal itself but also for the synthesis of microbial biomass in the rumen. As a result, a deficiency of P may limit the digestibility of plant cell walls, particularly the cellulose component, in the rumen (Durand and Komisarczuk, 1988).

Absorption of Dietary P by Ruminant Animals

The absorption of P from the diet occurs mainly in the small intestine, with little, if any, absorption in the rumen (Little, 1981; Yano et al., 1991). Absorption appears to be an active process (National Research Council, Subcommittee, 1989). Phosphate, like N, is recycled to the rumen through saliva, and is incorporated into the rumen microorganisms. When the microorganisms are partially digested further along the digestive tract, some of the microbial P is released in soluble form, but that occurring in the cell walls has a limited digestibility. The recycling of P in saliva is a characteristic of ruminant animals and is a major factor in the homeostatic control of P (Horst, 1986; National Research Council, Subcommittee, 1989). In cattle, the saliva normally contains between 370 and 720 mg P l^{-1}, depending on the concentration in the diet, but, even at the lower end of the range, the concentration is considerably higher than that in blood plasma, 40–80 mg l^{-1} (Yano et al., 1991). The proportion of the P not absorbed from the diet plus saliva is excreted in the faeces, together with additional endogenous P, which provides further homeostatic control (National Research Council, Subcommittee, 1989).

The wide variation in the proportion of endogenous P in the faeces makes it difficult to assess the true availability of dietary P, even in balance trials with ^{32}P. Most recent estimates of the availability of the P in grass and legume herbage have ranged from about 64 to 86% (Underwood and Suttle, 1999). However, the absorption of P is influenced by the genetic characteristics and age of individual animals, by the amount of P consumed and by interactions with other constituents of the diet (Suttle, 1987). In general, the absorption of P declines with the age of the animal, for example from about 90% in young calves to 55% in fully mature cattle (National Research Council, Subcommittee, 1989). Usually, for adult cattle consuming forage diets, it is 75–85% (Ternouth et al., 1996). Dietary constituents that influence the absorption of P include divalent and trivalent cations, which tend to curtail absorption by forming insoluble phosphates, and vitamin D, which promotes absorption (McDowell, 1992). Although, in non-ruminant animals, the availability of P in feeds such as cereal grains tends to be decreased by the presence of phytic acid, the effect is negligible in ruminants, due to the production of microbial phytase in the rumen (McDowell, 1992).

Nutritional Requirements for P in Ruminant Animals

The optimum concentration of P in the diet of ruminant animals varies with the type and productivity of the animal and with the overall quality of the diet. The daily requirement for P of a lactating dairy cow is much greater than that of a beef animal, though the difference is less in terms of dietary concentration, since the dairy cow consumes more food. Nevertheless, in early lactation, highly productive cows are unable to obtain sufficient P from their diet, whatever the dietary concentration, and so utilize reserves from the skeleton. It is therefore important that the concentration of P in the diet during the later stage of lactation and after lactation has ceased should be high enough for the P in the skeleton to be replenished (Reid, 1980).

The UK and US recommendations for dietary concentrations of P for dairy cows are shown in Table 6.10 and for growing bullocks in Table 6.11. The recommendations are in broad agreement between the two countries, though those from the UK are a little higher. However, work in Australia has suggested that maintenance requirements may be met by dietary concentrations 40–50% less than those recommended in the UK (Ternouth et al., 1996). Differences between countries in recommended dietary concentrations arise mainly from differences in estimated maintenance requirements, particularly those to replace endogenous losses, and differences in the estimated absorption of dietary P (Tamminga, 1992). The UK recommendations shown in Tables 6.10 and 6.11 are based on a herbage diet of typical quality; with a diet of higher quality (i.e. a higher proportion of the gross energy being metabolized), the recommended dietary concentrations would be somewhat lower. Also, for dairy cattle, a gain in body weight has the effect that a lower dietary P concentration is adequate, while a loss in weight increases the concentration required. For growing bullocks and lambs, the initial weight to which the particular amount of gain is added has a substantial effect, with smaller animals requiring a higher concentration for a particular weight gain (Agricultural and Food Research Council, 1991).

A dietary Ca : P ratio of between 1 : 1 and 2 : 1 is often considered to be ideal for growth and bone formation, since this is the approximate ratio in bone. However, although the ratio is important when P is deficient, it is less important when P is adequate, and animals will then tolerate wide ratios of Ca : P (National Research Council, Subcommittee, 1989; Underwood and Suttle, 1999). A wide ratio of Ca : P in the diet results most commonly from a large proportion of legume herbage, since legumes often have a ratio of between 6 : 1 and 10 : 1. A plentiful supply of vitamin D also reduces the importance of the Ca : P ratio, and enables the animal to make the best use of a limited intake of P.

The range of dietary concentration of P recommended for productive animals, 0.25–0.40%, is similar to the range of critical concentration reported for the growth of young herbage, but is higher than the concentrations that often occur in mature herbage. Even when grown with adequate P, mature

Table 6.10. Recommendations for dietary P concentration (% in DM) of dairy cow (600 kg live weight) as influenced by milk yield.

	Milk yield (kg day^{-1})				
	5	10	15	20	30
Concentration of P recommended in the UK (Agricultural and Food Research Council, 1991)	0.30	0.35	0.39	0.41	0.44
Concentration of P recommended in the USA (National Research Council, Subcommittee, 1989)	0.28	0.30	0.34	0.37	0.41

Table 6.11. Recommendations for dietary P concentration (% in DM) of bullock (400 kg live weight) as influenced by rate of live-weight gain.

	Live-weight gain (kg day^{-1})				
	0.00	0.25	0.50	0.75	1.00
Concentration of P recommended in the UK (Agricultural and Food Research Council, 1991)	0.19	0.25	0.30	0.33	0.35
Concentration of P recommended in the USA[a] (National Research Council, Subcommittee, 1984)	0.16	0.18	0.22	0.26	0.30

[a]Values obtained by interpolation.

herbage and hay often contain less than 0.25% P. Deficiencies of P in livestock are therefore widespread in many parts of the world, and especially so in areas where grassland receives no fertilizer P. However, there is some evidence that grazing animals are able to mitigate this problem to some extent by selecting herbage with a higher concentration of P than the average available. Thus, sheep given a choice between different plots of subterranean clover, ranging in P concentration from 0.2% to 0.4%, preferred the highest concentration, though they were less selective when their diet consisted of dry herbage (Ozanne and Howes, 1971b).

Effects of P Deficiency on Ruminant Animals

A deficiency of P is more likely to occur in cattle than in sheep. However, with both types of animal, it is most likely to occur in the latter part of the growing season, when the herbage is mature, especially after a prolonged dry period, and in situations where the soil is either markedly acid or calcareous, or has a high capacity for retaining P in insoluble forms (Cornforth, 1984). The symptoms of P deficiency in animals include reduced feed intake, slower growth, decreased milk production, impaired reproduction, lethargy and unthriftiness. Animals with chronic P deficiency sometimes develop lameness, due to stiff joints and muscular weakness. However, the concentration of P in milk shows little change. When the deficiency is severe, there may be skeletal abnormalities, such as rickets in young animals and osteomalacia in older animals. Animals deficient in P may also show 'pica', or depraved appetite, and chew wood, bones and other unusual materials (National Research Council, Subcommittee, 1989; McDonald et al., 1995; Underwood and Suttle, 1999). Phosphorus deficiency is more common than Ca deficiency, which induces similar symptoms, partly because the concentration of P in herbage declines to a greater extent than that of Ca with increasing maturity.

The P status of ruminant animals can be assessed approximately from the concentration of inorganic P in the blood plasma. The normal concentration is between 40 and 60 mg l^{-1} for cows and between 60 and 80 mg l^{-1} for calves less than 1 year old (Miller, 1979), and a concentration appreciably below these limits suggests that P is deficient. However, there is some uncertainty in the interpretation of marginal values, since the concentration of P in plasma is influenced by other factors, such as the pattern of feeding, stress and exercise (Langlands, 1987). Where P is deficient, it can be provided to grazing animals in salt-licks or blocks containing a source of P, such as $CaHPO_4$, in the water-supply or in supplementary feeds (Underwood and Suttle, 1999).

Excretion of P by Ruminant Animals

The proportion of the dietary P incorporated into livestock products ranges from about 10%, for live-weight gain and wool growth, to about 36%, for intensive milk production (Gillingham, 1987). The remainder of the P is excreted. Usually about 95% of the excreted P occurs in the faeces and only trace amounts occur in the urine (see Table 4.9, p. 91; Barrow and Lambourne, 1962; Sommers and Sutton, 1980; Betteridge et al., 1986; Tamminga, 1992). However, the amount in the urine increases slightly if the intake of P is increased – for example, by feeding a supplement of cereal grain – and it may become substantial if the diet is deficient in Ca (Braithwaite, 1976). Excretion in the urine appears to occur only when the concentration of P in the blood plasma exceeds a renal threshold, which is usually between 60 and 90 mg P l^{-1} (Challa et al., 1989). The concentration of P in the urine of grazing steers in Australia was in the range 4–100 mg l^{-1} (Betteridge et al., 1986), while the mean concentration in urine from lactating cows on seven North Carolina dairy farms was 190 mg l^{-1} (Safley et al., 1984). The latter were probably receiving supplementary feed.

In general, the concentration of P in the faeces is related to the concentration in the diet, and is typically within the range 0.5–1.5% in the DM (Safley et al., 1984; Holechek et al., 1985; Van Faassen and van Dijk, 1987; Kirchmann and Witter, 1992; Chiy and Phillips, 1993). When the supply of P is plentiful, much of the P in faeces is derived from saliva, which is a major route for the excretion of any surplus (Underwood and Suttle, 1999). In a study with sheep, the concentration of P in the faeces was higher in winter and spring than in summer, showing a similar seasonal pattern to herbage P; and was increased by the application of fertilizer P to the sward (Rowarth *et al.*, 1988). The concentration of P in the faeces of grazing animals is normally greater than that in herbage being consumed. With sheep grazing poor quality pasture, the concentration in faeces (0.4–0.7% P) was between two and four times greater than the concentration in herbage (Floate, 1970). Some of the faecal P is inorganic (mostly dicalcium phosphate) and some is organic; and the proportion in inorganic form increases with increasing concentration of P in the diet (Haynes and Williams, 1992a). At a dietary concentration of 0.24% P, about 75% of the P in the faeces is inorganic (Barrow, 1987), reflecting considerable conversion of organic to inorganic P during passage through the animal. There appears to be little effect of the rate of fertilizer application or season of the year on the amount of organic P in the faeces (Haynes and Williams, 1993).

The average concentration of P in cattle slurry is usually about 1% on a DM basis, which is equivalent to 1 g l^{-1} in slurry with 10% DM (see Table 2.7; Westerman *et al.*, 1985; Steenvoorden, 1986; Stevens *et al.*, 1995). During the storage of slurry, there is some mineralization of organic P, and the proportion of the total P in inorganic form often increases to about 80%. Although most of the inorganic P is present in various precipitated or adsorbed forms (Gerritse and Vriesema, 1984), it appears to be utilized by plants almost as effectively as fertilizer P (Tamminga, 1992).

When the pollution of watercourses by slurry or fertilizer results in eutrophication, it is often the concentration of P in the water that has the greatest impact. Concern about this effect has led to a number of studies aimed at reducing the amounts of P excreted by livestock – for example by reducing the amount of P in the diet while maintaining productivity (Morse *et al.*, 1992; Tamminga, 1992; Valk *et al.*, 2000). In one such study, reducing the daily intake of dairy cows from 82 g to 60 g P reduced the amount excreted in the faeces by 23%, whereas increasing the intake from 82 to 112 g P day^{-1} increased the amount in the faeces by 49% (Morse *et al.*, 1992). The movement of P from a field application of slurry to watercourses can also be minimized by ensuring that the slurry is applied at rates and at times that minimize runoff (e.g. by avoiding periods when the soil is at field capacity or is frozen and when heavy rain is forecast). In addition, no slurry should be applied within a distance of several metres of the banks of rivers and streams (Withers and Jarvis, 1998).

Transformations of P in Grassland Systems

The major forms of P in the soil, plant and animal components of grassland systems are shown in Fig. 6.4. In relation to the productivity of the grassland,

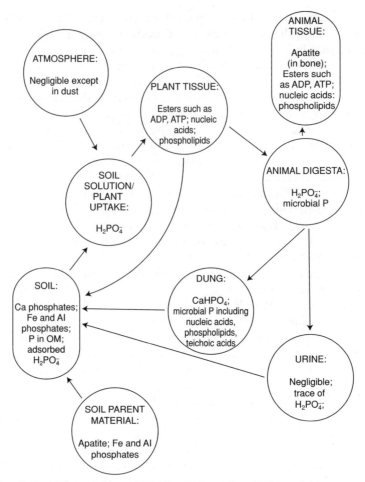

Fig. 6.4.   Outline of the main forms of P involved in the cycling of P in grassland.

the most important transformations are those that result in the release of $H_2PO_4^-$ from insoluble inorganic forms, and the microbial processes of mineralization and immobilization. Phosphorus is a macronutrient element for both plants and animals, and some of its functions (e.g. in ADP/ATP and nucleic acids) are similar in plants and animals. However, it has an additional role in animals as a major constituent of bone, and this is reflected in its concentration in whole-body tissue being five to ten times greater than in its concentration in grass herbage (see Table 4.6, p. 80).

Quantitative Balances of P in Grassland Systems

Estimated balances of P for three systems of grassland management are shown in Table 6.12. The amounts of P involved are much less than those of

Table 6.12. Estimated P balances (kg P ha⁻¹ year⁻¹) typical of three systems of grassland management: intensively managed grass, moderately intensive grass–clover and extensive grass, all grazed by cattle (except during the winter) in UK conditions (based on data from Parfitt, 1980; Aarts *et al.*, 1992; Haygarth *et al.*, 1998).

	Intensive grass (dairy cows)	Grass–clover (beef cattle)	Extensive grass (beef cattle)
Inputs			
Fertilizer	25	15	0
Atmosphere	1	1	1
Aspects of recycling			
Uptake into herbage	55	30	6
Consumption of herbage by animals	44	20	3
Dead stubble and herbage to soil	20	20	5
Dead roots to soil	9	9	2
Excreta to soil of grazed area	27	18	2.7
Outputs			
Milk/live-weight gain	12	2	0.8
Leaching/runoff	1	0.2	0.1
Loss through excretion off sward	5	0.0	0.0
Gain (loss) to soil	8	12	0.1

N, and losses of P from the system, apart from those in livestock products, are negligible, due to the retention of P in the soil. Where there is an input of P from fertilizers or from concentrate feeds supplied to livestock, there is generally an increase in the amount of P in the soil, particularly in the top few centimetres. Thus, in P balances calculated for dairy farming systems in the UK and The Netherlands, the additional inputs of P from concentrate feeds increased the amount of P in milk or live-weight gain, and also the amount of P in the soil (Aarts *et al.*, 1992; Haygarth *et al.*, 1998). In New Zealand, productive grass/clover swards are often given relatively high rates of fertilizer P (30–40 kg P ha⁻¹ year⁻¹), and P balances for such systems also show a greater accumulation of P in the soil, compared with the grass–clover sward in Table 6.12 (Parfitt, 1980; Nguyen and Goh, 1992a).

Chapter 7

Sulphur

Natural Sources and Concentrations of S in Soils

The S in igneous rocks occurs mainly in the form of metal sulphides and, when these are exposed to air during weathering, they are slowly oxidized to sulphates. However, the sulphates may be reduced back to sulphides if conditions become anaerobic. Although many sulphate salts are susceptible to leaching, some of the sulphate released by weathering is retained in sediments, as a result of precipitation or adsorption. Sediments containing clay minerals and hydrous Fe and Al oxides have the greatest capacity to retain sulphate by adsorption, while sandstones have little capacity for retention. Sediments that are formed in relatively arid conditions may contain precipitated sulphates, such as those of Ca and Mg, which would be leached with a higher rainfall. When sedimentary rocks, such as shales, are formed under chemically reducing conditions, insoluble sulphides, such as pyrites (FeS_2), are precipitated, and the concentrations of total S in shales may be higher than those in igneous rocks (see Table 2.3, p. 21).

In humid regions, it is common for the concentration of inorganic S in the soil to be less than that in the corresponding parent material, due to the loss of sulphate by leaching. Nevertheless, some sulphate is retained through adsorption by Fe and Al oxides and by clay minerals, especially under acid conditions. In soils that are anaerobic, some inorganic S is present in the form of insoluble metal sulphides, especially iron sulphides, which may be derived partly from the parent material and partly from precipitation *in situ*. In addition to the inorganic forms of S, many soils contain relatively large amounts of organic S and, in long-term grassland soils, organic S may account for as much as 90% of the total. Although much of the organic S will have been derived ultimately from the weathering of rocks, some will result from the assimilation of atmospheric or fertilizer inputs.

Differences in parent material, in the extent of leaching and in the amount of soil OM all contribute to the wide variation that occurs in the concentration of total S in soils. In leached sandy soils, there may be

< 200 mg S kg^{-1}, while, in soils from reclaimed tidal marshes and saline soils of arid and semi-arid regions, there may be > 3000 mg S kg^{-1} (Stevenson, 1986; Syers *et al.*, 1987). Grassland soils in humid temperate regions usually have a concentration within the range 200–1000 mg S kg^{-1} soil (Syers *et al.*, 1987; Nguyen and Goh, 1994a); and long-term grassland soils typically have higher concentrations of S than do comparable arable soils (Janzen and Ellert, 1998). When most of the S is in organic forms, the amount of total S in the soil tends to decline with depth (Syers *et al.*, 1987).

Atmospheric and Agricultural Inputs of S to Soils

All soils receive some input of sulphur from the atmosphere through wet and dry deposition, and many soils receive additional inputs through the addition of fertilizers or waste materials.

The atmosphere contains a number of S compounds, including sulphur dioxide (SO_2) and sulphate aerosols, hydrogen sulphide (H_2S), carbon disulphide (CS_2) and dimethyl sulphide ($(CH_3)_2S$). These compounds are derived partly from industrial activities and the burning of fossil fuels, and partly from natural sources. The SO_2 and sulphate aerosols are derived largely from human activities, and contribute most of the S deposited on the land surface. However, the naturally occurring gases, of which dimethyl sulphide accounts for about 75% (Janzen and Ellert, 1998), are slowly oxidized in the atmosphere to SO_2 and sulphate, and these are more susceptible than the sulphides to deposition. The amount of S deposited from the atmosphere varies widely, from about 1 kg S ha^{-1} year^{-1}, in areas that are remote from industrial and urban activities and are also remote from the influence of sea spray, to more than 100 kg ha^{-1} year^{-1}, in heavily polluted urban and industrial areas (Paul and Clark, 1996). Regions with the lowest inputs of S from the atmosphere include inland parts of New Zealand (Ledgard and Upsdell, 1991) and much of Australia (Till, 1975). In the UK, the amount deposited annually ranges from about 5 to > 70 kg S ha^{-1} (Sinclair *et al.*, 1993), while, in south-eastern USA, the amount is often 20–30 kg ha^{-1} year^{-1} (Kamprath and Jones, 1986). The distribution of the atmospheric input between wet and dry deposition varies with factors such as rainfall and the concentrations of S compounds in the atmosphere. The proportion of the total present in wet deposition may be < 30% in areas where rainfall is low and atmospheric pollution is high (e.g. south-east England), but is generally > 50% in areas of moderate to high rainfall and only slight pollution (Fowler, 1986).

In regions with large industrial and urban centres, such as north-west Europe and much of the USA, a large proportion of the S deposited from the atmosphere is contributed by man-made sources, mainly SO_2 released to the atmosphere from the combustion of coal, oil and natural gas. Sulphur dioxide has an average lifetime in the atmosphere of only a few days: some is absorbed directly by plant leaves, some is dissolved in rain and deposited as H_2SO_3 and H_2SO_4 and some reacts with other constituents of the atmosphere to form aerosols. In particular, SO_2 reacts with gaseous ammonia to form ammonium sulphate aerosol, and this type of reaction tends to prolong the time that the S spends in the atmosphere and to increase the distance over which it is transported.

Since the 1970s, many industrialized countries, particularly those in Europe and North America, have reduced their emissions of SO_2, often by between 20 and 60%, and this trend is continuing (Ceccotti, 1996; Department of the Environment, 1997). The amounts of S deposited from the atmosphere in these countries have therefore declined substantially. For example, at Rothamsted Experimental Station, the average amount was estimated at about 70 kg ha^{-1} year^{-1} in the mid-1970s and only 20 kg ha^{-1} year^{-1} in 1988 (McGrath and Goulding, 1990).

Some of the S that is deposited, either in rainfall or by dry deposition, is retained in the soil through the adsorption of sulphate or through immobilization in the OM. However, the fate of the remainder varies to some extent with the time of year. During the growing season, much of the S from the atmosphere is taken up by plants, either directly as SO_2 through the leaves (see p. 163) or as sulphate via the roots; but, during the winter, when there is little if any uptake, a higher proportion is leached (Syers et al., 1987). Of the S that is absorbed by plants, most reaches the soil eventually through the death and decay of plant material.

During the past 100 years, the soils in many areas have also received S incidentally from N, P and/or K fertilizers containing sulphate. In general, such additions were at a maximum during the period 1950–1965 and have declined markedly since then. The decline has been due mainly to a reduction in the use of two fertilizers that contain substantial amounts of sulphate, namely ammonium sulphate and superphosphate (Table 7.1). Ammonium sulphate has been partly replaced by ammonium nitrate and urea; single superphosphate has been partly replaced by triple superphosphate; and both have been partly replaced by compound NPK fertilizers. Although these compound fertilizers may contain S, their concentration varies from almost nil to about 9% S (Watson and Stevens, 1986; Syers et al., 1987). There is a

Table 7.1. Fertilizers and fertilizer components containing sulphur (data from Watson and Stevens, 1986; Syers et al., 1987).

Material	Formula	% S
Single superphosphate	$Ca(H_2PO_4)_2.H_2O + CaSO_4$	10–12
Triple superphosphate	$Ca(H_2PO_4)_2$	< 1.5
Ammonium sulphate	$(NH_4)_2SO_4$	24
Potassium sulphate	K_2SO_4	17–18
Gypsum	$CaSO_4$	15–18
Compound NPK fertilizers	Varied	0.1–9.1

tendency for the concentration of S in compound fertilizers to be high when the concentration of N is low, but there is considerable variation, even with the same NPK ratio (Watson and Stevens, 1986). This is because the compound NPK fertilizers are often based on mono- and/or diammonium phosphate, with other components being added to supply K and to adjust the NPK ratio. Only a few of such components contain S, and the choice is usually governed by price and availability. Thus ammonium nitrate, potassium nitrate or ammonium sulphate may be added to provide additional N, while K is provided by potassium nitrate, potassium chloride or potassium sulphate. Fertilizers that are intended for use in S-deficient areas often contain S that is added deliberately, either as a sulphate salt or as elemental S or ammonium thiosulphate (Hagstrom, 1986). The need for the deliberate addition of S to grassland has become more widespread in recent years, due partly to the reductions that have occurred in inputs from the atmosphere and fertilizers, and partly to increases in the yields of herbage.

An additional source of S in some areas is the application of sewage sludge, which usually contains between 0.6 and 1.5% S on a dry weight basis (Table 2.8, p. 32), a concentration range considerably higher than that in soils. Much of the S in sludge is in organic forms and becomes available slowly.

Recycling of S through the Decomposition of Organic Residues

The S in unharvested herbage and in roots is returned eventually to the soil. The amount returned in dead herbage is influenced by the extent to which the herbage is removed by cutting or grazing, but is likely to range between 1.5 and 20 kg ha^{-1} year^{-1}. For grazed grassland in New Zealand, the amount was estimated to be in the range 3–11 kg S ha^{-1} year^{-1} (Nguyen and Goh, 1994a). The amount of S returned in root residues is probably similar. A turnover of 5000 kg of root DM ha^{-1}, with a concentration of S of 0.15%, would provide 7.5 kg S ha^{-1} year^{-1}. However, 11 kg S ha^{-1} year^{-1} was considered a typical amount for root turnover in intensively managed grassland in New Zealand (Nguyen and Goh, 1994a).

The release of sulphate S during the decomposition of dead herbage and root material occurs in two main phases. The first phase, which takes place during the first few weeks, involves the release of any sulphate that is already water-soluble, together with the mineralization of labile organic S, while the second phase, which takes place over a period of many months, involves the mineralization of the more complex organic S compounds (Nguyen and Goh, 1994a). In a glasshouse study of the recycling of the S in clover leaf litter labelled with ^{35}S, about 9% of the S had been reincorporated into grass (shoots plus roots) within 2 months, with a day/night temperature regime of 15°/10°C, while 15% had been incorporated at 28°/22°C (Till and Blair, 1978).

When grass herbage is fed to livestock, less than 25% of the S consumed is converted into milk or body tissue, and the remainder is excreted. For example, with animals grazing intensively managed grassland, herbage consumption may amount to 10,000 kg DM ha^{-1} year^{-1}, and this would contain about 30 kg S, of which at least 22 kg would be excreted. The distribution of the excreted S between urine and faeces can vary widely (see Table 4.9, p. 91), but usually more than 50% of the S occurs in the urine, with sulphate as the major form. The sulphate is readily available for plant uptake, though it is also susceptible to leaching and incorporation into soil OM. In a study in New Zealand, in which urine was applied in both summer and winter, about 77% of the sulphate in urine applied in the summer was recovered in the herbage within 65 days, while the corresponding figure for application in winter was only 27% (Williams and Haynes, 1993a). The sulphate in urine is particularly susceptible to leaching because the amount added in the urine patch is often equivalent to about 40–50 kg S ha^{-1} in a single application, and is more than plants can take up during a period of several weeks. However, the susceptibility to leaching declines with time, due to continued uptake by plants and to the sulphate being slowly immobilized in organic forms and/or adsorbed by inorganic components of the soil (Nguyen and Goh, 1993).

The concentration of S in dung is usually 0.25–0.35% in the DM (see p. 178), and the higher the concentration, the more rapid the mineralization of S (Barrow, 1987). In dung from sheep fed grass–clover herbage, about 20% of the S was initially present as sulphate and 80% as C-bonded organic S, and, following deposition on the sward, the organic S was largely resistant to mineralization during a 9-month period (Williams and Haynes, 1993a). However, the mineralization of S from dung is greater in moist than in dry conditions, and is increased where earthworms and dung beetles are active (Saggar *et al.*, 1998). The extent to which the S from both urine and dung are taken up by plants is, of course, influenced by the pattern of their distribution within the grazing area.

When livestock excreta are collected and stored as slurries, there is little, if any, loss of S by volatilization (Beard and Guenzi, 1983), and the application of slurry is therefore an effective means of recycling S. As not all of the S in slurry is readily available, some contributes to soil reserves. The availability of the S in slurry appears to be about half that of the S in gypsum, as shown by measurements of recovery in herbage. When slurry containing 0.35 kg S m^{-2} was applied to a sward of tall fescue, the recovery of S in the herbage, during the first year, was 22%, compared with 45% for gypsum (Lloyd, 1994).

The ratios of C : S and N : S are important in the balance between the mineralization and immobilization of S, and typical values for these ratios in plant residues, animal excreta and components of the microbial biomass are shown in Table 7.2. In practice, ratios based on the amounts of C, N and S in the readily available fractions may be more important than ratios based on

Table 7.2. Some typical C : S and N : S ratios of plant residues, excreta of ruminant animals and the biomass of soil microorganisms decomposing in grassland soils (based on values for % in DM).

	C (%)	N (%)	S (%)	C : S	N : S
Dead grass herbage	48	1.8	0.15	320 : 1	12 : 1
Dead clover herbage	48	2.7	0.18	270 : 1	15 : 1
Grass roots	49	1.4	0.15	330 : 1	9 : 1
Clover roots	50	3.8	0.35	140 : 1	10 : 1
Faeces	48	2.4	0.3	160 : 1	8 : 1
Urine	43	11	0.65	66 : 1	17 : 1
Bacteria	50	15	1.1	45 : 1	14 : 1
Fungi	44	3.4	0.4	110 : 1	8.5 : 1

the total amounts of C, N and S (Ghani *et al.*, 1992), but they are not easily determined.

Forms and Availability of S in Soils

The distribution of soil S between organic and inorganic forms varies with factors such as the total amount of OM in the soil and the soil's pH, drainage status and mineralogical composition. In the top 10 cm of many grassland soils, more than 90% of the total S is organic. The organic S is derived partly from plant residues and animal excreta and partly from inorganic sulphate assimilated by the soil microorganisms. The application of fertilizer S increases the amount of organic S, as well as sulphate, in the soil, as shown following the repeated application of superphosphate to a grass–clover sward (Haynes and Williams, 1992b). In general, the amount of organic S in a soil is related to the amounts of C and N, and the ratio for C : N : organic S in grassland soils is typically about 120 : 10 : 1.3, equivalent to a C : S ratio of 92 : 1 and N : S ratio of 7.7 : 1 (Perrott and Sarathchandra, 1987; Nguyen and Goh, 1994a). However, the ratios between C, N and organic S vary to some extent with factors such as temperature, moisture status and the parent material of the soil and the extent to which excreta are returned (Nguyen and Goh, 1994a). The N : S ratio shows less variation than the C : S ratio, suggesting that the stabilization of S in humified OM is similar to that of N.

The soil organic S can be categorized into a labile pool, amounting to 10–50% of the total, and a more inert pool. The labile pool consists of the readily mineralizable S in plant litter, animal excreta and microbial biomass, whereas the inert pool consists mainly of humified OM (Nguyen and Goh, 1994a). Microbial S constitutes 2–3% of the total organic S (Paul and Clark, 1996), though the amount fluctuates in response to soil moisture and temperature (Ghani *et al.*, 1990). Although the soil organic S has not been fully characterized in chemical terms, much is carbon-bonded, as in the S-amino acids. However, some organic S occurs in the form of organic sulphates, including polysaccharide sulphate and choline sulphate (Paul and Clark, 1996), the latter being a component of microbial cell walls (Scott, 1985).

The mineralization of organic S to inorganic sulphate is primarily a microbiological process and, to that extent, is analogous to the mineralization of C and N. Although C, N and S often mineralize at rather different rates (Scott, 1985), the effects of temperature and moisture status are similar. Thus, the mineralization of S generally increases with increasing temperature over the range 5–30°C and is at a maximum when soil moisture status is close to field capacity and when the pH is near to neutral (Swift, 1983; Syers *et al.*, 1987). The decomposition of plant residues with a C : S ratio of > 400 : 1 ($\equiv 0.1\%$ S) generally results in the immobilization of S, whereas with a C : S ratio of < 200 : 1 ($\equiv 0.2\%$ S) there is fairly rapid net mineralization (Paul and Clark, 1996). Differences in the mineralization rates of N and S in soil OM may be due partly to some of the N and S occurring in different components of the OM and partly to recently added plant residues differing in N : S ratio from the bulk of the soil. In addition, some organic sulphates mineralize readily as a result of repeated drying and wetting, without microorganisms being involved (Freney, 1986).

In soils that are aerobic, the inorganic fraction of the soil S is almost entirely sulphate, and, in soils that are subject to leaching, most of the sulphate present is adsorbed by hydrous Fe and Al oxides and clay minerals. The Fe oxides are especially important in adsorption and, of the clay minerals, kaolinite adsorbs more than montmorillonite. Adsorption occurs most strongly under acid conditions and declines with increasing soil pH over the range 4.0–6.5, with little adsorption occurring when the pH is above 6.5 (Metson, 1979; Syers *et al.*, 1987; Marsh *et al.*, 1992). In grassland soils, adsorbed sulphate usually amounts to between 3 and 30 mg S kg^{-1}, depending on the quantity and nature of the clay fraction and the soil pH (Nguyen and Goh, 1994a). The amount of adsorbed sulphate often increases at lower depths in the soil, reflecting factors such as more acid conditions and/or larger amounts of hydrous Fe and Al oxides. Sulphate is adsorbed less strongly than phosphate and, as a consequence, fertilizer phosphate tends to displace sulphate already adsorbed (Nguyen and Goh, 1994a) and to reduce the capacity of soils to adsorb newly applied sulphate (Janzen and Ellert, 1998). Liming also tends to displace adsorbed sulphate and reduce the capacity for adsorption. Although in humid regions only a small proportion of the sulphate in soils is water-soluble, the proportion is much higher in semi-arid soils (Freney, 1986). Semi-arid soils that are calcareous often contain substantial amounts of S as calcium sulphate, a compound that is removed by prolonged, but not occasional, leaching.

In soils that are anaerobic, the dominant form of inorganic S is sulphide, and any sulphate present is progressively reduced to sulphide by microbial activity. The initial product may be H_2S, but this generally reacts with soil constituents, such as Fe oxides, to form FeS_2 or FeS (Metson, 1979). As a result, there is normally no volatilization of H_2S from soils (Freney, 1986), though, when soils are completely waterlogged, trace amounts of carbon disulphide, dimethyl sulphide and possibly other organic S compounds produced by microorganisms, may volatilize (Farwell *et al.*, 1979; Brown, 1982). If soils that are normally waterlogged are drained, the Fe sulphides are slowly oxidized to sulphates, partly through the action of soil micro-organisms, especially *Thiobacillus* spp.

The continued uptake of S by plant roots depends on the soil's supply of water-soluble sulphate being maintained. Some sulphate is supplied via deposition from the atmosphere and some by the application of fertilizers, but the mineralization of organic S and the return of sulphate in animal excreta are major sources for grassland. Since the mineralization of S is largely a microbial process, there is inevitably some immobilization of S at the same time, especially if the soil microbial population is increasing and if the OM being decomposed is low in S.

Losses of S from Soils

The main route for the loss of S from soils is the leaching of sulphate, though trace amounts of S may be lost in volatile forms from waterlogged soils. Leaching losses are greatest when there is a high concentration of sulphate, due to the application of sulphate in fertilizers or its return in urine. The prolonged deposition of sulphate from the atmosphere and the mineralization of S from soil OM can also contribute appreciably to the amount of sulphate subject to leaching (Sakadeven *et al.*, 1993a). Whatever the main input, the extent of leaching of sulphate varies with the adsorption capacity of the soil, and hence with soil type. It also varies with the amount and timing of rainfall (Heng *et al.*, 1991). As an example of the effect of soil type, the sulphate leached from four contrasting soils, sown to ryegrass in monolith lysimeters, ranged from the equivalent of 12 kg S ha^{-1} for a clay soil to 30 kg S ha^{-1} for a sandy soil in the first year of measurement (Bristow and Garwood, 1984). In this study, there was little input of S from fertilizer (3 kg ha^{-1} year^{-1}) but a substantial input from the atmosphere, though this declined from 49 to 33 kg S ha^{-1} year^{-1} during the measurement period (1979–1982). Over the 4-year period, the amount of sulphate leached was approximately equal to that removed in cut herbage for the sandy soil, though it was less than half that removed in the herbage from the clay soil. When large amounts of sulphate are applied in fertilizer, there is normally an increase in the amount of sulphate leached. However, when superphosphate, at a rate providing 43 kg S ha^{-1}, was applied to a grazed grassland in New Zealand, the increase was only small, from 11 to 15 kg ha^{-1} year^{-1} (Smith *et al.*, 1983). The extent of leaching of the sulphate returned in urine also varies considerably, and may be as much as 70% on soils with a low sulphate retention capacity (Hogg, 1981; Nguyen and Goh, 1993). Factors such as liming and the application of phosphate, which reduce the capacity of a soil to adsorb sulphate, increase its susceptibility to leaching (Bolan et al., 1988). When phosphate and sulphate are applied together, as in super-phosphate, the phosphate is often adsorbed mainly in the topsoil, whereas the sulphate moves downwards and tends to be adsorbed in the subsoil (Syers et al., 1987).

Although various volatile S compounds have been identified in soils or decomposing organic materials (see p. 161), only trace amounts are actually volatilized from soils, even under waterlogged conditions (Farwell et al., 1979; Freney, 1986). Volatilization from dung and urine also appears to be negligible (Kennedy and Till, 1981).

Assessment of Plant-available S in Soils

Plant-available S is often assessed by extracting the soil with a solution containing an anion, such as phosphate, bicarbonate, chloride or acetate (Jones, 1986; Saunders et al., 1988; Kowalenko, 1993). Extraction with water alone is sometimes used, but this removes only the small amount of sulphate in the soil solution, an amount that is liable to fluctuate from day to day (Nguyen and Goh, 1994a). Extraction with a solution containing an anion, such as phosphate, displaces at least some of the adsorbed sulphate, and therefore extracts a larger and less variable amount. However, in many investigations, the amount of extractable sulphate has shown only a poor correlation with the response to fertilizer S (Syers et al., 1987; Saunders et al., 1988; Nguyen and Goh, 1994a), probably due mainly to one of two factors. One factor is that, in some areas, the soil may contain an accumulation of sulphate in the subsoil, below the depth of sampling, which can be utilized by grassland plants (Metson, 1979; Syers et al., 1987; Saunders et al., 1988). A second factor is that more than 90% of the S in the topsoil of many grassland areas is organic, and the measurement of extractable sulphate does not take into account mineralization during the growing season. This may amount to about 20 kg S ha^{-1} year^{-1}, as reported for intensively managed grassland in New Zealand (Nguyen and Goh, 1992b). Although mineralization can be taken into account by incubating the soil, with or without the addition of non-sulphate S, this type of procedure is time-consuming. Nevertheless, the sulphate (extractable by 0.01 M CaCl$_2$) produced during the incubation of a soil amended with either cysteine or elemental S produced a better correlation with the yield response of grass to fertilizer S on ten soils than did several other methods, including extraction with Ca(H$_2$PO$_4$)$_2$ (Hoque et al., 1987).

As an alternative to soil analysis, herbage analysis can be used to assess the need for fertilizer S (see p. 169). Another approach to assessing the need for fertilizer S is to estimate the net loss from the system by calculating a mass balance, an approach that has been developed in New Zealand (Nguyen and Goh, 1993, 1994a; Goh and Nguyen, 1997).

Uptake of S by Herbage Plants

The main route for the uptake of S is through the roots, in the form of sulphate, but plants may also obtain appreciable amounts of S from the SO_2 in the atmosphere, via the stomata in their leaves. In contrast to phosphate, which moves in the soil mainly by diffusion, sulphate moves mainly by mass flow (Syers *et al.*, 1987). The difference is due to sulphate being adsorbed by soils less strongly than phosphate and to the concentration of sulphate ions in the soil solution being up to 100 times greater than that of phosphate. When there is a substantial amount of adsorbed sulphate in the subsoil, a significant proportion of total plant S may be taken up from a depth greater than 20 cm, as shown for ryegrass and white clover by the use of sulphate labelled with ^{35}S (Gregg *et al.*, 1977).

After being taken into the root, sulphate moves upwards in the xylem and is converted to organic forms in the leaves. The sulphate is first activated by reaction with ATP to form adenosine phosphosulphate. Much of this is then chemically reduced, with the S being assimilated into cysteine, some of which is then converted into cystine and methionine. A large proportion of each of these three S-amino acids is incorporated into proteins (Thompson *et al.*, 1986). A smaller proportion of the adenosine phosphosulphate provides sulphate for transfer to sulphate esters, and thus for the synthesis of sulpholipids and sulphated polysaccharides. If sulphate is taken up in amounts that exceed the needs for assimilation into organic compounds, it tends to accumulate in the plant in inorganic form. As plants mature, there may be some remobilization of S, and, when S moves in the phloem, it is mainly in the form of cysteine contained in the tripeptide glutathione (Thompson *et al.*, 1986).

When grass and clover are grown together, the grass has a greater ability than the clover to obtain sulphate from the soil (Nguyen and Goh, 1994a). Grasses also have a greater tendency than clovers to accumulate sulphate in their tissues (Metson, 1979). When grown in soils that are seriously deficient in S, but in areas with a high concentration of SO_2 in the atmosphere, plants may absorb as much as half their total S content as SO_2 through their leaves (Siman and Jansson, 1976). Even with smaller amounts of SO_2 in the atmosphere, such absorption may be sufficient to prevent plants with a marginal supply of S from the soil becoming deficient, especially when high rates of fertilizer N are applied (Cowling and Lockyer, 1978). The extent to which atmospheric SO_2 is absorbed appears to vary between species: for example, adsorption per unit area of leaf surface was found to be greater for lucerne than for ryegrass (Garsed and Read, 1977). With plants that are not deficient in S, the absorption of atmospheric SO_2 increases their concentration of S, and much of additional S may be in the form of sulphate (Lockyer, 1985).

Functions of S in Herbage Plants

Sulphur is essential as a constituent of plant proteins and also of several compounds involved in redox reactions. In plant proteins, the ratio of N : S is fairly constant at about 36 : 1 in terms of the number of atoms, which is equivalent to 15.7 : 1 on a weight basis. This is close to the ratio of total N to total S that occurs in plants with adequate amounts of each element but no appreciable excess of either. An important role of S in the structure of proteins is to link together adjacent polypeptide chains through the disulphide bonds of two cysteine units (Fig. 7.1). The position of such bonds is an important factor in the structural configuration of any particular protein, and therefore in its enzymic or other properties. Since nearly all plant enzymes are proteins, a deficiency of S tends to impair the functioning of many biochemical processes. When S is deficient, the concentrations of the S-containing amino acids and several other amino acids in plant tissues are reduced, but, as found with ryegrass, the concentration of asparagine may be increased (Millard *et al.*, 1985).

Of the non-protein S in plants, some is present in coenzymes, including biotin, thiamine and coenzyme A. About 1–2% of the organic S in the plant is present in glutathione, a water-soluble tripeptide (Fig. 7.1), which has a number of functions in plants. Being a powerful antioxidant, glutathione protects plant tissues against damage from oxidation. It also provides a temporary store of reduced S, and is a precursor of several compounds that are involved in the detoxification of heavy metals (Marschner, 1995). The sulpholipids, which contain sulphate groups linked to sugars, such as glucose, by an ester bond (Fig. 7.1), are structural constituents of all biological membranes, and are thought to be involved in the regulation of ion transport across membranes (Marschner, 1995). Methionine, as well as being a constituent of many proteins, has an important role in the biosynthesis of lignin, pectin, chlorophyll and flavonoids, through the formation of S-adenosyl methionine, which is a methyl group donor (Stevenson, 1986).

In legumes, a deficiency of S restricts N_2 fixation and, consequently, S-deficient legumes are relatively low in total N (Sinclair *et al.*, 1996b). For example, white clover containing < 0.13% S had a concentration of N of only 2%, compared with the normal 4.5–5.0% N (McNaught and Christoffels, 1961).

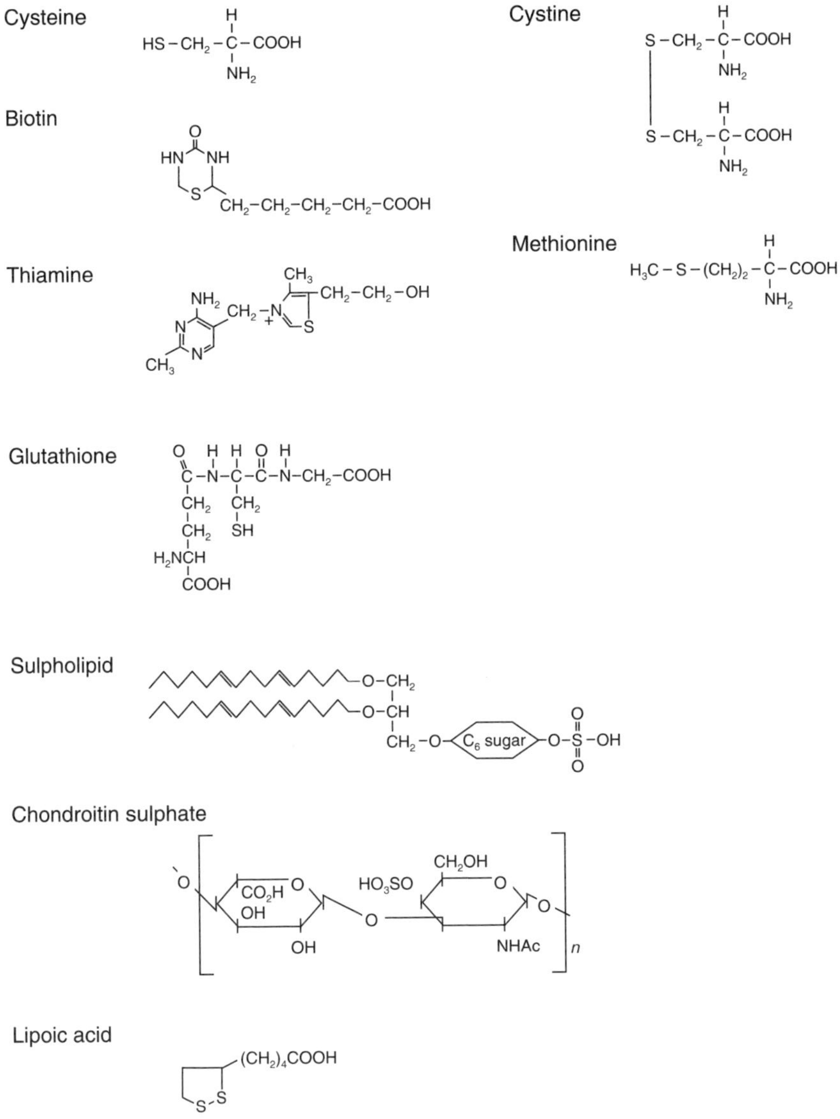

Cysteine

$$HS-CH_2-\underset{\underset{NH_2}{|}}{\overset{\overset{H}{|}}{C}}-COOH$$

Cystine

$$S-CH_2-\underset{\underset{NH_2}{|}}{\overset{\overset{H}{|}}{C}}-COOH$$

$$S-CH_2-\underset{\underset{NH_2}{|}}{\overset{\overset{H}{|}}{C}}-COOH$$

Biotin

Methionine

$$H_3C-S-(CH_2)_2-\underset{\underset{NH_2}{|}}{\overset{\overset{H}{|}}{C}}-COOH$$

Thiamine

Glutathione

Sulpholipid

Chondroitin sulphate

Lipoic acid

Fig. 7.1. Chemical structure of some plant and animal constituents containing S.

Influence of Fertilizer S on Yields of Grassland Herbage

The response of grassland to fertilizer S is usually greatest on soils of light to medium texture, in areas with little atmospheric pollution. As noted on p. 155, regions with large industrial and urban centres may receive substantial amounts of plant-available S from SO_2 in the atmosphere, but, in many countries, these emissions have declined in recent years and S deficiency has become more widespread. Two other factors that have contributed to an increased incidence of S deficiency are the reduction in the use of ammonium sulphate and superphosphate as fertilizers and the trend towards higher yields from both grassland and arable crops.

Where supplies of S are marginal or deficient, the response to fertilizer S is usually greater from intensively managed grass, especially cut grass, than from most arable crops (Syers *et al.*, 1987). On many soil types, grassland that is cut repeatedly is likely to show a response to fertilizer S when the deposition from the atmosphere is < 30 kg S ha^{-1} year^{-1} (Sinclair *et al.*, 1993), though responses are most likely on sandy soils containing little OM (Murphy and Boggan, 1988). The response tends to increase as the season progresses and, with grass cut for silage, fertilizer S may be needed only for the second and subsequent cuts (Dampney and Unwin, 1993; Richards, 1999). However, the likelihood of grassland responding to fertilizer S is increased when high rates of N are also applied, as shown in Fig. 7.2.

Responses to fertilizer S are common in New Zealand (Nguyen and Goh, 1992b), in Ireland (Murphy and Boggan, 1988) and in parts of the USA (Lamond *et al.*, 1995) and Canada (Beaton and Soper, 1986). In New Zealand, there is often a strong interaction between applications of S and P, and clover shows a greater response than grass to fertilizer S. Increasing the rate of application of fertilizer S from nil to 30 kg S ha^{-1} increased the average proportion of clover in a grass–clover sward from 30 to 46% (Sinclair *et al.*, 1996b). In areas showing a response, the recommended rates of application of S for grassland are usually within the range 10–40 kg ha^{-1} year^{-1} (Syers *et al.*, 1987; MAFF, 1994).

Fertilizer S is usually applied as a sulphate salt in conjunction with N, P or K. Gypsum (CaSO$_4$.2H$_2$O) is sometimes used as a source of S, but potassium sulphate, ammonium sulphate and ammonium thiosulphate are alternatives in which the S is more readily available (Lamond *et al.*, 1995). Elemental S may be incorporated into other fertilizers, and is particularly suitable for grassland, as the S becomes available more slowly than that from sulphate (Boswell and Gregg, 1998). It is unusual for S to be applied alone as a fertilizer nutrient, but, when this is necessary, possible sources include gypsum and elemental S (Syers *et al.*, 1987).

Distribution of S within Herbage Plants

In grasses, the concentration of S is generally lower in the roots than in the shoots (Table 7.3; Wilkinson and Lowrey, 1973; Gilbert and Robson, 1984b). This pattern also occurred in lucerne (Rominger *et al.*, 1975), but in clovers the concentration of S has been found to be higher in the roots than in the shoots (Table 7.3; Gilbert and Robson, 1984a, b). Concentrations of S in the roots of grasses are usually in the range 0.13 to 0.22% and are often increased by high rates of fertilizer N (Whitehead, 1970; Wilkinson and

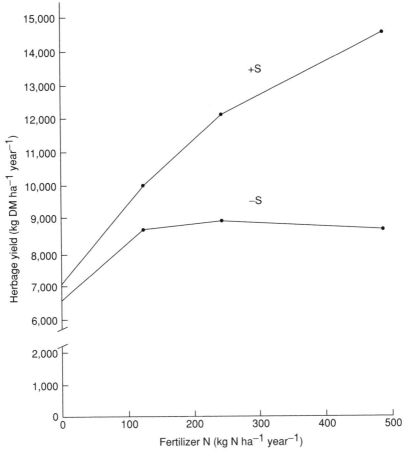

Fig. 7.2. The effect of the application of fertilizer S on the response of perennial ryegrass to fertilizer N (S applied at 25 kg ha⁻¹ to nil-N plots and at 0.18 times the rate of N to other plots) at a site in Ireland (data from Murphy and Quirke, 1997).

Lowrey, 1973; Nguyen and Goh, 1994a). Concentrations in the roots of white clover and red clover are generally higher than those of grasses, and show a wider range of up to 0.60% S (Whitehead, 1970; Gilbert and Robson, 1984b). However, root concentrations appear to the lower in lucerne than in clovers (Whitehead, 1970; Rominger *et al.*, 1975). Root concentrations increase with increasing S supply, as shown for a grass–clover sward receiving superphosphate (Nguyen and Goh, 1992b).

Within the shoots of grasses, the concentration of S is much higher in the leaf blades than in the leaf sheath and stem (see Table 3.3, p. 53). Similarly, in subterranean clover, the concentration of S in the leaf laminae was about twice that in the petioles (Spencer *et al.*, 1977). Although only small amounts of S are involved, volatile compounds, such as dimethyl sulphide and hydrogen sulphide, may sometimes be released from the leaves of growing plants (Janzen and Ellert, 1998).

Range of Concentrations and Critical Concentrations of S in Herbage Plants

The concentration of S in grassland herbage is usually within the range 0.15–0.60% (Table 7.4). However, in S-deficient areas, concentrations as low as 0.04–0.06% have been recorded – for example, in grass–legume mixtures in Pennsylvania, USA (Adams, 1975). With a plentiful supply of S, concentrations may be as high as 0.75% S, as found, for example, in Italian ryegrass in the UK (Cunningham, 1964) and in grass–clover herbage in New Zealand (Smith and Cornforth, 1982).

The concentration of S in grasses declines steadily with advancing maturity (see p. 171), and this has an effect on the critical concentration. However, the effect of the stage of maturity can be taken into account to some extent by analysing for N and/or fibre in addition to S. Some reported values for the critical concentration of S in grasses and legumes, mainly at a young stage of growth, are shown in Table 7.5, together with some critical values for the N : S ratio. These and other studies suggest that

Table 7.3. Distribution of S between shoots and roots of ryegrass and white clover grown in a mixed sward and sampled 76 days after the application of gypsum at 45 kg S ha^{-1} (from Goh and Gregg, 1980).

	Shoots	Roots
Ryegrass	0.57	0.16
White clover	0.19	0.22

Table 7.4. Concentrations of S in grassland herbage analysed for survey or advisory purposes in several regions (% in DM).

Region	No. of samples	S concentration Range	Mean	Reference
UK	101	0.13–0.75		Cunningham, 1964
Scotland	126	0.12–0.48	c. 0.24	McLaren, 1975
Norway	c. 130	0.09–0.41		Frøslie and Norheim, 1983
The Netherlands	> 700	0.28–0.56[a]	0.42	Kemp and Geurink, 1978
Pennsylvania, USA	235	0.04–0.43	0.23	Adams, 1975
North Island, New Zealand	> 2500	0.12–0.93	0.34	Smith and Cornforth, 1982

[a]Based on 2 × SD.

Table 7.5. Some reported values for the critical concentration of S and critical N : S ratio in some grasses and legumes.

Species	Portion of plant	% S	N : S ratio	Reference
Ryegrass	Young herbage	0.2	14–15	Cowling and Jones, 1978
Ryegrass	Young herbage	0.1	20	Gilbert and Robson, 1984a
Ryegrass	Young herbage	0.18–0.25		Smith *et al.*, 1985
Kentucky bluegrass	4–6 weeks' regrowth	0.16–0.24		Kelling and Matocha, 1990
White clover	Young herbage	0.25		McNaught and Christoffels, 1961
White clover	Young herbage	0.25–0.30	17–19	McNaught, 1970
White clover	Pre-flowering	0.18	22	Andrew, 1977
Subterranean clover	Young herbage	0.18–0.21	18–21	Jones *et al.*, 1980
Subterranean clover	Young herbage	0.18		Gilbert and Robson, 1984a
Lucerne	Pre-flowering	0.2	18–21	Andrew, 1977
Lucerne	Whole shoots	0.10–0.25		Kelling and Matocha, 1990
Lucerne	Whole shoots	0.22	11 : 1	Pumphrey and Moore, 1965b
Lucerne	Early bloom	0.20	17 : 1	Westermann, 1975

the critical S concentration in young grass herbage is about 0.20–0.21% and the critical N : S ratio about 14 : 1–15 : 1 (Syers *et al.*, 1987). However, the N : S ratio alone does not provide a good indication of the S status of the plant, as it tends to decline with increasing maturity (Salette, 1978; Whitehead and Jones, 1978). Also, for any specific N : S ratio, it is possible for both N and S to be excessive, optimal or deficient (Stevens, 1985). Therefore, in assessing S status, it is best to take into account both the total S concentration and the N : S ratio. Despite these complications, herbage analysis is often regarded as being more effective than soil analysis in assessing the S status of grassland (e.g. MAFF, 1994).

In legumes, especially white clover, the critical S concentration shows less change than in grasses with increasing maturity. Also, the use of the N : S ratio alone may be more valid for legumes than for grasses, because they generally receive no fertilizer N and are therefore unlikely to contain excessive N. On the other hand, a deficiency of S tends to reduce the concentration of N in clover, though it does increase the N : S ratio to some extent (Sinclair *et al.*, 1997). The critical S concentration for clovers appears to be about 0.20–0.25% and, as with grasses, the critical N : S ratio is about 15 : 1 (Syers *et al.*, 1987).

Another means of assessing S status is to measure the concentration of sulphate in the herbage, since this concentration is always low when S is deficient, but it increases when the supply is more than adequate. For ryegrass, the critical concentration of sulphate S appears to be about 300–320 mg kg^{-1} (Dijkshoorn *et al.*, 1960; Cowling and Jones, 1978). In lucerne, a critical concentration of about 150 mg kg^{-1} SO$_4$-S in the third leaf from the top of the stem has been proposed (Ulrich *et al.*, 1967). Although the concentration of sulphate S tends to decline with maturity, the ratio of sulphate S : total S appears to be more stable and may therefore provide a better indication of S status. A value of about 0.1 for this ratio has been proposed as critical for subterranean clover (Jones *et al.*, 1980).

Influence of Plant Species and Variety on Concentrations of S in Herbage

Herbage concentrations of S are generally similar in grasses and legumes (Walker and Adams, 1958; McNaught and Christoffels, 1961; Nguyen *et al.*, 1989; Mackay *et al.*, 1995). When grasses and legumes are grown separately, the legumes may have higher concentrations than the grasses (Mayland and Wilkinson, 1996), but, in mixed swards, the grasses often contain more than the legumes. With grass and clover growing together in mixed swards at seven sites in New Zealand and analysed at monthly intervals, the concentrations in both species were in the range 0.2–0.4% and, in any direct comparison, the concentration of S was almost always greater in the grass (Metson and Sanders, 1978b). With mixed swards in the UK, there was on average no difference between the grass and the clover at the first harvest in May, but the grass contained more than the clover in July and September (Table 7.6). The difference between grasses and legumes tends to increase as the supply of S is increased (Metson and Saunders, 1978b; Nguyen *et al.*, 1989).

There are few comparative data on concentrations of S in individual grass species, but differences generally appear to be small (Whitehead and Jones, 1978). However, tall fescue was reported to have a higher concentration than cocksfoot at the same herbage yield (Salette, 1978); and tall fescue and perennial ryegrass had higher concentrations than did cocksfoot and smooth brome-grass (Powell *et al.*, 1978). In another study, reed canary-grass contained much more than brome-grass, timothy and cocksfoot (Smith *et al.*,

Table 7.6. Total S and sulphate S (% in DM) in ryegrass and white clover from mixed swards, with and without fertilizer N as ammonium nitrate at 200 kg N ha^{-1} year^{-1}, at alternate harvests taken in May, July and September: mean values for 12 sites (from Whitehead *et al.*, 1983).

	May		July		September	
	Total S	Sulphate S	Total S	Sulphate S	Total S	Sulphate S
Ryegrass – N	0.30	0.13	0.38	0.21	0.49	0.27
Ryegrass + N	0.31	0.14	0.34	0.15	0.36	0.16
Clover – N	0.30	0.07	0.24	0.04	0.25	0.02
Clover + N	0.30	0.08	0.24	0.04	0.24	0.02

1974). With the legumes, appreciable differences between species have been reported, with mean values of 0.29% for white clover, 0.21% for red clover, 0.27% for lucerne and 0.24% for sainfoin in the UK (Whitehead and Jones, 1969). A similar difference between red clover and lucerne was found in New Zealand (McNaught and Christoffels, 1961) and in USA (Smith *et al.*, 1974).

Influence of Stage of Maturity on Concentrations of S in Herbage

In grasses, the concentration of S declines with advancing maturity, though to a lesser extent than the concentration of N, and so the N : S ratio also declines. The concentration of S in mature plants is often rather more than half that in young herbage in early spring (Fig. 7.3). The concentration of S may decline further when herbage is dead; and, in prairie grassland, the average concentration in dead herbage was 0.06% S compared with 0.14% S in young herbage (Clark and Woodmansee, 1992).

In legumes, the change in concentration with advancing maturity is less consistent, especially with white clover. However, there is an appreciable decline in red clover and sainfoin (Whitehead and Jones, 1969) and sometimes in lucerne (Pumphrey and Moore, 1965a). In subterranean clover, the concentration of S in both young and old laminae and petioles declined with the age of the plant, up to 20 weeks, and the declines were greater when fertilizer S was applied (Spencer *et al.*, 1977).

As plants senesce, a small amount of S (generally less than 1 kg ha^{-1} year^{-1}) may be released to the atmosphere (Mikkelsen and Camberato, 1995), but this process makes a negligible contribution to the decline in concentration with maturity.

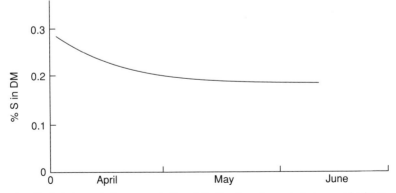

Fig. 7.3. Typical change in the concentration of S with advancing maturity in grass herbage (based on data from Salette, 1978; Whitehead and Jones, 1978).

Influence of Season of the Year and Weather Factors on Concentrations of S in Herbage

Seasonal changes in herbage S concentration are often small or inconsistent. Both grass and clover in New Zealand showed a tendency for the concentration of S to be lowest in summer and highest in late autumn and winter (Metson and Saunders, 1978b), and a similar pattern was observed for subterranean clover in Australia, with peak concentrations in winter (Spencer, 1978). In the UK, the concentration of S in the herbage of a grazed grass sward increased during the season, from May to September (Leech and Thornton, 1987), and also in the grass, though not the clover, component of grass–clover swards grown without fertilizer N (Whitehead et al., 1983). However, there was no consistent trend during the season in ryegrass grown alone with high rates of N and K at a range of sites in England and Wales (Whitehead and Jones, 1978).

A relatively low concentration in summer is consistent with the finding that, in both ryegrass and subterranean clover, the concentration of S in the herbage declined as the soil temperature was increased over the range 7–19°C (Gilbert and Robson, 1984a). However, when subterranean clover was grown in pots under controlled conditions, there was no difference in herbage S concentration between soil temperatures of 12°C and 22°C or between different soil moisture contents (Reddy et al., 1981).

Influence of Fertilizer N on Concentrations of S in Herbage

The application of fertilizer N tends to reduce the concentration of S in grass herbage if supplies of S are limited (e.g. Goh and Kee, 1978), but may increase the concentration if supplies of S are plentiful (e.g. Salette, 1978). Sometimes there is little or no effect (e.g. Rahman et al., 1960; Whitehead et al., 1978; Burmester et al., 1981). In the UK, fertilizer N has been found to increase herbage S at the first harvest of the season, but to reduce it later (Table 7.6; McLaren, 1976; Hopkins et al., 1994), possibly due to supplies of S from the atmosphere and/or fertilizer being less during the summer than during the spring.

When fertilizer N is applied in the form of ammonium sulphate, the concentration of S in the herbage is, of course, increased. The application of 470 kg S ha^{-1} year^{-1} as ammonium sulphate increased herbage S in a grass–clover mixture from 0.36 to 0.59% (Conroy, 1961); and the application of 180 kg S ha^{-1} year^{-1}, also to a grass–clover sward, increased herbage S from 0.27 to 0.46% in the third year (Jones, 1960). In an S-deficient area, the application of S as ammonium sulphate or ammonium thiosulphate, at 34 kg S ha^{-1}, increased the average concentration of S in brome-grass from 0.08% to 0.12% (Lamond et al., 1995).

Influence of Fertilizer P, K and S on Concentrations of S in Herbage

Fertilizers providing P and K, other than those containing S, generally have only a small effect on the concentration of S in herbage. However, fertilizer P as triple superphosphate reduced the concentration of S in white clover (Bailey and Laidlaw, 1999) and, in a trial with a grass–clover sward, a significant reduction was caused by KCl providing 78 kg K ha^{-1}, possibly due to a dilution effect (McLaren, 1976). Fertilizers that contain S as an incidental constituent are likely to increase the concentration of S in herbage. For example, superphosphate supplying about 60 kg S ha^{-1} increased the herbage S concentration of long-term grassland from an average of 0.30% to 0.36% S (McLaren, 1976).

Sulphur applied deliberately, for example as gypsum, potassium sulphate or elemental S, normally increases the concentration of S in the herbage (Nguyen *et al.*, 1989; Chiy *et al.*, 1999; Morton *et al.*, 1999). Thus, total S was increased from 0.33 to 0.40% in tall fescue, and from 0.29 to 0.37% in cocksfoot, when gypsum providing 132 kg S ha^{-1} was applied: the increase was mainly in non-protein S (Spears *et al.*, 1985). There was also an increase in brome-grass from 0.09% to 0.26% S following the application of potassium sulphate supplying 76 kg S ha^{-1} year^{-1} (Laughlin *et al.*, 1973). In white clover grown with ryegrass, the application of 60 kg S ha^{-1} increased the concentration of S from 0.18 to 0.26% (Sinclair *et al.*, 1996a).

Influence of Livestock Excreta on Concentrations of S in Herbage

Urine has been reported to reduce the concentration of S in ryegrass herbage (Joblin and Pritchard, 1983) and to have no significant effect with a grass–clover pasture (Saunders, 1984). These results are somewhat surprising since most excreted S occurs as sulphate in the urine (see p. 177). However, the N : S ratio of urine is somewhat wider than the N : S ratio of herbage (Table 7.2; Barrow, 1987), and so a dilution effect due to the N is possible.

Chemical Forms of Herbage S

Usually, about 90% of the S in herbage is organic, and is present mainly as the S-containing amino acids (cysteine, cystine and methionine) in various proteins. Some organic sulphur is also present in sulpholipids and in a number of other compounds, including glutathione, thiamine, biotin, ferredoxin, lipoic acid and coenzyme A (Stevenson, 1986).

The inorganic S in herbage is present as sulphate and is usually at a concentration within the range 0.03–0.20% sulphate S. Sulphate tends to accumulate in plants in which protein synthesis is restricted by a deficiency of N (Goh and Kee, 1978) and when the supply of sulphate exceeds the requirement (Nguyen *et al.*, 1989). The concentration of sulphate S tends to be higher in the grass than in the clover component of grass–clover swards, even when the concentrations of total S in the two species are similar (Table 7.6). Although not harmful to the plant, high concentrations of sulphate may have adverse effects in livestock (see p. 177).

Occurrence and Metabolic Functions of S in Ruminant Animals

Sulphur has several important roles in the metabolism of ruminant animals. It is essential for the synthesis of microbial biomass in the rumen and is therefore important in digestion. The rumen microorganisms are able to utilize both organic and inorganic forms of S in herbage, and both forms are converted to sulphide and then assimilated into the S amino acids of microbial protein. When S is deficient, the synthesis of microbial biomass may be curtailed, and this, in turn, may limit the degradation of fibre in the rumen and sometimes decrease feed intake (Spears *et al.*, 1985). The other metabolic roles of S reflect the requirements of the animal itself, for the synthesis of body tissue and of milk, wool, etc. Body tissue as a whole contains an average of 0.15% S (see Table 4.1, p. 73; National Research Council, Subcommittee, 1989), most of it in the form of the S-amino acids in proteins. Individual proteins may range in S concentration from about 0.3 to 1.6%. As in plant proteins, cysteine has the important role of forming disulphide bridges, thus stabilizing the structure of the proteins. The individual S amino acids also have a number of other functions. Again as in plants, cysteine occurs in the peptide glutathione, which is involved in oxidation–reduction processes through the reversible formation of a disulphide bridge (Underwood and Suttle, 1999). Cysteine is also a precursor of two vitamins, biotin and thiamine, of the hormone insulin and of coenzyme A (McDonald *et al.*, 1995), as well as being a component of molecules such as metallothionein, which are involved in the detoxification of heavy metal ions when these are present in excess (Underwood and Suttle, 1999). Methionine has an important role in transmethylation reactions. In addition to being present in the proteins of body tissues, S occurs in the structure of body tissues in sulpholipids and in chondroitin sulphate (Fig. 7.1), which is a component of cartilage, tendons and the walls of blood-vessels. Sulphur is also a constituent of haemoglobin and of cytochrome (McDonald *et al.*, 1995) and is present in lipoic acid (Fig. 7.1) which is a cofactor in several enzyme systems (Ammermann and Goodrich, 1983). Another metabolic function of S is to form sulphate conjugates with compounds such as phenols and indole, compounds which are derived from bacterial action in the digestive system and would otherwise be toxic to the animal. After conjugation with sulphate, these compounds are less toxic, and they are excreted in this form in the urine (Fontenot and Church, 1979).

Milk contains about 0.03% S, most of which is present in S-containing amino acids in proteins (National Research Council, Subcommittee, 1989). Wool and hair have high concentrations of S and generally contain between 2.5% and 3.8% S on a dry weight basis (see Table 4.5, p. 79; Langlands and Sutherland, 1973). About 90% of the S in wool occurs in the form of cysteine/cystine, and the N : S ratio is about 5 : 1 (Underwood and Suttle, 1999).

Absorption of Dietary S by Ruminant Animals

Most of the S in the diet is converted to sulphide S in the rumen, and the sulphide is then assimilated into microbial protein. Any excess sulphide is absorbed into the blood, probably as HS^-, and subsequently converted to sulphate by the liver. Some of the sulphate is then returned to the rumen in saliva and some is excreted in the urine. The recycling of S in saliva enables S to be conserved, an important factor when the intake is low (Grace, 1983b). In this respect, the metabolism of S in ruminant animals is similar to that of N and P. As well as degrading plant proteins, the rumen microorganisms convert sulphate in the diet, plus the sulphate in saliva, to sulphide and then incorporate most of this into microbial protein.

Most of the S leaving the rumen is either in undigested dietary protein or in microbial protein, though there may be small amounts of sulphate or sulphide not assimilated by the microorganisms. In the second phase of the digestion process (see p. 70), the S-amino acids released by the digestion of proteins are absorbed from the small intestine (Goodrich and Garrett, 1986). Small amounts of S are also recycled into the small intestine via secretions from the pancreas and liver. Taurine, a compound secreted in bile, and other S compounds are subsequently degraded in the large intestine, with the formation of sulphide, some of which is absorbed into the bloodstream (Grace, 1983b).

Animals can readily synthesize cysteine/cystine from methionine in the diet, but, unlike plants, cannot synthesize methionine from cystine. However, ruminant animals normally receive an adequate supply of methionine, as a result of its synthesis by microorganisms in the rumen.

Nutritional Requirements for S in Ruminant Animals

In ruminant animals, the dietary requirement for S is closely related to that for N, since both elements are assimilated into microbial protein in the rumen and into the protein of body tissues. The supply of S for the synthesis of microbial protein is normally adequate when the N : S ratio of the diet is no wider than about 14 : 1–15 : 1, though, as with the critical N : S ratio for plant growth, this ratio is meaningful only if the N is not present in excess. However, although cattle appear to have sufficient S if the requirement of the rumen microorganisms is met, sheep may need a slightly higher dietary concentration than that required by the microorganisms, due to the relatively high concentration of S in wool. The N : S ratio in wool, at about 5 : 1, is considerably lower than that in microbial protein and, if the supply of S is inadequate, wool growth is reduced and the wool tends to become brittle.

In general, the requirement for S, like that for N, is influenced by the rate of live-weight gain and/or milk or wool production. As a result of the close relationship with N, requirements for S are often expressed in terms of the N : S ratio. In the UK, recommendations are that the ratio of total N : S (or ideally of the rumen-degradable N : S) should be no greater than 14.3 : 1, resulting in a minimum concentration of S in the diet of between 0.11% and 0.16% S, depending on the concentration of protein and digestibility (Agricultural Research Council, 1980). However, feeding trials involving the supplementation of diets with S have shown little effect on live-weight gain or milk yield when concentrations of S in the diet are more than about 0.1% (Agricultural Research Council, 1980). In the USA, the dietary concentration recommended for lactating dairy cattle is higher, at 0.20%, and the requirement for non-lactating animals is calculated from the minimum protein requirement, based on an N : S ratio of 12 : 1 (National Research Council, Subcommittee, 1989). For sheep, the recommended dietary concentration is 0.14–0.18% S for mature ewes, and rather more for young lambs (National Research Council, Subcommittee, 1985). For optimal wool growth, the requirement for S may well be higher, at about 0.25% S in the diet (Underwood and Suttle, 1999).

If animals are given non-protein N (e.g. urea) as additional N for the synthesis of microbial biomass, it may be necessary to provide additional S. However, if the amount of N exceeds the requirement for microbial synthesis, the addition of S to maintain the N : S ratio at less than 15 : 1 is unnecessary (Grace, 1983b).

Not all the S that is ingested is absorbed and, with diets consisting of concentrate feeds, hay and silage, the absorption of S is generally about 65%. There is a lack of data for diets of grazed herbage but, in assessing requirements using the factorial approach (see p. 83), 65% absorption has been assumed (Grace, 1983b).

Effects of S Deficiency and Excess on Ruminant Animals

A deficiency of S relative to N may limit the synthesis and/or the activity of the microbial biomass and thus curtail the rate of digestion in the rumen

(Millard *et al.*, 1987). Dietary N may also be poorly utilized (Spears *et al.*, 1985; Goodrich and Garrett, 1986). An increase in digestibility was observed, in a study with sheep, when supplementary S was provided which reduced the N : S ratio of the diet from 34 : 1 to about 11 : 1 (Hume and Bird, 1970). Since the initial effects of deficiency are on digestibilty and feed intake, symptoms are not specific but include reduced live-weight gain (Table 7.7) and reduced production of milk and/or wool (Miller, 1979; Goodrich and Garrett, 1986).

With S, the dietary concentration that is potentially harmful is only two to three times greater that the optimal concentration, a rather small difference. Reductions in appetite and growth rate have been observed in animals consuming diets with 0.3–0.4% S, possibly due to sulphide toxicity (Underwood and Suttle, 1999). Excessively high concentrations of S in the diet, and therefore high concentrations of sulphide in the digestive tract, may reduce the motility of the rumen and impair respiration (McDonald *et al.*, 1995). They can also adversely affect the absorption and metabolism of other elements, especially Cu and Se, as discussed in Chapters 9 and 11.

Excretion of S by Ruminant Animals

Usually, more S is excreted in the urine than in the faeces, but the distribution varies with the concentration of S in the diet. Thus, with sheep, the proportion of the excreted S in the urine ranged from 6% with herbage containing

Table 7.7. Influence of the application of S, at 45 and 90 kg S ha^{-1} year^{-1}, on herbage concentrations of N and S and N : S ratio, and on feed consumption and daily weight gain of lambs: means of 2 years (from Jones *et al.*, 1982).

	Addition of S (kg ha^{-1})		
	0	45	90
Grass–clover			
Herbage N (%)	1.66	2.03	2.04
Herbage S (%)	0.14	0.18	0.20
N : S	12.3	11.3	10.3
Feed consumption (kg day^{-1})	1.57	1.65	1.71
Live-weight gain (g day^{-1})	141	180	207
Ryegrass			
Herbage N (%)	1.12	1.17	1.24
Herbage S (%)	0.09	0.17	0.21
N : S	12.4	7.1	6.0
Feed consumption (kg day^{-1})	1.30	1.27	1.58
Live-weight gain (g day^{-1})	32	88	113

< 0.1% S to 90% with herbage containing 0.5% S (Barrow, 1987). With sheep grazing grass–clover pastures in Australia and New Zealand, it is usual for 50–70% of the excreted S to be in the urine (Haynes and Williams, 1993; Nguyen and Goh, 1994a). The amount of S excreted in the faeces is related to DM intake and, with sheep, is about 0.11 g S 100 g^{-1} of feed DM, though it does increase somewhat with increasing S intake (Haynes and Williams, 1993; Nguyen and Goh, 1994a).

Reported average concentrations of S in urine include 0.7 g l^{-1} for cattle grazing a fertilized grass–clover sward (Ledgard *et al.*, 1982) and 1.3 g l^{-1} for sheep consuming a similar diet (Nguyen and Goh, 1994b). However, in another study, urine from sheep on a commercial farm contained, on average, only 0.3 g l^{-1} (Williams and Haynes, 1993b). Urine contains both inorganic sulphate and organic S. The proportion of urine S in the form of inorganic sulphate increases with increasing S intake and, for animals consuming grassland herbage, is often 50–80% of the total (Nguyen and Goh, 1994a). The organic forms of S in urine include ester sulphate, which may account for about 11% of the total S (Nguyen and Goh, 1994b), and sulphate conjugated by compounds such as phenol and indole (see p. 174). In addition, cystine, methyl sulphide, ethyl sulphide, thiocyanate and taurine derivatives may be present (Nguyen and Goh, 1994a).

The average concentration of S in the faeces of grazing sheep and cattle is around 0.3% in DM (Kennedy and Till, 1981; Chiy and Phillips, 1993; Nguyen and Goh, 1994a), which is consistent with an excretion of 0.11 g S 100 g^{-1} feed DM if the digestibility of the herbage consumed is about 73%. The concentration of S in the faeces tends to increase with increasing digestibility of the diet. Most of the S in faeces is organic, but some inorganic sulphate is also present. The organic S consists of undigested protein, microbial protein, ester sulphate and compounds such as indican, which are insoluble in water and rather resistant to bacterial breakdown (Nguyen and Goh, 1994a). Between 10 and 25% of the faecal S is readily soluble in water (Nguyen and Goh, 1993), and most of this is probably sulphate S (Williams and Haynes, 1993b).

Neither urine nor faeces has the same N : S ratio as the feed. Thus, when sheep consumed herbage with an N : S ratio of about 15 : 1, the N : S ratio of the urine was about 20 : 1 and that of the faeces about 7 : 1 (Barrow, 1987).

The concentration of S in slurry is variable, and is influenced by the degree of dilution of the slurry and the period of storage. For cattle slurries in south-west England and in Northern Ireland, the average concentration of S was 0.35 kg S m^{-3} (Watson and Stevens, 1986; Lloyd, 1994). However, in undiluted cattle slurry (10% DM), the average concentration was higher, at 0.7 kg S m^{-3}, and in undiluted pig slurry it was 1.4 kg S m^{-3} (Watson and Stevens, 1986). Small amounts of S can volatilize from slurry, mainly as hydrogen sulphide and methyl sulphide (Beard and Guenzi, 1983). In farmyard manure from cattle, the concentration of S is typically about 0.07% on a fresh weight basis (Syers *et al.*, 1987).

Transformations of S in Grassland Systems

The main chemical forms of S involved in the soil, plant and animal components of grassland systems are shown in Fig. 7.4. The release of sulphate during the mineralization of S from organic residues is often a major source of S for new plant growth, and soil microorganisms have an important role in the balance between mineralization and immobilization. Microorganisms are also involved in the digestion of plant S within ruminant animals, mainly through breaking down plant proteins, with the release of S-amino acids and sulphide. The presence of grazing animals increases the rate at which S is recycled within a grassland system, because passage through the animal accelerates decomposition of the herbage and much of the S in excreta occurs

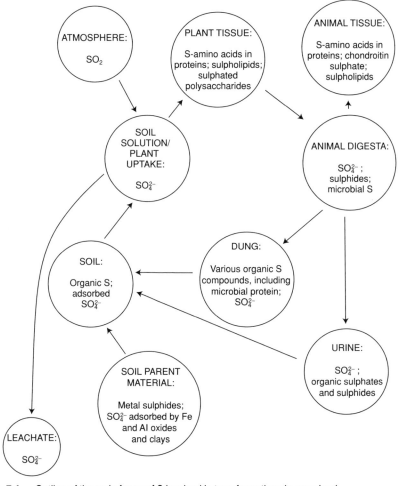

Fig. 7.4. Outline of the main forms of S involved in transformations in grassland.

as sulphate, which is readily plant-available (Saggar *et al.*, 1998). However, sulphate that is not taken up is retained less strongly than phosphate in the soil and, particularly on soils with pH > 6.5 and low in Fe and Al oxides, is susceptible to leaching.

Quantitative Balances of S in Grassland Systems

The estimated S balances in Table 7.8 show small gains in the amounts of S in the soil, but, with some soils, losses by leaching would be greater and gains non-existent. Thus, in a balance for grass grazed by cattle in the USA, the loss by leaching was estimated at 40 kg ha^{-1} year^{-1}, despite no S being added in fertilizer, and the net loss from the soil was estimated at 36 kg ha^{-1} year^{-1} (Wilkinson and Lowrey, 1973). Although this pattern is feasible for a soil in which sulphate is being released by weathering or by a decline in the content of OM, it clearly could not be maintained in the long term. In balances assessed for grazed grass–clover swards in New Zealand, fertilizer inputs of 12–42 kg S ha^{-1} year^{-1} were considered to be approximately balanced by leaching losses, with little change in the amount of S in the soil (Nguyen and Goh, 1992b).

Table 7.8. Estimated sulphur balances (kg S ha^{-1} year^{-1}) typical of three systems of grassland management: intensively managed grass, moderately intensive grass–clover and extensive grass, all grazed by cattle (except during the winter) (based partly on data from Nguyen and Goh, 1992; Haynes and Williams, 1993).

	Intensive grass (dairy cows)	Grass–clover (beef cattle)	Extensive grass (beef cattle)
Inputs			
Fertilizer	20	0	0
Atmosphere	20	20	15
Aspects of recycling			
Uptake into herbage	40	25	6
Consumption of herbage by animals	32	16	3
Dead herbage to soil	15	14	5
Dead roots to soil	8	8	3
Excreta to soil of grazed area	25	15	2.8
Outputs			
Animal products	3	0.8	0.16
Leaching/runoff	30	16	8
Loss through excreta off sward	4	0.0	0.0
Gain (loss) to soil	3	4	6

Chapter 8

Macronutrient Cations: Potassium, Sodium, Calcium and Magnesium

Sources and Concentrations of K, Na, Ca and Mg in Soils

In igneous rocks, K occurs mainly in the muscovite and biotite micas and the orthoclase feldspars. Sodium occurs mainly in the plagioclase feldspars and amphiboles, with smaller amounts in the micas and pyroxenes. Calcium is distributed in a wide range of minerals, but concentrations are highest in the pyroxenes and amphiboles, while Mg is present mainly in the olivines, pyroxenes, amphiboles and biotite micas (see Table 2.1, p. 16).

When these various minerals are subject to weathering, there is usually a partial loss of all four cations by leaching. Potassium and Na are leached most readily, though they are retained to some extent by incorporation into secondary minerals and by cation exchange. Potassium is released from biotite more readily than from muscovite, and more readily from muscovite than from orthoclase or microcline. However, during the weathering sequence from micas to vermiculite and smectite (see Fig. 2.1, p. 17), an initial K concentration of about 10% may well be reduced to less than 1% (Kirkman *et al.*, 1994). In arid conditions, K and Na may be retained through being reprecipitated in salts, such as sulphates and chlorides. In general, Ca and Mg are leached less readily than are K and Na; and Ca and Mg are more likely to be retained in sedimentary materials. One factor that contributes to the retention of Ca and Mg is that their carbonates, phosphates and sulphates have a lower solubility than those of K and Na. Some limestone sediments are composed almost entirely of $CaCO_3$, but others contain appreciable amounts of $MgCO_3$ and/or non-carbonate deposits. Some Mg may also be retained during the formation of clay minerals, by partially substituting for Al in the lattice structure, particularly of smectite and illite (Metson, 1974). All rock types vary in their contents of the major cations, but typical concentrations in several igneous and sedimentary rocks are shown in Table 2.2 (p. 20).

Since weathering is usually accompanied by leaching, most soils have rather lower concentrations than their parent materials of all four major cations. However, there is considerable retention in soils, partly through the presence of the cations in unweathered rock fragments, partly through their incorporation into secondary minerals and partly through adsorption on to cation exchange sites. In most non-acid soils, Ca is the dominant cation, with Mg the second most abundant, though the actual and relative concentrations show wide variations (Table 8.1). The relative amounts of the four cations are strongly influenced by the nature of the parent material and by the severity of leaching. For example, soils derived from limestone have high concentrations of Ca, often between 3% and 25% of dry weight, and soils rich in ferromagnesian minerals have relatively high concentrations of Mg, unless they have been strongly leached. Strongly leached soils are usually low in K and Na.

Agricultural and Atmospheric Inputs of K, Na, Ca and Mg

Of the four major cations, only K is applied routinely as fertilizer and, in intensively managed grassland, the rate of application is sometimes more than 400 kg K ha^{-1} year^{-1}. However, there is often little effect on the total soil content, as most of the fertilizer K is either taken up into the herbage or is lost by leaching. In soils containing illite or similar clay minerals, there may be some retention of fertilizer K, due to fixation (see p. 18), and retention appears to be increased where K is recycled by intensive grazing. The other three major cations are applied much less widely. Calcium is applied, as lime, only to acid soils, and the main function of lime is to raise the soil pH rather than provide Ca. The lime requirement depends not only on the initial soil

Table 8.1. Ranges of concentration of K, Na, Ca and Mg in soils (% in dry weight).

		K	Na	Ca	Mg	Reference
Temperate regions	Typical[a]	2.1	–	0.44	0.33	Brady and Weil, 1999
Temperate regions	Typical range	0.5–2.5	0.1–1.0	0.7–25.0	0.1–4.0	Tisdale *et al.*, 1985
England and Wales	Usual range	0.3–0.5	0.018–0.033	0.19–0.70	0.17–0.47	McGrath and Loveland, 1992
South Dakota, USA	Prairie soil	1.5	0.94	1.15	0.55	Jackson, 1964
Georgia, USA	Grassland soil	0.51	–	0.26	0.19	Wilkinson and Lowrey, 1973
Scotland	Four soils	0.5–2.5	0.6–2.0	1.0–2.5	0.2–1.5	Berrow and Mitchell, 1980

[a]Assuming soil to 15 cm depth weighs 1800 t ha^{-1}.

pH but also on the buffering capacity of the soil, and it increases with contents of clay and OM. When lime is applied, its effects are more long-lasting than those of fertilizers and, typically, an application of 4–6 t $CaCO_3$ ha^{-1} might be repeated at intervals of 3–5 years, providing an average input of about 500 kg Ca ha^{-1} year^{-1}. Although Na and Mg are applied occasionally to grassland, such applications are intended to benefit livestock rather than plant growth, and the amounts added are insufficient to have an appreciable effect on soil content.

Where sewage sludge is applied, there is usually a small net addition of the major cations, as these are generally present in larger concentrations in sludge than in soils. Calcium is usually present in the greatest amount (see Table 2.8, p. 32), and an application of 1 t of sludge ha^{-1} would supply approximately 40 kg Ca ha^{-1}.

Inputs of K and Ca from the atmosphere are usually negligible in relation to the amounts in soil, but the input of Na and, to a lesser extent, of Mg is sometimes sufficient to make an appreciable contribution (Table 8.2). With both Na and Mg, the input is greatest in areas near the coast (Metson, 1974; Smith and Middleton, 1978). Thus, in Germany, the amount of wet deposition of Na was found to decline from about 28 kg Na ha^{-1} year^{-1} at the coast, in the north of the country, to 3–4 kg Na in the south: the corresponding amounts of Mg were 3.4 kg in the north and about 1 kg ha^{-1} year^{-1} in the south (Ulrich, 1991).

Recycling of K, Na, Ca and Mg through the Decomposition of Organic Residues

Few assessments have been made of the amounts of the major cations recycled to grassland soils through the death and decomposition of plant material. Measurements are difficult because the cations, particularly K and Na, are liable to be leached from senescent and dead material before

Table 8.2. Amounts of K, Na, Ca and Mg deposited from the atmosphere to the land surface (kg ha^{-1} year^{-1}) reported in several investigations.

	K	Na	Ca	Mg	Reference
Europe (central)	2–4	4–8	10–20	3–6	Mengel and Kirkby, 1987
UK (three sites)	9.2–20.0	21–41	15–19	5.4–10.5	Cawse, 1987
UK (three lowland sites)	2–8	8–35	5–25	2–6	Goulding et al., 1986
UK (upland)	1.3–1.6	2.2–2.3	3.8–6.5	2.9–3.0	Taylor et al., 1999
UK (Scotland)	–	10–70	–	–	Sinclair et al., 1993
Germany (Berlin)	3.4	7.8	22.9	2.5	Marschner, 1995
USA (south-east)	3.6–5.2	–	4.9–12.0	1.3–2.4	Lowrance et al., 1985
USA (Ohio)	3.6	4.0	9.1	2.3	Owens et al., 1998

decomposition begins. However, the estimates for K in Table 8.3, which include transfers by leaching from senescent and dead material, are similar to those made for a grazed tall fescue pasture in Georgia, USA, namely 64 kg ha^{-1} year^{-1} in unharvested herbage and 34 kg ha^{-1} in roots (Wilkinson and Lowrey, 1973).

Typical amounts of the cations returned in the excreta of dairy cattle, grazing at a density of 700 cow-days ha^{-1} year^{-1}, are estimated in Table 8.4. However, distribution of the excreta, and therefore of the excreted cations, is extremely uneven (see p. 29). Of the four major cations, potassium is of most significance for grassland productivity, and the impact of its return in urine has been examined in several studies. Although a large proportion may be taken up by the grass, the amount of K in a urine patch generally exceeds the uptake capacity of the grass, and much of the remainder is susceptible to leaching (Williams *et al.*, 1989, 1990), especially with high rainfall and well-drained soils. The proportion of the K that is taken up by plants is also influenced by the extent to which the N in the urine is nitrified, since the uptake of K is increased when N is taken up as nitrate (Carran, 1988). In some soils, the K returned in urine is subject to fixation by clay minerals.

While K and Na are returned mainly in urine, the return of Ca and Mg occurs mainly in dung. Less attention has been given to the cations in dung, but the rate at which they move into the soil tends to increase with increasing

Table 8.3. Typical amounts of K, Na, Ca and Mg recycled to soil through the death and decay of herbage and roots from a moderately intensive sward (kg ha^{-1} year^{-1}), assuming 7000 kg dead herbage and 5000 kg dead roots ha^{-1} year^{-1}.

	K	Na	Ca	Mg
Assumed concentration in dead herbage (%)	1.2	0.24	0.50	0.17
Return in herbage (kg ha^{-1} year^{-1})	84	16	35	12
Assumed concentration in dead roots (%)	0.58	0.12	0.50	0.18
Return in roots (kg ha^{-1} year^{-1})	29	6	25	9
Total return in herbage + roots (kg ha^{-1} year^{-1})	113	22	60	21

Table 8.4. Estimated amounts (kg ha^{-1} year^{-1}) of K, Na, Ca and Mg returned to the soil in the excreta of dairy cattle grazing at a density of 700 cow-days ha^{-1} year^{-1}, assuming 25 l urine and 3.8 kg faecal DM cow^{-1} day^{-1}, and typical concentrations derived from Tables 8.21 and 8.22.

	K	Na	Ca	Mg
Assumed concentration in urine (g l^{-1})	6.0	0.98	0.27	0.51
Return in urine (kg ha^{-1})	105	17	4.7	8.9
Assumed concentration in dung (% in DM)	0.86	0.20	1.65	0.55
Return in dung (kg ha^{-1})	23	5.3	44	14.6
Total return in urine + dung (kg ha^{-1})	128	22	49	24

rainfall (Dickinson and Craig, 1990). Potassium is leached from dung more quickly than are Ca and Mg (Weeda, 1977); in the absence of prolonged heavy rainfall, the loss of Ca and Mg from the dung may occur no more rapidly than the loss of OM (Underhay and Dickinson, 1978).

The concentrations of the major cations in slurries vary widely, depending partly on the composition of the excreta and partly on the degree of dilution. Some typical concentrations are shown in Table 8.5 and in Table 2.7 (p. 31).

Losses of K, Na, Ca and Mg from Soils

All four of the major cations are subject to loss by leaching, but there are no volatile losses. Leaching losses vary with the particular cation and with the soil type, and also with fertilizer application and sward management. The monovalent cations are leached more readily than the divalent cations but, when Ca^{2+} is the dominant cation in the soil, it may be leached in the greatest amount. Leaching occurs more readily from sandy soils than from loam or clay soils, due to the lower cation exchange capacity of sandy soils. The leaching of all four cations is increased when fertilizer K is applied, reflecting some displacement of Ca, Mg and Na from cation exchange sites. Leaching is also increased when swards are grazed rather than cut, the losses being much greater from urine than from dung (Hogg, 1981). The cations in urine are liable to move downwards rapidly in macropores (Goh and Williams, 1999). However, the extent of leaching from grassland, especially of K, is reduced when the ammonium derived from urine is nitrified rapidly, as this favours the uptake of cations by plants (Carran, 1988).

Reported amounts of the four cations leached from grazed grassland soils are shown in Table 8.6. In some situations, greater losses of K may occur: for example, up to 50 kg K ha^{-1} may be lost from grazed pastures in New Zealand (Goh and Williams, 1999) and up to 140 kg K ha^{-1} year^{-1} from grazed fescue pasture in Georgia, USA (Wilkinson and Lowrey, 1973). However, when urine was applied to grassland on a silt loam soil in New Zealand, most of the K was either taken up by plants or retained in the top 5 cm of soil, while substantial amounts of Ca and Mg were leached, partly due to their displacement by K (Early et al., 1998). With Na, leaching losses are influenced by current inputs from the atmosphere, and amounts similar

Table 8.5. Typical concentrations of K, Na, Ca and Mg in cattle slurries (g l^{-1}).

Source	K	Na	Ca	Mg	Reference
The Netherlands, mean, 9 years	4.6–5.4	0.6–0.9	1.4–1.5	0.7–0.8	Schroder and Dilz, 1987
Denmark, mean, four slurries	5.5	0.6	1.6	0.5	Sommer and Husted, 1995
UK, mean, 48 slurries	3.3	0.44	1.05	0.39	Stevens et al., 1995

Table 8.6. Leaching losses of K, Na, Ca and Mg from grazed grassland (kg ha^{-1} year^{-1}).

Country	Grazed by	Fertilized with	K	Na	Ca	Mg	Reference
New Zealand	Steers	–	14	71	154	32	Steele *et al.*, 1984
New Zealand	Steers	N	21	86	216	44	Steele *et al.*, 1984
USA	Cattle	N, P, K	10	25	86	51	Owens *et al.*, 1998
New Zealand	Sheep	K, P, S	3–11	31–65	31–53	9–15	Heng *et al.*, 1991

to those reported from New Zealand, namely 30–50 kg Na ha^{-1} year^{-1} (Table 8.6), have been suggested as typical for the UK (Sinclair *et al.*, 1993). There is often considerable variation in the leaching of the cations from year to year, reflecting differences in inputs and in rainfall (Heng *et al.*, 1991; Owens *et al.*, 1998). In some areas, the amounts of Ca, Mg and Na lost from grassland by leaching may well be greater than the amounts taken up into harvested herbage, as shown by comparing Table 8.6 with Table 3.5 (p. 55).

When moderate to high rates of fertilizer are applied to sloping fields of grassland, and especially if they are grazed, there may be an appreciable loss of cations in surface runoff. In a study of this type of situation in Ohio, USA, the average amounts in runoff, in soluble form, were about 20 kg K, 1.5 kg Na, 9 kg Ca and 5 kg Mg ha^{-1} year^{-1} (Owens *et al.*, 1998).

Forms and Availability of K, Na, Ca and Mg in Soils

The major cations in soils are present partly in the structure of primary and secondary minerals, partly in exchangeable form on cation exchange sites and partly as soluble ions in the soil solution. With potassium, there is an additional category, that 'fixed' within the lattice structure of clay minerals, such as illite and vermiculite, in a form that is not readily accessible for exchange with other cations (see p. 18). Potassium is more susceptible than Na, Ca and Mg to fixation, because, in ionic form, it is less hydrated and therefore causes less swelling of the clay lattice structure. In calcareous soils and in some recently limed soils, much of the soil Ca occurs in the form of calcium carbonate and, in some arid and semi-arid soils, Ca also occurs as gypsum.

In most soils, a much larger proportion of each cation is present in unweathered minerals than in exchangeable form, and a much larger proportion is in exchangeable form than in soluble form. With K, the amount of exchangeable K is usually between ten and 100 times larger than the amount in solution. However, in arid and semi-arid regions, there may be an accumulation of the cations in the form of potentially soluble salts, such as sulphates and carbonates. When the cations are in soluble form in the soil solution, K and Na occur only as the free ions, but some Ca and, to a lesser extent, Mg may occur in the form of complexes.

Although the fraction of any cation that is immediately available for plant uptake is that in the soil solution, the most important fraction, in terms of supply for the whole growing season, is the exchangeable fraction. This provides the source from which the soluble fraction is replenished. Of the exchangeable cations, Ca is usually dominant, often amounting to between 60 and 85% of the total in non-acid soils. In the soils of humid regions, Mg normally accounts for 5–30% of the total exchangeable cations, K for between 2 and 6% and Na also for between 2 and 6% (Bohn *et al.*, 1979; Mengel and Kirkby, 1987).

The cations tend to move from one soil category to another, as the amount in the soluble fraction is either depleted or augmented. With K, equilibration between the soluble and exchangeable fractions is usually rapid, but equilibration between the exchangeable fraction and that fixed by clay minerals is much slower, and requires days or even months (Mengel and Kirkby, 1987). Nevertheless, the actual rate at which the exchangeable K is replenished may have a large influence on the plant-available supply, especially when large amounts of K are removed in silage or hay. Although many soils have large reserves of total K (for example, a soil with 1% K contains about 30,000 kg K ha^{-1} to a depth of 20 cm), the rate at which the reserve is released to exchangeable and soluble forms varies widely. In a pot experiment involving 20 UK soils, the amount of K made available for uptake by ryegrass, which was cut repeatedly, varied from the equivalent of < 50 kg to > 1000 kg ha^{-1} year^{-1}, with about a quarter of the soils providing more than the 200 kg K ha^{-1} year^{-1} needed by a productive grass sward (Arnold and Close, 1961). In a longer-term study with a sandy loam soil sown to ryegrass, more than 200 kg K ha^{-1} year^{-1} was available for uptake in the first year, but the exchangeable K was depleted within 3 years, and subsequently K was released from non-exchangeable sources at a constant rate of about 90 kg ha^{-1} year^{-1} (Clement and Hopper, 1968). The rate at which available K is replenished depends mainly on the types and amounts of micas and clay minerals present in the soil. Soils that contain substantial amounts of muscovite and biotite micas, illites and vermiculites release much more K than do soils whose main types of clay are smectite and kaolinite (Kirkman *et al.*, 1994).

Assessment of Plant-available K, Na, Ca and Mg in Soils

Assessment of the plant-available fraction of any of the four cations is often based on measurement of the exchangeable fraction. Ammonium acetate is a widely used extractant for this purpose. For grassland to produce a maximum yield of herbage, it has been suggested that the amount of exchangeable K in the soil should be more than 114 mg K kg^{-1} dry soil (Barraclough and Leigh, 1993). However, although the exchangeable fraction of the soil K provides a good assessment for some soils, for others there is an improvement by including a proportion of the non-exchangeable K. Extraction procedures that have been proposed for this combined assessment include: (i) boiling

with 1 M HNO$_3$; and (ii) continuous leaching with 0.01 M HCl (Kirkman
et al., 1994). In a study involving a number of UK soils, differences in the rate
of release of K in potentially plant-available forms were related quite closely
to the amount dissolved by two complementary extractions, with a Ca-resin
and with concentrated HCl (Goulding and Loveland, 1986).

Uptake of K, Na, Ca and Mg by Herbage Plants

All four cations move within the soil by a combination of mass flow and
diffusion. Potassium and Na move mainly by diffusion, while Ca and Mg
move mainly by mass flow. Each of the four cations is absorbed by roots as
the free ion, and the relative amounts taken up are influenced by the absolute
and relative concentrations of the ions in solution. Potassium is normally
taken up in the largest amount (see Table 3.5). Over the range of K and Ca
concentrations that occur in soil solutions, K uptake appears to compete with
Ca uptake (Evangelou et al., 1994), and K also competes with Mg (Mayland
and Wilkinson, 1989) and with Na (Smith, 1975). Although Na is thought
not to be essential for plants, it is taken up quite readily if available in the soil.
As well as being influenced by competition between the cations, the relative
amounts taken up are influenced by the cation exchange capacity of the plant
roots. As shown in a comparison of 20 grass species, the higher the cation
exchange capacity of the roots, the higher the proportion of divalent cations
taken up and the lower the proportion of monovalent cations (Wacquant,
1977).

Detailed studies of the absorption of cations, carried out mainly with K,
have shown that the rate of uptake generally depends on the concentration
close to the root surface. The actual uptake of K into the root appears to
involve two types of mechanism. When the concentration in the soil solution
is less than about 1 mM, K uptake across the plasma membrane is thought to
be due to a carrier protein, which is transported across the plasma membrane
and releases K inside the cell. This type of absorption occurs against an
electrochemical potential gradient, and therefore consumes energy and is
metabolically 'active'. At higher concentrations of K in the soil solution, K
is transported increasingly by a relatively 'passive' process, which may
involve ion channels in the plasma membrane (Kochian and Lucas, 1988).

When the supply of K is plentiful, grass readily takes up more than is
needed for maximum yield, sometimes twice as much, resulting in herbage
concentrations of > 4% (e.g. Hopper and Clement, 1967). When the supply is
low and grasses and legumes are grown together, the grasses compete more
successfully for soil K and the legumes are more liable to become deficient,
probably due mainly to the difference in rooting pattern (see p. 42). When
ryegrass and red clover were grown together on a soil low in exchangeable K,
the grass appeared to be the more effective at obtaining non-exchangeable K
(Mengel and Steffens, 1985).

Functions of K, Na, Ca and Mg in Herbage Plants

All four cations, but especially K, contribute to the maintenance of osmotic potential and, as a result, plants that are deficient in K are more susceptible to drought. The cations are also involved in maintaining the pH of the plant by neutralizing organic acids, including the acidic groups of various polymer molecules, such as pectin. However, in addition to maintaining osmotic potential and pH, each of the major cations, except Na, has a number of individual functions.

Potassium has a specific role in the opening and closing of stomata, through its influence on the osmotic potential and turgor of the guard cells. Opening occurs in response to an increase in the osmotic potential of the guard cells, due to a temporary accumulation of K, together with Cl and/or malate. Potassium is also involved in the transport of photosynthate from the leaves, and is an activator of a large number of enzymes, including some involved in protein synthesis (Blevins, 1994).

Much of the plant Ca is bound to the galacturonic acid component of pectin, and the resulting calcium pectate is an important constituent of the middle lamella of cell walls. Since Ca is divalent, it is able to link adjacent polymer molecules that contain acidic groups, and this role probably explains why Ca is important in the growth of meristems, and particularly in the growth and functioning of root tips. Calcium is also involved in maintaining the integrity and selectivity of membranes, and is required by a number of enzymes, including amylase and some nucleases (Clarkson and Hanson, 1980; Blevins, 1994).

Magnesium is a constituent of chlorophyll (Fig. 8.1) and therefore has a major role in photosynthesis. It is also involved in the functioning of ribosomes, and thus in the synthesis of protein. In addition, Mg acts as a cofactor of many enzymes, especially those involved in the transfer of energy through substrates such as ATP (Blevins, 1994).

Distribution of K, Na, Ca and Mg in Herbage Plants

The relative concentrations of K, Ca and Mg in the roots and shoots of two grasses and two legumes are shown in Table 8.7. Concentrations of these three cations were either higher in shoots than in roots or the same in shoots and roots for ryegrass, tall fescue and lucerne grown in soil. However, the concentrations were higher in roots than in shoots for white clover grown in solution culture (Table 8.7). The relative concentrations in roots and shoots depend, to some extent, on supply. For example, in subterranean clover grown in solution culture, Mg was generally higher in the roots than in the shoots at low to moderate supply, but was higher in the shoots when a plentiful supply had been maintained for several weeks (Scott and Robson, 1990). With Na, the relationship between root and shoot concentrations

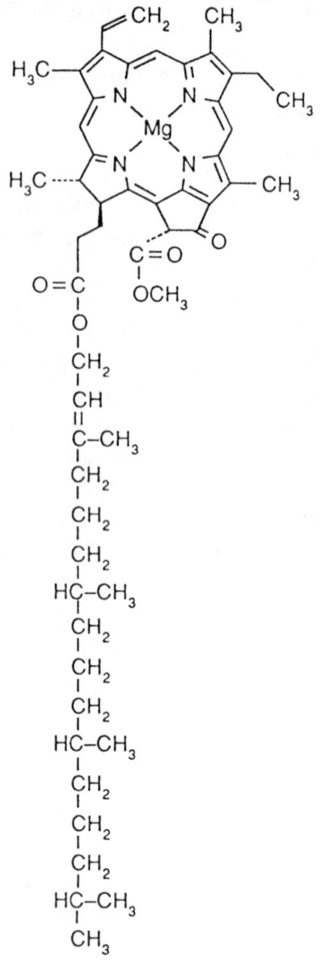

Fig. 8.1. Structure of chlorophyll *a*.

depends largely on plant species (see p. 198), though it is influenced by supply. Thus, in perennial ryegrass and white clover, Na is readily translocated to the shoots if the supply is plentiful, whereas, in timothy and *Trifolium hybridum*, Na accumulates mainly in the roots (Marschner, 1995).

The distribution of the cations between leaves and stems is also variable. In two studies involving grasses, there were higher concentrations of K, Ca and Mg in the leaves than in the stems (Pritchard *et al.*, 1964; Smith and Rominger, 1974). In another study, with perennial ryegrass, there were higher concentrations of K, Na and Ca in the leaves than in the stems, but lower concentrations of Mg (Laredo and Minson, 1975). However, sometimes, the concentration of K is higher in stems than in leaves (Thomas, 1967) and, sometimes, the concentrations of all four cations are slightly higher in

Table 8.7. Concentrations of K, Na, Ca and Mg (% in DM) in shoots, compared with roots, of grass and legume species as reported in various investigations.

	Growth medium	K	Na	Ca	Mg	Reference
Ryegrass	Fertilized soil					
Shoots		4.70	–	0.77	0.28	Goh *et al.*, 1979
Roots		0.79	–	0.50	0.14	
Tall fescue	Soil					
Shoots		2.84	–	0.30	0.20	Wilkinson and Lowrey, 1973
Roots		1.50	–	0.30	0.20	
White clover	Culture solution					
Shoots[a]		3.60	–	1.06	0.25	Wilkinson and Gross, 1967
Roots		3.81	–	1.38	0.55	
Lucerne	Soil					
Shoots		2.57	0.03	1.10	0.33	Rominger *et al.*, 1975
Roots		1.11	0.67	0.20	0.21	

[a]Assuming equal amounts of leaflet, petiole and stolon.

stems than in leaves (Chiy and Phillips, 1996a). In four grasses and red clover, leaf and stem were similar in K, while the concentration of Ca was at least twice as high in leaf as in stem, and the concentration of Mg was also considerably greater in the leaf (Fleming, 1963). With lucerne, one study showed the leaves to be higher than the stems in Ca, and the stems to be higher than the leaves in K, Na and Mg (Rominger *et al.*, 1975), but in another the leaves were higher than the stems in K (James *et al.*, 1995). In subterranean clover, the concentration of Mg was higher in leaves than in petioles (Scott and Robson, 1990). While these contrasting results may be due partly to differences in nutrient supply, they may also be influenced by the stage of growth: for example, with red clover, the concentration of K was greater in stem than in leaf in young growth but not in more mature herbage (McNaught, 1958).

Potassium, Na and Mg are all mobile in plant tissues, but Ca has a low mobility and cannot be moved readily from old to new leaves (see Table 3.2, p. 51; Mengel and Kirkby, 1987; Scott and Robson, 1990).

Ranges of Concentration and Critical Concentrations of K, Na, Ca and Mg in Herbage Plants

The average concentrations of K, Na, Ca and Mg reported in surveys of grassland herbage from various countries (Table 8.8) are, with the exception of the data from Finland, surprisingly close. However, with each element, it is possible for herbage concentrations to vary widely. Thus, with K, concentrations as low as 0.4% (Hemingway, 1963) and as high as 5.2% (Smith and

Table 8.8. Mean concentrations of K, Na, Ca and Mg (% in DM) in grassland herbage analysed for survey or advisory purposes in different regions.

Region	Number of samples	K	Na	Ca	Mg	Reference
England and Wales	> 2000	2.59	0.25	0.60	0.17	MAFF, 1975
The Netherlands	> 700[a]	3.02	0.37	0.61	0.23	Kemp and Geurink, 1978
Finland	> 2000[b]	2.36	0.005	0.26	0.13	Kahari and Nissinen, 1978
Pennsylvania, USA	> 9000	2.06	0.02	0.86	0.20	Adams, 1975
New Zealand	> 2500	2.89	0.20	0.73	0.24	Smith and Middleton, 1978; Smith and Cornforth, 1982

[a]All from intensively managed farms.
[b]All timothy.

Middleton, 1978) have been reported, though the usual range is between 1% and 3.5%. With Na, herbage concentrations may range from a trace to more than 2% (Cunningham, 1964). Concentrations of Ca in herbage are usually within the range 0.1% to 2.6%, the upper part of the range reflecting legume rather than grass species (see Table 3.7, p. 58), and concentrations of Mg are usually between 0.10 and 0.30% though they may be as low as 0.05% (Drysdale *et al.*, 1980).

The total cation content of herbage, expressed in terms of mEq 100 g^{-1}, varies less widely than the content of individual cations. For ryegrass grown with various fertilizer treatments at different sites in the UK and harvested at various stages of maturity, the total content of cations ranged from 80 to 200 mEq 100 g^{-1} DM (Cunningham, 1964). Expressed in terms of mEq 100 g^{-1}, herbage with typical concentrations of 2.5% K, 0.5% Na, 0.6% Ca and 0.25% Mg would contain 136 mEq 100 g^{-1}. The total cation content of herbage was found to be positively correlated with contents of total anions and of N (Cunningham, 1964).

Some reported values for critical concentrations of K, Ca and Mg in the young vegetative herbage of grasses and legumes are shown in Table 8.9. As it is rare for grasses to be deficient in either Ca or Mg, there is little information on their critical concentrations, but there is rather more information for legumes, which are more susceptible to deficiency. Factors that contribute to differences in critical concentration include differences in the stage of maturity of the herbage and whether the concentrations are assessed on the basis of maximum yield or on 90% or 95% of maximum yield (see p. 56). As an example of the effect of stage of maturity, a critical concentration of 2.25% K was proposed for young grass, compared with 1.65% K for older herbage with < 1.6% N (Knauer, 1966). In this context, it has been proposed that the N : K ratio of the herbage might provide a better indication of whether K is deficient than the concentration of K alone, and an N : K ratio of more than 1.3 : 1 has been suggested as indicating K deficiency (Dampney, 1992). In relation to whether the critical concentration is based on maximum yield

Table 8.9. Critical concentrations of K, Ca and Mg reported for ryegrass and white clover in young vegetative herbage (% in DM).

K

Temperate grasses	1.2–1.6	Mayland and Wilkinson, 1996
Perennial ryegrass	1.8	Dampney, 1992
Perennial ryegrass	2.0	Clement and Hopper, 1968
Kentucky bluegrass	2.0–2.4	Kelling and Matocha, 1990
White clover	1.0	Andrew and Robins, 1969
White clover	1.2	Rangeley and Newbould, 1985
White clover	1.4	Whitehead, 1982
White clover	1.8–2.3	McNaught, 1970
White clover	1.7	Evans *et al.*, 1986
Lucerne	2.0	Lanyon and Smith, 1985

Ca

Perennial ryegrass	0.1	De Wit *et al.*, 1963
Perennial ryegrass	0.2–0.3	McNaught, 1970
White clover	1.0	Andrew, 1960
White clover	1.3	Whitehead, 1982
White clover	2.0	Rangeley and Newbould, 1985

Mg

Perennial ryegrass	0.06	De Wit *et al.*, 1963
Perennial ryegrass	0.10–0.13	McNaught, 1970
Perennial ryegrass	0.07–0.10	Smith *et al.*, 1985
Grass	0.13	Mayland and Wilkinson, 1989
White clover	0.12–0.14	McNaught, 1970

or on a percentage of maximum, estimates for K in ryegrass were as high as 3.8% K for 99% of maximum yield, compared with 2.8% for 90% of maximum yield (Smith *et al.*, 1985), though both of these values are high in comparison with other assessments. Another factor that may influence the critical concentration of K is the supply of Na : the critical concentration of K in the youngest, fully open leaf blade of Italian ryegrass was reported to be as low as 0.8% K when the concentration of Na was > 2.0%, but as high as 3.5% K when the concentration of Na was < 0.3% (Hylton *et al.*, 1967).

As an alternative to expressing critical concentrations in terms of dry weight, it is possible to express them in terms of tissue water (fresh weight minus dry weight). A study to assess, for K, whether this might result in a more consistent value for the critical concentration concluded that the critical concentration in grass herbage was about 160 mM K, but that there was little, if any, improvement in precision compared with a dry weight basis (Barraclough and Leigh, 1993). The lack of improvement was probably due mainly to other cations being able to substitute partially for K, in its main role of maintaining osmotic pressure.

Influence of Fertilization with K, Na, Ca and Mg on Yields of Grassland Herbage

Potassium is the only macronutrient cation to be applied more or less routinely in fertilizers, and its application can have a large effect on herbage yield. The soil supply of available K varies widely from soil to soil and, for any one soil, it may change from one year to the next, particularly if the exchangeable fraction is depleted by the removal of cut herbage. The extent to which grassland responds to fertilizer K therefore varies widely with differences in the soil supply of K and in sward management (Simpson et al., 1988; Webb et al., 1990; Dampney, 1992). In the UK, recommended rates of application for fertilizer K depend on the K status of the soil at the beginning of the season and on whether the sward is to be cut or grazed. Soil analysis is used to allocate the soil to one of five categories and, when a need for fertilizer K is indicated, the recommended rate is much greater for cut than for grazed swards (Table 8.10). Fertilizer K is often applied only once during the growing season, but, if a large amount is required, splitting the total amongst several successive applications is more effective. Ideally, the rate of application should be just sufficient to ensure that the yield of herbage is as high as possible, while avoiding excessive uptake (Dampney, 1992). When too much fertilizer K is applied, grass readily takes up excessive amounts which, as well as being inefficient, can impair the absorption of Mg by animals (see p. 211). Too much fertilizer K also incurs the risk of a substantial loss of K by leaching. For these reasons, it is often impossible to maintain the K status of the soil when grass is cut repeatedly for silage or hay, and there is

Table 8.10. Amounts of fertilizer K (kg K ha^{-1}) recommended for grassland in the UK (from MAFF, 1994).

	Extractable K in soil (mg K l^{-1} soil)[a]				
	0–60	61–120	121–240	241–400	> 400
Grazed grass and grass–clover	50[b]	25	0	0	0
Cut grass–clover					
1st cut	115	80	50	25	0
2nd cut	100	80	65	30	0
Cut grass for silage					
1st cut	115	80	50	25	0
2nd cut	100	80	65	30	0
3rd cut	65	50	30	15	0
4th cut	65	50	30	15	0

[a]Extracted by 1 M ammonium acetate.
[b]Annual amount, best divided equally between the first and third grazings.

no point in aiming to do so if the yield of herbage is satisfactory (Hopper and Clement, 1967). Although there is less depletion of the soil K when grass is grazed rather than cut, depletion may occur in patches that escape the return of urine for 2–3 years; and grass growth may be restricted in the patches that are depleted unless adequate fertilizer is applied. Evidence of this type of effect, based on variation in herbage K concentration in a grazed sward, has been reported (Tallowin and Brookman, 1988).

Although Na is thought not to be essential for higher plants, the addition of Na to grass sometimes produces a yield response when there is a slight to moderate deficiency of K (e.g. Mundy, 1984; Webb *et al.*, 1990). The effect may be due partly to the Na displacing K from the roots to the shoots (McNaught and Karlovsky, 1964) and partly to the Na substituting for K in its role of maintaining osmotic pressure (Chiy and Phillips, 1995a). However, when Na is applied to grassland, it is usually to improve the nutritional quality of the herbage for animals rather than for any effect on yield (Chiy and Phillips, 1995b).

The application of lime is usually confined to fields where the soil pH is < 5.5, and the aim is to bring the pH to about 6.0. On grassland soils that have a tendency to become acid, regular liming is necessary to counteract the acidifying effects of nitrification and of acidic inputs from the atmosphere and to offset the leaching losses of Ca. There is a greater tendency towards acidification, due to nitrification, when high rates of ammonium or urea fertilizer are applied (Chambers and Garwood, 1998). However, the adverse effects of acidity on plant growth are usually not due to a lack of Ca *per se*, but to a reduced availability of P or to toxicity from increased amounts of soluble Al or Mn in the soil. Consequently, when liming increases herbage yield (e.g. Stevens and Laughlin, 1996), it is likely to be due to the alleviation of one of these effects or to an increase in the mineralization of N (Wheeler, 1998).

Magnesium deficiency in grass or clover is rare, though it has been reported from New Zealand (Hogg and Karlovsky, 1968). However, Mg deficiency in livestock is quite widespread, and Mg is sometimes applied to grassland to prevent this problem. Kainite, which contains KCl, NaCl and $MgSO_4$ (10% K, 18% Na, 3.6% Mg) is used quite widely for this purpose in Germany and The Netherlands (Mengel and Kirkby, 1987).

In addition to their direct effects on herbage yield, both fertilizer K and lime may influence the species composition of mixed swards, especially when applications are continued for many years. Such effects are illustrated by the Park Grass experiment at Rothamsted, UK. For example, on plots that had received fertilizer K annually for more than 100 years and had been limed at intervals, dandelion was many times more abundant than it was on plots that received neither lime nor fertilizer K (Tilman *et al.*, 1994). With grass–legume mixtures, the application of fertilizer K generally tends to increase the ratio of legume to grass, and such a change will have an indirect influence on the yield and nutritional quality of the herbage.

Influence of Species and Variety on Concentrations of K, Na, Ca and Mg in Herbage

Grasses and legumes grown under the same conditions are usually similar in K concentration, whereas the concentration of Ca is often two to three times higher in legumes than in grasses, and the concentration of Mg is often about 20% greater in legumes. Although grasses and legumes are usually similar in K, some comparisons have shown the legume to contain more than the grass (Adams, 1975; Drysdale *et al.*, 1980; Mayland and Wilkinson, 1996; Stockdale, 1999) whereas others, involving grass and legumes grown together in mixed swards, have shown the grass to contain more than the legume (Blaser and Kimbrough, 1968; Whitehead *et al.*, 1983). A higher concentration in the grass than in the legume is likely when the supply of K is limited (Blaser and Kimbrough, 1968; Follett and Wilkinson, 1995). With Na, it is impossible to make general comparisons between grasses and legumes, as there are large species differences within both.

For each of the four cations, typical ranges of concentration in perennial ryegrass and white clover are shown in Table 3.7. In a more specific comparison of perennial ryegrass and white clover grown together in mixed swards, the concentration of K was higher, and the concentrations of Ca and Mg lower in the grass than in the clover (Table 8.11). However, in a study in Victoria, Australia, concentrations of K, Ca and Mg were higher in white clover than in perennial ryegrass, while Na was higher in the ryegrass (Stockdale, 1999).

In general, individual grass species are broadly similar in their concentrations of K, Ca and Mg, though meadow fescue appears to be relatively high in Ca, and timothy relatively low in Mg (Table 8.12). In the USA, cocksfoot and tall fescue were 20–40% higher in Mg than were brome-grass and couch grass (Fairbourn and Batchelder, 1980), and tall fescue was higher in K and Mg than was Kentucky bluegrass (Hojjati *et al.*, 1977). Substantial differences in K and Mg were also found amongst nine cool-season grass species, including cocksfoot, Kentucky bluegrass and smooth brome-grass (Thill and George, 1975). Tall fescue had a higher concentration of Mg, and smooth brome-grass

Table 8.11. K, Na, Ca and Mg in perennial ryegrass and white clover grown together in mixed swards: mean and range of concentrations in herbage from alternate harvests taken in May, July and September at 12 sites in the UK; % in DM (from Whitehead *et al.*, 1983).

	K	Na	Ca	Mg
Perennial ryegrass	3.38	0.14	0.65	0.19
	(1.91–4.91)	(0.06–0.38)	(0.33–1.07)	(0.09–0.34)
White clover	2.98	0.13	1.73	0.28
	(1.37–4.49)	(0.04–0.41)	(1.21–2.54)	(0.15–0.54)

a lower concentration than did perennial ryegrass and cocksfoot (Powell *et al.*, 1978). However, differences between species in K, Ca and Mg are also influenced by the choice of cultivars (see below).

Amongst the legume species, white clover and red clover are generally higher in K and Mg than are lucerne and sainfoin, while lucerne is often similar to the clovers in Ca and higher in Ca than sainfoin (Table 8.13). However, in comparison with other species, lucerne can be either low or high

Table 8.12. Herbage concentrations of K, Na, Ca and Mg in four grass species: mean values from three investigations (% in DM).

	K	Na	Ca	Mg	Reference
Perennial ryegrass	2.35	0.57	0.65	0.25	Copponet, 1964
Timothy	2.30	0.11	0.59	0.18	
Cocksfoot	2.64	0.46	0.64	0.26	
Meadow fescue	4.01	0.07	0.73	0.23	
Perennial ryegrass	2.35	0.48	0.46	0.15	Whitehead, 1966
Timothy	2.74	0.08	0.38	0.13	
Cocksfoot	2.59	0.40	0.36	0.17	
Meadow fescue	2.79	0.08	0.47	0.15	
Perennial ryegrass	3.00	0.05	0.62	0.21	Baker and Reid, 1977
Timothy	2.16	0.004	0.33	0.08	
Cocksfoot	2.84	0.02	0.38	0.19	

Table 8.13. Herbage concentrations of K, Na, Ca and Mg in four legume species: mean values from four investigations (% in DM).

	K	Na	Ca	Mg	Reference
White clover	3.11	0.28	1.36	0.19	Davies *et al.*, 1966
Red clover	3.42	0.06	1.25	0.22	
Lucerne	2.55	0.14	1.05	0.17	
White clover	2.26	0.12	1.47	0.27	Whitehead and Jones, 1969
Red clover	2.07	0.04	1.84	0.21	
Lucerne	1.65	0.06	1.82	0.15	
Sainfoin	1.73	0.01	0.94	0.17	
Red clover	2.4	0.02	1.5	0.5	Smith *et al.*, 1974
Lucerne	1.9	0.06	1.5	0.4	
Sainfoin	2.1	0.02	0.9	0.3	
White clover	2.48	0.04	1.47	0.27	Baker and Reid, 1977
Red clover	1.89	0.01	1.50	0.30	
Lucerne	2.47	0.02	1.54	0.19	

in Mg, depending on the supply (Grunes, 1983) and on cultivar (Gross and Jung, 1978). *Lotus* spp. and subterranean clover have been reported to be high in Ca (Sherrell, 1978). When both red clover and white clover were grown with grass in a mixed sward, the concentration of Mg in the red clover was about 50% higher than in the white clover (Jones, 1963).

Some, but not all, dicotyledenous forbs are much higher than grasses in one or more of the major cations (Thomas *et al.*, 1952; Forbes and Gelman, 1981; Greene *et al.*, 1987; Wilman and Derrick, 1994). For example, yarrow and chickweed appear to be particularly rich in K, plantain and dandelion particularly rich in Ca, and burnet and dandelion particularly rich in Mg (see Table 3.10, p. 61; Forbes and Gelman, 1981).

With Na, both grasses and legumes show potentially large differences in concentration between different species. Some species, such as perennial ryegrass, cocksfoot and white clover, have the potential to take up large amounts of Na if the supply is plentiful, whereas others, such as timothy, have consistently low concentrations in the herbage whatever the supply (Ap Griffith and Walters, 1966). The extent of the variation in a potentially high-Na species is illustrated by the range from a trace to 2.12% Na reported for Italian ryegrass in the UK (Cunningham, 1964). In general, and based on a number of comparisons, the concentration of Na in ten common grassland species appears to be in the order: Yorkshire fog > perennial ryegrass > white clover > cocksfoot > *Agrostis tenuis* > Italian ryegrass > tall fescue > meadow fescue > timothy and *Poa pratensis* (Ap Griffith *et al.*, 1965; Sherrell, 1978; Chiy and Phillips, 1995a). Amongst the legume species, white clover often contains about twice as much Na as lucerne, and lucerne about twice as much as red clover (Table 8.13). Sainfoin also appears to be low in Na (Thomas *et al.*, 1952; Whitehead and Jones, 1969). Of the non-legume forb species, dandelion is high in Na while plantain is low (Forbes and Gelman, 1981). The differences amongst plant species and cultivars in their potential concentration of Na appear to reflect differences in the extent of retention in the roots. Plants that translocate only small amounts of Na to their leaves tend to retain large amounts in their roots and, in some instances, in their stems (Smith *et al.*, 1978; Chiy and Phillips, 1995a).

With each of the four major cations, there are substantial differences between varieties or cultivars of both grass and legume species. For example, in a comparison of K in several varieties of each of the species perennial ryegrass, cocksfoot, meadow fescue and tall fescue, there were only small differences between the species in mean concentration, but, within each species (except meadow fescue), there was an approximately twofold range between varieties (Ap Griffith *et al.*, 1965). Differences between varieties in Na were even greater, and there was a general tendency for Na and K to show an inverse relationship (Ap Griffith *et al.*, 1965). Varietal differences in Na were, of course, much greater in species with potentially high concentrations than in 'low-Na' species (Ap Griffith and Walters, 1966). Substantial differences between cultivars of grass species have also been reported for

Ca (Forbes and Gelman, 1981) and Mg (Brown and Sleper, 1980). In New Zealand, no evidence was found of consistent differences in Mg between ryegrass cultivars, but there were large differences within cultivars (Crush, 1983). With prairie grass, there were differences between cultivars in K, Na and Ca (Rumball *et al.*, 1972).

Within the legumes, varietal differences have been reported in K, Ca and Mg for red clover (Hunt *et al.*, 1976), in Ca for both red and white clover (Davies *et al.*, 1968) and in K and Na for lucerne (James *et al.*, 1995).

Evidence of substantial differences between cultivars has prompted attempts to breed varieties with consistently high or low concentrations of the major nutrient cations. Most attention has been given to Mg, as a productive variety high in Mg would be valuable in reducing the risk of hypomagnesaemia in livestock. However, the major problem appears to be combining a high concentration of Mg with high yield. For example, although a cultivar of Italian ryegrass bred specifically for high Mg concentration was found to reduce the incidence of hypomagnesaemia in ewes from 21% to 2.5%, its yield of DM was 8–10% lower than that of high-yielding cultivars (Moseley and Baker, 1991).

Influence of Stage of Maturity on Concentrations of K, Na, Ca and Mg in Herbage

Changes in the concentrations of K, Na, Ca and Mg with advancing maturity are less consistent than changes in N, P and S. In grasses, there is sometimes a decrease with maturity in all four elements, as reported for K (Fleming and Coulter, 1963; Ap Griffith and Walters, 1966; Baker and Reid, 1977), for Na (Pritchard *et al.*, 1964; Ap Griffith *et al.*, 1965), for Ca (Fleming and Coulter, 1963; Whitehead, 1966; Fleming and Murphy, 1968) and for Mg (Pritchard *et al.*, 1964). However in other studies, there was little change in Na (Fleming and Murphy, 1968), Ca (Baker and Reid, 1977) or Mg (e.g. Wilson and McCarrick, 1967; Fleming and Murphy, 1968; Baker and Reid, 1977). Sometimes K and Na increase during the first few weeks of growth in spring and then decrease (Ap Griffith and Walters, 1966). Part of the inconsistency, particularly with Mg, may be due to differences in the portion of the plant material sampled, as there may be a greater decline in the stem than in the leaves (Pritchard *et al.*, 1964). In a study involving fertilized ryegrass, concentrations of Na, Ca and Mg in the leaves actually increased until the stage of senescence (5–40% of leaf area yellow or brown), while K showed little change, though all four cations then declined in dead leaves (Chiy and Phillips, 1997). The dead herbage of prairie grassland also had much lower concentrations than living herbage of K and Mg, but there was little difference in Ca (Greene *et al.*, 1987), and dead leaf tissue of tall fescue, sampled in early spring, contained only 0.5% K, while new growth contained 4.5% K (Wilkinson and Mays, 1979).

In legumes, there is often no consistent trend in K, Ca or Mg with advancing maturity (Fleming and Coulter, 1963; Davies *et al.*, 1968; Baker and Reid, 1977). However, K sometimes declines (Whitehead and Jones, 1969; Baker and Reid, 1977; Wilman *et al.*, 1994) and Ca sometimes increases (Wilkinson and Gross, 1967; Wilman *et al.*, 1994). With Na, there is generally either an increase (Thomas *et al.*, 1952; Fleming and Coulter, 1963; Wilman *et al.*, 1994) or no consistent change (Whitehead and Jones, 1969).

Influence of Season of the Year and Weather Factors on Concentrations of K, Na, Ca and Mg in Herbage

When swards undergo frequent defoliation and are adequately supplied with nutrients, there is often no consistent seasonal change in concentrations of K and Na (e.g. Thomas, 1967; Metson and Saunders, 1978a). However, if the supply of K is low or is not maintained, there may be a progressive fall in herbage K concentration during the season (e.g. Hemingway, 1963; Fleming and Murphy, 1968; Kappel *et al.*, 1985). On the other hand, if fertilizer K is applied at intervals throughout the growing season, herbage K may increase (e.g. Whitehead *et al.*, 1983). With Na, although some studies have shown no consistent seasonal trend (e.g. Thomas, 1967; Forbes and Gelman, 1981; Whitehead *et al.*, 1983), others have shown an increase during the season (Hemingway, 1961; Reith *et al.*, 1964; Reay and Marsh, 1976; Thompson and Warren, 1979). There is evidence that herbage Na tends to decline during cool wet weather (Fleming and Murphy, 1968).

Reports of seasonal variation in Ca are contradictory. Although the concentration of Ca in grasses is often lower in spring and summer than in autumn and winter (Fleming, 1973; Reid and Jung, 1974; Thompson and Warren, 1979; Whitehead *et al.*, 1983), relatively high concentrations of Ca sometimes occur during periods of active growth in summer (Reay and Marsh, 1976; Metson and Saunders, 1978a; Edmeades *et al.*, 1983). On the other hand, low concentrations of Ca were noted during cool wet periods in summer (Fleming and Murphy, 1968).

Concentrations of Mg in grasses are generally at their lowest in early spring and increase considerably as the season advances towards autumn (Hemingway, 1961; Jones, 1963; Reith, 1963; Thomas, 1967; Fleming and Murphy, 1968; McIntosh *et al.*, 1973a; Reay and Marsh, 1976; Metson and Saunders, 1978a; Grunes, 1983; Whitehead *et al.*, 1983). However, as with Ca, temporarily low concentrations of Mg may occur in grasses during cool wet periods in summer (Fleming and Murphy, 1968). In the legumes, there is less change in Mg concentration during the course of the season, and the difference between grasses and clovers therefore diminishes during late summer and autumn (e.g. Reith, 1963; Whitehead *et al.*, 1983).

The seasonal trends are probably due, at least partly, to changes in temperature, but changes in soil water appear to have little effect. Increasing temperature, over the range of about 10–30°C, tends to increase the concentrations of K, Ca and Mg in grasses (Parks and Fisher, 1958; Nielsen and Cunningham, 1964; Grunes *et al.*, 1970; Grunes and Welch, 1989), but, over a similar temperature range, appears to have little, if any, effect on Na (Nielsen and Cunningham, 1964), though temporarily low concentrations of Na did occur during cool wet periods in summer (Fleming and Murphy, 1968). There was an increase in herbage Mg with increasing temperature in several investigations (Elkins and Hoveland, 1977; Fairbourn and Batchelder, 1980; Sleper *et al.*, 1980). However, it has been suggested that one reason why hypomagnesaemia in animals is most prevalent when warm weather follows a period of cool weather is that both the K concentration and the K : (Ca + Mg) ratio are increased by the increase in temperature (Grunes *et al.*, 1970). In lucerne and red clover, increasing temperature generally had little effect on Ca and Mg, but K was lower at a day/night temperature of 15/10°C than at 21/15°C and above (Smith, 1970).

Drought was found to have small and inconsistent effects on concentrations of K, Na, Ca and Mg in eight grass and legume species (Kilmer *et al.*, 1960). However, a high soil moisture content (equivalent to flooding) caused an increase in herbage Ca and a decrease in herbage K in cocksfoot (Currier *et al.*, 1983).

Influence of Soil Type on Concentrations of K, Na, Ca and Mg in Herbage

In a comparison of the concentrations of K, Na, Ca and Mg in timothy grown on three types of soil, there were only small differences between sandy soils, clay and silt soils, and organic soils (Kähäri and Nissinen, 1978). In a comparison of cation concentrations in grasses and legumes grown on 16 soil series in Virginia, USA, there were at least twofold differences between soil series in K, Ca and Mg and a much larger difference in Na (Baker and Reid, 1977), but the number of samples was insufficient to show that the concentrations were related closely to soil series. With K in particular, soils differ widely in their ability to release plant-available forms from reserves: some release sufficient K to maintain optimum plant growth for many years, whereas others are rapidly depleted (Goulding and Loveland, 1986). In practice, the supply of K from the soil is often dominated by that applied in the form of fertilizer. With Na, the effect of soil parent material is often less than that of the atmospheric input, and hence of proximity to the coast (Smith and Middleton, 1978). The concentration of Ca in herbage tends to increase with increasing soil pH and with an increasing proportion of Ca in the total exchangeable cations of the soil (Gross and Jung, 1981). Herbage Mg is most likely to be low on soils that are either sandy or acid or that have high contents of Ca or K. Calcareous soils often induce low herbage concentrations of Mg (Stewart and McConaghy, 1963; Whitehead *et al.*, 1983), while high concentrations of Mg are associated with soils derived largely from dolomite or serpentine (Kubota *et al.*, 1980).

Influence of Fertilizer N on Concentrations of K, Na, Ca and Mg in Herbage

Fertilizer N has inconsistent effects on herbage concentrations of K and Ca, but generally increases concentrations of Na and Mg. The inconsistencies are due partly to differences between soils in the amounts of the cations in plant-available form and partly to the form in which the fertilizer N is taken up. When supplies of the cations, especially K and Ca, are limited, the increase in growth caused by fertilizer N generally tends to reduce their concentrations in herbage. However, when supplies of the cations are plentiful, and when N is taken up mainly as nitrate, herbage concentrations may increase, due to the synergistic effect between nitrate and cation uptake (see p. 46).

With K, it has been suggested that fertilizer N tends to increase the herbage concentration when the initial concentration is more than about 2%, and to decrease it when the initial concentration is below about 2% (Kemp, 1983). However, other studies have shown increases with rather low initial concentrations (e.g. Mudd, 1970a; Follett *et al.*, 1977) and decreases at initial concentrations of more than 2% (Stewart and Holmes, 1953; Mortensen *et al.*, 1964; MacLeod, 1965). In some studies, fertilizer N has had little or no effect (e.g. Reith *et al.*, 1964; Adams, 1973; Reid and Strachan, 1974; Whitehead *et al.*, 1983; Hopkins *et al.*, 1994). In investigations in Finland, fertilizer N increased herbage K markedly in the first year, had little effect in the second year and decreased herbage K in the third year, by which time soil K on the control plots was low (Rinne *et al.*, 1974a). The tendency of ammonium N to depress herbage K was shown following the application of ammonium sulphate to a grass–clover sward (Hemingway, 1961). Also, the herbage concentration of K was lower when N was applied as anhydrous ammonia than when the same amount of N was applied as calcium ammonium nitrate (Van Burg and van Brakel, 1965).

The concentration of Na in those species that tend to accumulate Na is generally increased by fertilizer N, at least when this is added in the absence of K (Stewart and Holmes, 1953; Reith *et al.*, 1964; Kershaw and Banton, 1965; Mudd, 1970a; Wilman and Mzamane, 1982; Whitehead *et al.*, 1983; Hopkins *et al.*, 1994). Although herbage Na is increased more by nitrate than by ammonium (e.g. Kershaw and Banton, 1965), the application of ammonium sulphate may also cause a large increase in Na (Hemingway, 1961). However, with species that are inherently low in Na, such as timothy and meadow fescue, the application of fertilizer N has little effect (Ap Griffith *et al.*, 1965).

Fertilizer N has an inconsistent effect on herbage Ca, due partly to the fact that some N fertilizers contain Ca and partly to the form in which the N is taken up. Fertilizers that do not contain Ca often decrease herbage Ca (Gardner *et al.*, 1960; Hopkins *et al.*, 1994) though, in some instances, there is little effect (Table 8.14; Reid *et al.*, 1967b; Mudd, 1970a). With fertilizers, such as calcium ammonium nitrate, that do contain Ca, there may be little effect on herbage Ca or even a decrease at low rates of application, but an increase at higher rates (Stewart and Holmes, 1953; Whitehead, 1966; Rinne *et al.*, 1974a; Wilman and Mzamane, 1982). When fertilizer N is supplied in the form of ammonium or urea, herbage Ca is lower than when the N is supplied as nitrate (Van Burg and van Brakel, 1965; Reid, 1966; Follett *et al.*, 1977; Adams, 1984). With grass–clover swards, fertilizer N generally reduces the proportion of clover in the sward and this, in turn, reduces the average concentration of Ca in the herbage (Rodger, 1982), even though there may be no significant effect on the concentration of Ca in either component (Rahman *et al.*, 1960; Hemingway, 1961; Whitehead *et al.*, 1983).

The influence of fertilizer N on herbage Mg also depends on the form of N. Herbage Mg is often increased by N taken up as nitrate (Stewart and Holmes, 1953; Reith *et al.*, 1964; Mudd, 1970a; McIntosh *et al.*, 1973b; Follett *et al.*, 1977; Hojjati *et al.*, 1977; Wilman and Mzamane, 1982; Hopkins *et al.*, 1994), though sometimes there is little or no effect (Gardner *et al.*, 1960; Mortensen *et al.*, 1964; Whitehead *et al.*, 1983). When fertilizer N is applied in the form of ammonium, the effects on herbage Mg are inconsistent, probably due to differences in the rate of nitrification. However, N applied as ammonium sulphate produced a lower concentration of Mg in herbage than did N applied as calcium nitrate (Kershaw and Banton, 1965; Follett *et al.*, 1977). Nitrogen applied as anhydrous ammonia produced herbage with a lower concentration of Mg than did an equal amount of N applied as calcium ammonium nitrate (Van Burg and van Brakel, 1965). Nevertheless, the application of ammonium sulphate and of urea has sometimes increased herbage Mg (Rahman *et al.*, 1960; Hemingway, 1961; Adams, 1984), presumably because much of the N was taken up as nitrate.

Table 8.14. Concentrations of K, Na, Ca and Mg in perennial ryegrass as influenced by fertilizer N, applied as ammonium nitrate, in two investigations (% in DM).

Rate of fertilizer N (kg ha^{-1} year^{-1})	K	Na	Ca	Mg	Reference
Nil	3.35	0.12	0.66	0.19	Whitehead *et al.*, 1983[a]
200	3.41	0.15	0.64	0.19	
Nil	2.46	0.16	0.82	0.18	Hopkins *et al.*, 1994[b]
300	2.49	0.29	0.76	0.20	

[a]Mean values from three harvest dates, 12 sites, 1 year.
[b]Mean values from two harvest dates, 16 sites, 4 years.

In two multicentre investigations, the average effects of ammonium nitrate fertilizer on each of the four cations were relatively small (Table 8.14), though there was some variation between individual sites.

Influence of Fertilizer P on Concentrations of K, Na, Ca and Mg in Herbage

Fertilizer P usually has little effect on herbage concentrations of the four major cations. This has been shown for all four cations together (Hemingway, 1961; Reid and Jung, 1965; Balasko, 1977), for K, Ca and Mg (e.g. Gardner et al., 1960; Reith et al., 1964; Heddle and Crooks, 1967; Follett et al., 1977; Rodger, 1982) and for Na (Kemp, 1960). However, in two studies in the USA, fertilizer P increased herbage concentrations of Ca and Mg in tall fescue (Reinbott and Blevins, 1994; Sweeney et al., 1996).

Influence of Fertilizer K on Concentrations of K, Na, Ca and Mg in Herbage

Fertilizer K normally increases the concentration of K in herbage, but decreases the concentrations of the other three cations (Table 8.15; McNaught, 1959). A substantial increase in herbage K has been reported in many investigations (Table 8.15; Stewart and Holmes, 1953; Gardner et al., 1960; Hopper and Clement, 1967; Heddle and Crooks, 1967; Brockman et al., 1970; Schellberg et al., 1999), and the increase tends to be greater in clovers than in grass (Hemingway, 1961).

The application of fertilizer K generally causes a large reduction in herbage Na when the initial concentration of Na is relatively high (Hemingway, 1961; McNaught and Karlovsky, 1964; Smith, 1975; Evans et al., 1986), but, since added N and K have opposite effects on herbage Na

Table 8.15. Herbage concentrations of K, Ca and Mg in brome-grass (% in DM) as influenced by rate of fertilizer K, applied as KCl: mean values from two harvests in each of 2 years (from Laughlin et al., 1973).

Rate of fertilizer K (kg ha^{-1} year^{-1})	K	Ca	Mg
0	1.11	0.48	0.22
37	1.56	0.44	0.18
74	1.98	0.39	0.15
112	2.17	0.36	0.14
148	2.32	0.36	0.13
186	2.51	0.32	0.12

(Kemp, 1960), when both are added together there may be little effect or even a small increase (Rahman *et al.*, 1960; Whitehead *et al.*, 1978).

Fertilizer K also tends to decrease the concentration of Ca in herbage, as shown for both grasses (McNaught, 1959; Gardner *et al.*, 1960; Kemp, 1960; Mudd, 1970a; Laughlin *et al.*, 1973; Lightner *et al.*, 1983) and legumes (e.g. Smith, 1975). However, with grass–clover swards, fertilizer K may have little effect on the Ca concentration of the mixed herbage, because it tends to increase the proportion of clover in the swards (Gardner *et al.*, 1960; Hemingway, 1961; Reith *et al.*, 1964; Rodger, 1982).

Decreases in herbage Mg resulting from the application of fertilizer K have been noted in several investigations (Hemingway, 1961; Hunt *et al.*, 1964; Reith *et al.*, 1964; Laughlin *et al.*, 1973; Rodger, 1982; Lightner *et al.*, 1983; Evans *et al.*, 1986; Reinbott and Blevins, 1994). However, when the rate of K application is no greater than that required for maximum herbage yield, there may be little or no effect on herbage Mg (McConaghy *et al.*, 1963; Reith, 1963; Mudd, 1970a; Smith, 1975).

Influence of the Application of Na, Mg and Lime on Concentrations of K, Na, Ca and Mg in Herbage

Sodium is sometimes applied to grassland in order to increase the palatability or nutritional value of the herbage. The application may be in the form of NaCl or as low-grade potassium salts. An application, in spring, of 200 kg Na ha^{-1} as NaCl increased the concentration of Na in grass by 50% and the concentration in clover by 30%, while having little effect on K and Ca, but reducing Mg (Hemingway, 1961). However, data from several other experiments indicated that, as the concentration of Na in grass herbage increased over the range 0.02–0.4%, reflecting the supply of Na, there was a linear reduction in K, though Ca and Mg tended to increase (Chiy and Phillips, 1995b). In contrast to these results, the application of 100 kg Na ha^{-1} as NaCl increased K in grass–clover herbage when the concentration of K was above about 1.2%, though not when K was seriously deficient (McNaught and Karlovsky, 1964). In some situations, the application of 50–100 kg Na ha^{-1}, while increasing the concentration of Na, has little effect on K, Ca or Mg (Cushnahan *et al.*, 1996; Chiy and Phillips, 1998).

Liming grassland on acid soils generally increases herbage Ca, but often has little effect on concentrations of K, Na or Mg (Table 8.16; Edmeades *et al.*, 1983; Adams, 1984; Evans *et al.*, 1986). In some situations, liming causes a reduction in herbage Mg (John *et al.*, 1972; McNaught *et al.*, 1973; Jones and Sparrow, 1977; Evans *et al.*, 1986; Stevens and Laughlin, 1996; Wheeler, 1998), though it has also been reported to cause an increase (Fleming, 1973). Dolomitic limestone, in particular, tends to increase herbage Mg (Riggs *et al.*, 1995).

Table 8.16.   Herbage concentrations of K, Na, Ca and Mg in grass, mainly perennial ryegrass (% in DM), as influenced by the rate of application of lime at the beginning of a 10-year period: mean values for the period (from Stevens and Laughlin, 1996).

Rate of lime (t ha^{-1})	K	Na	Ca	Mg
0	1.81	0.46	0.50	0.18
4	1.78	0.46	0.56	0.18
8	1.76	0.47	0.61	0.17
12	1.78	0.46	0.64	0.17

Magnesium is sometimes applied to grassland with the aim of increasing its concentration in herbage and thus preventing hypomagnesaemia in livestock. Different sources of Mg vary in effectiveness, due partly to differences in solubility and partly to differences in particle size (Mayland and Wilkinson, 1989). Magnesium sulphate is generally an effective source, and Epsom salts ($MgSO_4.7H_2O$) is a more soluble (and expensive) form than kieserite ($MgSO_4.H_2O$) (Mengel and Kirkby, 1987). Of other possible sources of Mg, calcined magnesite (MgO) is generally more effective than dolomitic limestone (Jones and Sparrow, 1977). The latter is moderately effective on acid soils, but not on neutral and calcareous soils, in which its solubility is low. The application of kieserite at 260 kg Mg ha^{-1} increased the average concentration of Mg in grass–clover herbage from 0.14% to 0.19% (Hunt *et al.*, 1964); but applications of dolomite or calcined magnesite at rates up to 450 kg Mg ha^{-1} had only small effects on the concentration of Mg in herbage (McConaghy *et al.*, 1963). In practice, to prevent hypomagnesaemia, it may be more effective to adjust the rates of fertilizer K and N or to supply Mg in concentrate feeds than to apply Mg to the sward.

Influence of Livestock Excreta on Concentrations of K, Na, Ca and Mg in Herbage

Urine is generally high in K and low in Ca and Mg, and is variable in Na. It therefore tends to increase the herbage concentration of K and to reduce concentrations of Ca and Mg, as shown in Table 8.17. Increases in herbage K resulting from the application of urine have also been reported in several other investigations (e.g. Lotero *et al.*, 1966; Ledgard *et al.*, 1982; Williams and Haynes, 1994), as have small reductions in Ca and Mg (Hogg, 1981; Williams and Haynes, 1994). Sometimes urine appears to have no consistent effect on herbage Ca (Joblin, 1981). Dung has been reported to increase herbage K (Williams and Haynes, 1995) and sometimes also Mg (Weeda, 1977; Hogg, 1981). It has an inconsistent effect on Na, and may cause a short-term reduction in herbage Ca (Weeda, 1977).

Table 8.17. Herbage concentrations of K, Na, Ca and Mg in ryegrass (% in DM) as influenced by the application of sheep urine (from Joblin and Keogh, 1979).

	K	Na	Ca	Mg
Without urine	2.61	0.45	0.57	0.31
With urine	3.35	0.42	0.46	0.28

As retention of the major cations by grazing animals is relatively low, the amounts returned in the dung and urine are important in relation to the amounts taken up by grass swards, as shown by comparing Table 8.4 with Table 3.5. This is especially so for K, which is the cation most likely to be deficient. However, the actual uptake of urine K is influenced by the application of fertilizer K. In a field trial with plots of ryegrass, the uptake of excreted K into the herbage of a grazed sward when no fertilizer K was applied (based on comparison with a cut sward) was about 140 kg K ha^{-1} year^{-1}, but when fertilizer K was applied, at 125 kg K ha^{-1} year^{-1}, the uptake of excreted K was only 60 kg K ha^{-1} year^{-1} (Hopper and Clement, 1967). In another study, in which all four cations were examined, plots of a long-term upland sward that had been grazed for 25 years had much higher herbage concentrations of K and Na than did plots that had been cut, but were little different in Ca and Mg (Davies *et al.*, 1959). However, although it is clear that the K returned in excreta can be utilized for new plant growth, there is a major problem in combining it with fertilizer K, due to the uneven distribution of dung and urine (see p. 29).

When excreta are returned in the form of slurry, there is often an increase of about 10–20% in the concentrations of K and Na in herbage, and a smaller increase in the concentration of Mg (Adams, 1973). Slurry is more effective than excreta returned directly from grazing animals in reducing the need for fertilizer K, as it is applied more evenly over the sward. However, losses of K due to leaching and runoff are possible, especially if the slurry is applied in excessive amounts at a time when the grass is not growing (Smith and van Dijk, 1987).

Chemical Forms and Availability of K, Na, Ca and Mg in Herbage

The monovalent cations, K and Na are present in herbage almost entirely as free or readily exchangeable ions, whereas the divalent cations, Ca and Mg, are largely bound to organic plant constituents. Consequently it is usual for >85% of the K and Na in herbage to be soluble in water, while the solubility of Ca is 40–70%, and of Mg 55–85% (Makela, 1967; McIntosh *et al.*, 1973b; Grace *et al.*, 1977; Whitehead *et al.*, 1985; Marschner, 1995). Of the Ca which is not soluble in water, some is present as calcium pectate, some is bound to protein, and some is bound to other organic compounds containing carboxyl, phosphoryl and phenolic groups. In addition, some Ca may be precipitated as calcium phosphate and, in some species such as lucerne, as calcium oxalate (Butler and Jones, 1973). Lucerne contains about 20–30% of its Ca as Ca oxalate, and this appears to be largely unavailable to ruminant animals (Ward et al., 1979). The Mg in chlorophyll usually accounts for less than 25% of the total (Marschner, 1995), and may be only about 4% in mature herbage with a high proportion of stem (Grunes, 1983). Some Mg is combined with organic anions including oxalate and citrate, and some is associated with cell walls, probably in the form of pectate (Grunes, 1983).

The true availability of both K and Na to ruminant animals is probably at least 95% (Fontenot and Church, 1979; Agricultural Research Council, 1980; McDowell, 1992; Chiy and Phillips, 1995b, for Na) while the true availability of Ca is thought to be at least 68% (Agricultural and Food Research Council, 1991). The true availability of Mg is uncertain (Underwood and Suttle, 1999).

Occurrence and Metabolic Functions of K, Na, Ca and Mg in Ruminant Animals

Typical concentrations of K, Na, Ca and Mg in the body tissue, blood serum and milk of ruminant animals are shown in Table 8.18. Calcium is the second most abundant nutrient element (after N) in the animal body, comprising about 1.5% of live weight, mainly in bone. Potassium, Na and Mg each amount to < 0.25% of live weight. In milk, K is the major cation, followed closely by Ca, and then by Na and Mg (Table 8.18).

Potassium has numerous functions in animals. It is involved in maintaining the optimum amount of water in the tissues and in maintaining osmotic pressure and acid–base equilibrium. It also activates several enzymes involved in carbohydrate metabolism and protein synthesis, and plays a role in the electrical activation of nerve and muscle cells, including the contraction of muscles (Ammermann and Goodrich, 1983; Naylor, 1991). Within animal cells, the concentration of K is 20–30 times greater than the concentration of Na, whereas, in the extracellular fluid, the concentration of Na is much greater than that of K.

Table 8.18. Typical concentrations of K, Na, Ca and Mg in the body tissue, blood serum and milk of cattle and sheep.

	K	Na	Ca	Mg	Reference
Body tissue (% of live weight)	0.2	0.15	1.5	0.04	Naylor, 1991 + Table 4.1
Blood serum (g l^{-1})	0.19	3.3	2.1	0.25	Kincaid, 1988
Milk (mg l^{-1})	1.5	0.6	1.2	0.12	White and Davies, 1958 + Table 4.4

The functions of Na are similar to those of K: it is involved, together with K and Cl, in maintaining osmotic pressure and pH, and also in muscle contraction, the active transport of amino acids and glucose, and the absorption and transport of Ca (Ammermann and Goodrich, 1983; National Research Council, Subcommittee, 1989; McDowell, 1992; Underwood and Suttle, 1999). Sodium may also be involved in the production of milk, as its addition to low-Na diets has been found to increase the proportion of fat in milk and, when the diet is seriously deficient, to increase the actual yield of milk (Chiy and Phillips, 1995b). The concentration of Na in the body is controlled partly by the mechanisms that maintain a nearly constant Na : K ratio in the extracellular fluid (McDowell, 1992) and partly by the mechanisms of thirst and antidiuretic hormone that control water balance (Michell, 1985). When the intake of Na is low, there is also an increased production of the hormone aldosterone, which reduces the loss of Na in the urine (Kincaid, 1988). Some of the body Na, about 30–40% of the total, is present in the skeleton in an insoluble form (Ammermann and Goodrich, 1983), and some of this can be used during periods of deficiency.

About 98–99% of the Ca in the body is present in bones and teeth, mainly in the form of hydroxyapatite, while the other 1–2% is distributed in various soft tissues and extracellular fluids (Grace, 1983a; National Research Council, Subcommittee, 1989). The Ca : P ratio in bone is fairly constant at 2 : 1, and Ca accounts for about 8% of bone on a wet weight basis (Fontenot and Church, 1979). However, the small amount of Ca outside the skeleton has important metabolic functions. These include acting as a cofactor for various enzymes and binding to the protein calmodulin, which has a role in regulating the intermediary metabolism of cells (Kincaid, 1988; Naylor, 1991). Calcium is also involved in the transmission of nerve impulses, in the contraction of muscles and in the clotting of blood (National Research Council, Subcommittee, 1989). In particular, it is required at the neuromuscular junction where it regulates the transmission of nerve impulses from nerve to muscle. Calcium is an important constituent of milk, which contains about 1.2 mg l^{-1}. The skeleton provides a major reserve of Ca, and this can be utilized to maintain blood Ca when the dietary supply is low and when there is a particularly large demand for the production of milk during early lactation. Homeostatic mechanisms operate to maintain a constant concentration of Ca in plasma and extracellular fluids (at about 100 mg l^{-1}), whatever the fluctuations in supply and demand (Underwood and Suttle, 1999). The need for Ca increases by a factor of between two and five at the beginning of lactation (Naylor, 1991), and it is normal for some Ca from the skeleton to be utilized at this time, even when the dietary supply is plentiful (National Research Council, Subcommittee, 1989). Of the Ca in blood serum, between 50 and 70% is in ionic form, being balanced by anions such as phosphate and bicarbonate, while the remainder is bound to serum proteins (Underwood and Suttle, 1999).

Although Mg is a minor component of bone, about 60–70% of the total body Mg occurs in the skeleton (National Research Council, Subcommittee, 1989; Naylor, 1991; Underwood and Suttle, 1999). The ratio of Ca : Mg in bone is about 50 : 1. The remainder of the body Mg is distributed between muscle and other soft tissues, with most being in the muscle (Miller, 1979; Grace, 1983a). Within animal cells, Mg is the second most abundant cation, after K. It is involved in the functioning of membranes and in muscle contraction, where it tends to counteract the effects of Ca (Naylor, 1991; Underwood and Suttle, 1999). It is also involved in energy metabolism as a constituent or cofactor of numerous enzymes, including many of those utilizing ATP/ADP or catalysing the transfer of phosphate (Underwood and Suttle, 1999). Magnesium also plays a role in Ca metabolism; it facilitates the absorption of dietary Ca and promotes the release of parathyroid hormone from the parathyroid glands (Naylor, 1991). When the dietary intake of Mg is low, there is some mobilization from bone, but hypomagnesaemia (a concentration of Mg in the blood plasma that is below the normal range) may result if the rate of mobilization is insufficient to meet requirements. There may also be some mobilization of Mg from bone unnecessarily, when bone is utilized as a source of Ca or P. About 68% of the Mg in blood serum is in ionic form, while about 32% is bound to protein (Underwood and Suttle, 1999).

Absorption of Dietary K, Na, Ca and Mg by Ruminant Animals

Potassium and Na are absorbed mainly from the upper small intestine and, to a lesser extent, from the rumen, the omasum, the lower small intestine and the large intestine (Little, 1981; McDowell, 1992). Absorption of K is normally 80–95% (Underwood and Suttle, 1999). Sodium is absorbed from the lower parts of the intestines only if there is a metabolic need (Agricultural Research Council, 1980). Both elements, but especially Na, are recycled through the digestive system, mainly by secretion in saliva.

Although some Ca is absorbed from the rumen, most absorption occurs from the small intestine (National Research Council, Subcommittee, 1989). Both passive and active absorption occur the latter requiring vitamin D. The proportion of the dietary Ca that is absorbed from the small intestine is related to need, and is increased by factors such as milk production and growth. When a diet is relatively low in Ca, the proportion absorbed increases and, conversely, when a diet contains more Ca than is needed, the proportion absorbed decreases (Horst, 1986). The absorption of Ca tends to decrease with age and with a low intake of vitamin D. Absorption is also influenced by the extent to which Ca is released from the diet and by interactions with other components, which may render the Ca insoluble. Some experiments have indicated that the utilization of Ca is affected by the Ca : P ratio of the diet, and a ratio of between 1 : 1 and 2 : 1 is often considered ideal.

However, ratios up to 7 : 1 generally have no detrimental effect, unless one element is deficient (National Research Council, Subcommittee, 1989; Naylor, 1991). The absorption of Ca may also be improved by supplementation of the diet with Na (Chamberlain and Wilkinson, 1996). On the other hand, there is evidence that feeding a diet low in K in the period before calving is an effective means of preventing milk fever in dairy cows, possibly by reducing the ratio of total cations to total anions (Goff and Horst, 1997). However, the concentration of K suggested as ideal (< 2% K) is less than that needed for the optimum growth of young grass herbage. Within the animal, homeostatic control of Ca is brought about mainly by two hormones, the parathyroid hormone and calcitonin. When the concentration of Ca in the blood falls, parathyroid hormone is produced, and this stimulates the production by the kidney of a third hormone, the active form of vitamin D (1,25-dihydroxy vitamin D). This, in turn, promotes the formation of a Ca-binding protein in the intestinal mucosa, and thus increases the absorption of Ca through the wall of the intestine (Little, 1981; Naylor, 1991). On the other hand, when the intake of Ca exceeds requirements, the secretion of parathyroid hormone is reduced, thus reducing the production of active vitamin D and hence the absorption of Ca (Horst, 1986; National Research Council, Subcommittee, 1989). The active form of vitamin D, in conjunction with parathyroid hormone, also increases the mobilization of Ca from bone, whereas calcitonin, which is produced by the thyroid gland in response to a high concentration of Ca in the blood, suppresses mobilization and also promotes the excretion of Ca in the urine (Naylor, 1991). Short-term adjustments of blood Ca are largely due to calcitonin. The mobilization of Ca from bone enables the concentration of Ca in the milk to be maintained, whatever the concentration in the diet, but, if continued for a prolonged period, is liable to result in the bones becoming soft and easily broken. Although there is evidence that the true availability of Ca is at least 68% for most diets (Agricultural and Food Research Council, 1991), measurements of actual absorption have often been lower, even for animals with a substantial demand for Ca (Fontenot and Church, 1979; Underwood and Suttle, 1999).

The absorption of Mg occurs mainly from the rumen and omasum, and to a lesser extent from the small and large intestines (Underwood and Suttle, 1999). There appear to be both active and passive mechanisms for Mg absorption, the active mechanism being more important when intake is low. The extent of absorption of Mg appears to be restricted more by conditions in the rumen than by the form in which it is present in the diet. Absorption is reduced by a high concentration of K, probably because Na is essential for the transfer of Mg across the rumen wall, and the K tends to displace Na (Dua and Care, 1995; Underwood and Suttle, 1999). Similarly, a deficiency of Na may restrict the absorption of Mg (Martens *et al.*, 1987; Naylor, 1991; Chiy and Phillips, 1996b; Chamberlain and Wilkinson, 1996), while the application of Na to grassland may enhance Mg absorption (Cherney *et al.*, 1998). Another factor that tends to reduce the absorption of Mg is a high concentration of ammonia in the rumen, as this may reduce the solubility of the Mg (Kincaid, 1988). A combination of a high concentration of K with a high concentration of ammonia in the rumen is most likely to arise in animals consuming young grass herbage recently fertilized with N and K. The risk of hypomagnesaemia is therefore greatest with this type of diet. Although various organic compounds, including lipids and organic acids, have also been suggested as curtailing the absorption of Mg, these are no longer considered important (Underwood and Suttle, 1999). As with Ca, the actual absorption of Mg is usually less than the true availability, and has been reported to range from nil to 66% (Underwood and Suttle, 1999). Measurements of apparent absorption have usually been in the range 10–40% (Agricultural Research Council, 1980; Reid, 1983). The absorption of Mg and its excretion in the urine are both increased by vitamin D (Fontenot *et al.*, 1989), while homeostatic control results partly from the action of the parathyroid hormone, which decreases excretion in the urine and stimulates resorption from bone (Fontenot *et al.*, 1989).

Nutritional Requirements for K, Na, Ca and Mg in Ruminant Animals

The requirement in terms of dietary concentration for each of the four cations is greater for dairy cows than for beef cattle and sheep (Table 8.19), due mainly to the needs of lactation.

Potassium is required at a concentration of between 0.5% and 0.9% in the diet, depending mainly on the production of milk, which contains about 1.5 g K l^{-1} (see Table 4.4, p. 78; Ammermann and Goodrich, 1983; National Research Council, Subcommittee, 1989; Naylor, 1991). However, it is rare for the diet of ruminant animals to contain less than the 0.9% K at which a deficiency might arise.

The requirement for Na by non-lactating animals is low (Michell, 1985), but, as with K, it increases with milk yield. It also increases in conditions that induce sweating. Recommendations range from about 0.10% in the diet for most non-lactating animals to about 0.18–0.20% for animals producing large

Table 8.19. Recommendations for dietary concentrations of K, Na, Ca and Mg for cattle and sheep (% in DM).

	K	Na	Ca	Mg	Reference
Dairy cow yielding 20 kg milk day^{-1}	0.90	0.18	0.60	0.21	National Research Council, Subcommittee, 1989
Beef cattle, age > 1 year	0.65	0.10	0.29	0.16	National Research Council, 1984
Ewes, non-lactating	0.5	*c.* 0.10	0.20–0.32	0.12–0.15	National Research Council, 1985
Ewes, lactating	0.7–0.8	*c.* 0.10	0.32–0.39	0.18	National Research Council, 1985

amounts of milk, and also for working animals (Table 8.19; National Research Council, Subcommittee, 1989; Naylor, 1991). However, in the UK, the recommendation for dairy cattle producing up to 30 kg milk day^{-1} is only 0.12% Na (Agricultural Research Council, 1980). Deficiencies may develop in animals grazing on herbage consistently low in Na, and in some natural grassland systems the population of mammals such as moose may be limited by the supply of Na (Botkin *et al.*, 1973). However, when animals with an adequate supply initially encounter a low-Na diet, there is only a slow change to deficiency, because Na is conserved by restricting excretion in the urine and by reabsorption from saliva (Australian Agricultural Council, 1990). Although, in comparison with the other cations, the actual requirement for Na is relatively low, animals prefer to consume more if it is available to them. When additional Na is provided in the form of salt-licks, dairy cattle may consume up to 40 g day^{-1} (Chiy and Phillips, 1996b), equivalent to an additional 0.2% in the diet with a DM intake of 20 kg; and it has been suggested that animals may benefit from having 0.3% Na or more in their diet (Chiy and Phillips, 1995b; Chamberlain and Wilkinson, 1996). The benefit appears to be greatest when the additional Na is provided by application to the sward, and may then be due partly to secondary effects on the composition of the herbage. Thus, in a trial in which 32 kg Na ha^{-1} year^{-1} was applied, there was a marked increase in herbage Na, a slight increase in Ca and Mg and a decrease in K, while the content of water-soluble carbohydrate in the herbage also increased and non-protein N decreased. Dairy cattle consuming the herbage showed increases in Na in saliva and in rumen pH, together with a greater ratio of acetate to propionate in the rumen fluid, and increases in milk-fat percentage and in milk yield (Chamberlain and Wilkinson, 1996).

The requirement for Ca is influenced strongly by lactation and, to a lesser extent, by pregnancy, by the rate of live-weight gain and by the age of the animal. The maintenance requirement, which is predominantly to replace the endogenous loss of Ca in the faeces, is relatively low, at about 0.2–0.3% in the diet (Agricultural Research Council, 1980). For dairy cows producing 5 kg milk day^{-1}, the recommended dietary concentration is 0.30% Ca, whereas, with milk production of 30 kg day^{-1}, the recommendation is 0.41% (Table 8.20). When the demands of lactation cannot be met by a combination of dietary Ca intake and mobilization from bone, the animal may develop milk fever (see below). Paradoxically, one means of reducing milk fever is to reduce the intake of Ca during the last few weeks of pregnancy, so that the animal relies less on Ca absorption from the diet and is more able to mobilize Ca from bone at the onset of lactation (Braithwaite, 1976).

In general, the requirement for Mg by growing cattle and sheep is met when the diet contains 0.10% Mg, but dairy cattle require 0.18–0.20% during lactation (Table 8.19). When the concentration in the diet is less than this, hypomagnesaemia in lactating animals is a definite possibility, especially if the diet consists of young, fertilized grass with high contents of K and N.

Table 8.20. Recommendations, in the UK, for dietary Ca concentration (% in DM) of dairy cow (600 kg live weight) as influenced by milk yield (Agricultural and Food Research Council, 1991).

	Milk yield (kg day^{-1})				
	5	10	15	20	30
Ca in diet (% in DM)	0.30	0.35	0.37	0.39	0.41

Effects of Deficiencies of K, Na, Ca and Mg on Ruminant Animals

It is rare for the concentration of K in herbage to be deficient for livestock: an excess is more common. When K deficiency does occur, it is usually the result of an abnormal loss of K due, for example, to vomiting, diarrhoea or excessive sweating or to an excessive secretion of aldosterone causing additional excretion in the urine (Kincaid, 1988).

A deficiency of Na in animals is most likely to occur in areas far from the coast and on freely drained soils, and especially if the grassland is heavily fertilized with K. It is also rather more likely on swards dominated by low-Na species, such as timothy and fescue, than on potentially high-Na species, such as ryegrass, cocksfoot and white clover (Australian Agricultural Council, 1990). One of the first symptoms of Na deficiency in animals is the abnormal chewing or licking of soil, wood or sweat. Symptoms that may develop subsequently include a decrease in appetite, an unthrifty appearance, loss in weight and decreased milk production (Miller, 1979; Chiy and Phillips, 1995b). Sodium deficiency can be diagnosed from the Na : K ratio of saliva, since deficient animals usually have a ratio that is less than 10 : 1, and possibly as low as 1 : 1 (Morris, 1980; Naylor, 1991; Chiy and Phillips, 1995b). A deficiency of Na can also be diagnosed from the concentration of aldosterone, the hormone mainly responsible for the regulation of Na (McSweeney et al., 1988). The concentration of Na in urine provides another indication of deficiency, with a value < 70 mg l^{-1} being marginal, though subject to some variation due to differences in urine volume (Morris, 1980; Underwood and Suttle, 1999). An excretion in the urine of < 3 g Na day^{-1} for dairy cattle has also been suggested as marginal (Langlands, 1987).

Calcium deficiency is most likely to occur with all-grass herbage and in areas where the soil is acid and strongly leached. The risk is also increased by high rates of fertilizer N (especially as ammonium) and K (Cornforth, 1984). The symptoms of mild Ca deficiency are not easily recognizable. Thus, in young animals, there is a slowing of overall growth and development, partly due to reduced bone growth (National Research Council, Subcommittee, 1989). In older animals, homeostatic mechanisms result in the mobilization of Ca from bone, thus maintaining the concentration of Ca in the blood plasma and delaying the onset of visible symptoms. However, if the deficiency becomes severe, more serious effects may develop. Thus, symptoms of rickets may develop in growing animals, and symptoms of osteomalacia, with the possibility of bone fractures, in adult animals (Naylor, 1991). There may also be a reduction in milk yield, though there is generally no decrease in the concentration of Ca in the milk (National Research Council, Subcommittee, 1989). Muscle weakness may develop, so that the functioning of the rumen and heart is impaired (Naylor, 1991). A severe deficiency in lactating animals results in milk fever and often, if no treatment is given, in death. In dairy cows, milk fever is most common within 3 days of calving. The diagnosis of Ca deficiency cannot be made from the concentration of Ca in the blood, as this is maintained, until symptoms become severe, by mobilization, and even demineralization of the bone does not provide definite evidence of Ca deficiency, as deficiencies of P or vitamin D can have the same effect (Miller, 1979).

Magnesium deficiency is most likely to occur in early spring, when herbage is young and growing rapidly and has high concentrations of K and N. Deficiency results in hypomagnesaemia, an abnormally low concentration of Mg in the blood, which, in turn, induces the symptoms of grass tetany, namely stiffness, staggering, twitching of muscles and eventually, if untreated, convulsions, coma and death (Grunes *et al.*, 1970; National Research Council, Subcommittee, 1989; Underwood and Suttle, 1999). Tetany is most likely following the onset of lactation, and usually occurs when serum Mg falls below 12 mg l^{-1} in cattle or 5 mg l^{-1} in sheep (Underwood and Suttle, 1999). The risk is greatest where the soil is derived from sandstone or granite and where high rates of fertilizer N and K are applied (Cornforth, 1984; Robinson *et al.*, 1989). Hypomagnesaemia is more common in cattle than in sheep, and in older cows than in heifers, probably due to their decreased ability to mobilize Mg from bone (National Research Council, Subcommittee, 1989). As animals have relatively low reserves of available Mg (see Table 4.3, p. 77), a change from an adequate diet to one deficient in Mg can induce hypomagnesaemia within a few days. Although hypomagnesaemia *per se* is due to inadequate Mg, the symptoms may well be accentuated by a low concentration of Ca (Grunes, 1983). For dairy cows, an excretion in the urine of > 3 g Mg day^{-1} indicates an adequate supply, and < 1 g day^{-1} suggests that the supply is marginal in relation to hypomagnesaemia (Langlands, 1987).

Excretion of K, Na, Ca and Mg by Ruminant Animals

The monovalent ions K and Na are excreted mainly in the urine, whereas the divalent ions Ca and Mg are excreted mainly in the faeces. Also, in warm conditions, substantial amounts of the monovalent ions may be excreted through the skin in sweat (McDowell, 1992). Ignoring excretion by this route, the proportions of K and Na excreted in the urine are usually 60–90% of the total. However, with low intakes, the proportions in the urine may be less than 60% (Fontenot and Church, 1979; Chiy *et al.*, 1994) and, with Na, sometimes less than 50% (Michell, 1985; Van Vuuren and Smits, 1997). With Ca and Mg, generally more than 75%, and often more than 85%, of the amount excreted is in the faeces (Fontenot and Church, 1979).

The excretion of K in the faeces is related to DM intake and amounts, on average, to about 2.6 g K kg^{-1} DM intake in cattle, and rather less in sheep (Agricultural Research Council, 1980). Any excess is excreted in the urine. With steers grazing grass–clover swards for periods of 10–12 days, the faeces accounted for 8–21% of the K consumed and the urine for 33–76%, but surprisingly large proportions were unaccounted for (Betteridge *et al.*, 1986). With Na, the extent of excretion in the urine varies widely with intake, as the homeostatic control of Na status is achieved mainly via this route.

The Ca in the faeces is a combination of unabsorbed dietary Ca and endogenous Ca from intestinal secretions. In both cattle and sheep, the endogenous faecal Ca appears to be directly proportional to the live weight of the animal (15–16 mg Ca kg^{-1} live weight day^{-1}) and not related to the amount of Ca ingested or absorbed (Agricultural Research Council, 1980; National Research Council, Subcommittee, 1989). However, the unabsorbed Ca increases with the amount ingested. Calcium in the urine normally amounts to only 2–3% of total excretion (Kincaid, 1988). Most of the Mg in faeces is unabsorbed from the diet, though there is an endogenous loss of Mg in the faeces of about 3 mg kg^{-1} live weight day^{-1} (Agricultural Research Council, 1980). When the intake of Mg is no more than adequate, excretion in the urine is negligible, but any excess Mg from greater intakes is excreted mainly in the urine (National Research Council, Subcommittee, 1989; Underwood and Suttle, 1999). The amount of Mg in the urine therefore reflects the available supply of Mg, and it responds more rapidly than does serum to a change in supply.

The large proportions of dietary Ca and Mg that are excreted in the faeces result in the total amount of cations in the faeces exceeding the total of phosphate plus sulphate anions, and the excess of cations is balanced largely by carbonate (Barrow, 1975).

The actual concentrations of K, Na, Ca and Mg in the urine and, to a lesser extent, in the faeces of cattle and sheep vary widely, as a result of variation in diet composition and water intake. There are also differences between individual animals. Even the average concentrations reported from several investigations (Tables 8.21 and 8.22) show considerable variation. Concentrations in slurry also vary widely around the typical concentrations shown in Table 8.5.

Transformations of K, Na, Ca and Mg in Grassland Systems

The major forms of K, Na, Ca and Mg in the soil, plant and animal components of grassland systems are shown in Fig. 8.2. Once released by the

Table 8.21. Concentrations of K, Na, Ca and Mg in the urine of cattle and sheep (mg l⁻¹).

	Diet	K	Na	Ca	Mg	Reference
Cattle	Ryegrass	4,190	1,240	–	–	Richards and Wolton, 1976
Cattle	Varied	7,470	1,110	160	520	Safley *et al.*, 1984
Cattle	Ryegrass–clover	6,500	600	38	49	Williams and Haynes, 1994
Cattle	Grass–clover	6,300	200	60	40	Hogg, 1981
Sheep	Grass–clover	8,700	–	negl.	110	Herriott and Wells, 1963
Sheep	Grass–clover, spring	8,500	410	280	85	Ledgard *et al.*, 1982
Sheep	Grass–clover, winter	10,400	31	1,185	147	Ledgard *et al.*, 1982
Sheep	Grass–clover	1,900	–	38	76	Sakadeven *et al.*, 1993b

Table 8.22. Concentrations of K, Na, Ca and Mg in faeces of cattle and sheep (% in DM).

	Diet	K	Na	Ca	Mg	Reference
Cattle	Grass–clover	0.9–1.5	0.2–0.5	0.9–2.3	0.4–0.6	During and Weeda, 1973
Cattle	Grass–clover	0.91	0.36	1.14	0.33	Hogg, 1981
Cattle	Varied	0.84	0.22	1.28	0.63	Safley *et al.*, 1984
Cattle	Grass–clover	0.73	0.001	1.96	0.60	Kirchmann and Witter, 1992
Cattle	Ryegrass	0.67	0.22	1.76	0.45	Chiy and Phillips, 1993
Cattle	Grass–clover	1.8	–	1.0	0.83	Williams and Haynes, 1995
Sheep	Grass–clover	1.33	–	3.07	0.65	Herriott and Wells, 1963
Sheep	Grass–clover	0.7–1.2	–	1.2–1.6	0.3–0.6	Haynes and Williams, 1992a
Sheep	Grass–clover	0.89	–	1.25	0.32	Sakadeven *et al.*, 1993b
Sheep	Grass–clover	1.5	–	1.3	0.64	Williams and Haynes, 1995

weathering of soil minerals, the cations are retained to a large extent on cation exchange sites in the soil, in equilibrium with soil solution. They are augmented to a small extent by inputs from the atmosphere, and sometimes to a large extent by fertilizers and lime. When taken up by plant roots, K and Na remain almost entirely in ionic form, while Ca and Mg are largely bound by organic compounds in the plant. During the decomposition of plant tissue, the cations are released to the soil again in ionic forms, Na and K rapidly, and Ca and Mg more slowly. When plant material is consumed by animals, large proportions of the ingested cations are excreted, the K and Na mainly in ionic form in the urine and the Ca and Mg mainly in organically bound forms in the faeces. With grassland that is not managed intensively, the cycling of all four cations may continue indefinitely without appreciable losses, but, as management becomes more intensive, the losses of cations by leaching, particularly from urine patches, become greater.

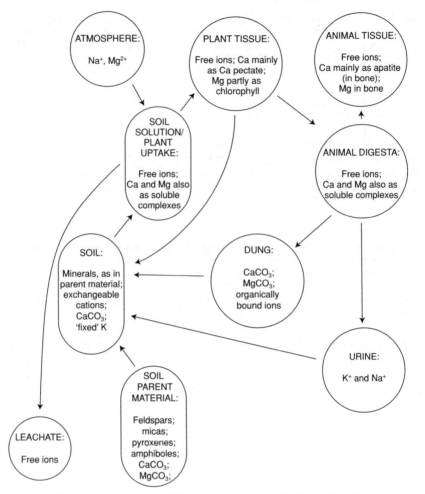

Fig. 8.2. Outline of the main forms of K, Na, Ca and Mg involved in the cycling of these elements in grassland.

Quantitative Balances of K, Na, Ca and Mg in Grassland Systems

The estimated balances for K, Na, Ca and Mg shown in Table 8.23 suggest that, with intensive management, there are substantial losses of Ca and Mg, and also of K unless it is provided as fertilizer. In some situations, the losses are accounted for by the weathering of soil minerals, and with some soils this mechanism may be sufficient to maintain an adequate supply of plant-available K. Thus, in a grazed pasture in Georgia, USA, with no fertilizer input, leaching losses were considered to be about 140 kg K, 170 kg Ca and 75 kg Mg ha^{-1} year^{-1} (Wilkinson and Lowrey, 1973). In New Zealand, many

Table 8.23. Estimated balances of K, Na, Ca and Mg (kg ha^{-1} year^{-1}) typical of two systems of grassland management, intensively managed grass and extensively managed grass–clover, both grazed by cattle (except during the winter) in UK conditions.

	Intensive grass (dairy cows)				Extensive grass–clover (beef cattle)			
	K	Na	Ca	Mg	K	Na	Ca	Mg
Inputs								
Fertilizer	60	0	0	0	0	0	0	0
Atmosphere	4	20	15	3	4	20	15	3
Aspects of recycling								
Uptake into herbage	250	40	120	30	25	3.7	12	3
Consumption of herbage by animals	200	32	96	24	15	2.2	7	1.8
Dead stubble + herbage to soil	100	16	48	12	30	4.5	15	3.6
Dead roots to soil	30	10	25	9	20	7	16	6
Excreta to soil of grazed area	155	23	71	20	14	1.7	6	1.7
Outputs								
Milk/live-weight gain	15	5	12	1	1.5	0.5	1.2	0.1
Leaching/runoff	15	20	150	40	2.5	2.0	30	4
Loss through excreta off sward	28	4	13	3	0	0	0	0
Gain (loss) to soil	6	(–9)	(–160)	(–41)	0	0	(–16)	(–1)

soils are well supplied with K and, although it is common for there to be a loss of up to 100 kg K ha^{-1} year^{-1} from dairy farms, deficiencies of K are uncommon (Goh and Williams, 1999). Leaching losses of Ca and Mg are greatly influenced by soil type, and may sometimes be much higher and sometimes much lower than indicated in Table 8.23.

Compared with the other macronutrient elements, K and Mg are incorporated to a rather small extent in milk and live-weight gain, and the proportions of the amounts in circulation that are present in excreta are therefore relatively high. As a consequence, losses from the system due to excreta being deposited off the grazed area are also potentially high. With the intensive grass system illustrated in Table 8.23, the loss of K due to 15% of the excreta being deposited off the sward represents almost half the amount applied in fertilizer.

Chapter 9

Micronutrient Cations: Iron, Manganese, Zinc, Copper and Cobalt

Sources and Concentrations of Fe, Mn, Zn, Cu and Co in Soils

For each of these five elements, the soil content is derived almost entirely from the parent material. Each of the elements occurs in a variety of minerals in both igneous and sedimentary rocks, with Fe being by far the most abundant and Co the least abundant (see Table 2.2, p. 20). However, the elements vary in the extent to which they persist during weathering, and their retention in sediments and soils is influenced by factors such as pH and redox conditions and by the severity of leaching.

Iron is present in igneous rocks in both ferric and ferrous forms. It is retained to some extent during the formation of clay minerals by substituting for aluminium in the lattice structure. It is also retained, except at low pH, by the precipitation of hydrous ferric oxides (Norrish, 1975). These hydrous oxides are insoluble at neutral or alkaline pH, but become increasingly soluble when conditions are acid. Under reducing conditions, ferric iron tends to be converted to the ferrous form, whose hydroxides are more soluble at neutral pH and therefore more susceptible to leaching.

The Mn in rocks is often associated with Fe, especially as a minor substituent in ferromagnesian minerals. Much of the Mn is retained during weathering, some in residual primary minerals, some as a substituent in clay minerals, such as smectite, some precipitated as oxides and hydroxides, such as MnO_2, and some adsorbed on to various mineral surfaces. The precipitated Mn may substitute for Fe^{3+} in secondary minerals, such as goethite (Gilkes and McKenzie, 1988).

The Zn in igneous rocks occurs partly as a minor constituent of minerals, such as the pyroxenes, amphiboles, biotite and iron oxides (Adriano, 1986).

It also occurs partly in the form of sulphides, such as sphalerite (ZnS), and partly as oxides, carbonates, phosphates and silicates (Barak and Helmke, 1993). In sedimentary rocks, small amounts of Zn may also occur as a substituent for Fe^{2+} and Mg^{2+} in clay minerals (Giordano and Mortvedt, 1980).

The Cu in igneous rocks is present almost entirely as sulphide and, in contrast to Zn, it does not substitute for other cations in silicate minerals (Norrish, 1975). Weathering of the sulphide minerals releases Cu in solution, as either the divalent cupric ion, Cu^{2+}, or the monovalent cuprous ion, Cu^+. However, the latter is unstable unless in the form of anionic complexes. The divalent Cu^{2+} tends to be adsorbed and occluded by Fe and Mn oxides, and also forms complexes with OM.

Cobalt is present in igneous rocks, mainly in ferromagnesian minerals as a substituent for Fe. Ultrabasic rocks, especially those containing olivine, are particularly rich in Co (Adriano, 1986). During weathering, Co tends to be precipitated as the oxide, hydroxide or carbonate, all of which are insoluble at neutral to alkaline pH, but become increasingly soluble with acidity. There is also a strong tendency for Co to be adsorbed and occluded by Mn oxides (Smith and Paterson, 1995).

Since the micronutrient cations in soils are derived almost entirely from the parent material, soils formed from basic rocks generally contain more of each of these elements than do soils formed from granite or sandstone. Of the sedimentary rocks, sandstones are normally low in all the micronutrient cations, whereas shales and clays may show some enrichment compared with igneous rocks (see Table 2.2). Shales that contain large amounts of OM may be particularly high in Cu (Stevenson, 1986). Limestones are normally low in all the micronutrient cations except Mn (see Table 2.2; Stevenson, 1986). Soils in temperate regions usually contain 1–5% Fe, 50–500 mg Mn kg^{-1}, 20–200 mg Zn kg^{-1}, 5–50 mg Cu kg^{-1} and 2–20 mg Co kg^{-1} (Table 9.1 plus other data). In general, the concentrations of all five elements are much lower in sandy soils than in loams and clays (e.g. Verloo and Willaert, 1990).

Table 9.1. Reported concentrations of Fe, Mn, Zn, Cu and Co in soils (mg kg^{-1} dry soil).

	Fe	Mn	Zn	Cu	Co	Reference
Temperate, normal range	–	100–4,000	10–300	2–100	1–40	Berrow and Burridge, 1980
Temperate, typical	30,000–40,000	500–600	50–100	c. 30	–	Mikkelsen and Camberato, 1995
California	12,000–100,000	40–1,400	28–130	2–90	4–71	Bradford et al., 1967
England and Wales	–	50–200	60–120	10–25	6–15	McGrath and Loveland, 1992
Scotland, ten soils	–	370–3,000	25–240	6–41	1.2–70	Ure et al., 1979
New Zealand, six soils	27,500–30,500	410–550	82–111	9–14	14–18	Haynes and Swift, 1984

Relatively large amounts of Zn occur in some alluvial soils, and there may be local enrichment in both Zn and Cu due to industrial activities, such as mining, smelting and the tipping of wastes (McGrath and Loveland, 1992). Serpentine soils are rich in Co, and those in New Zealand may contain up to 1000 mg Co kg⁻¹ with a mean value of about 400 mg kg⁻¹ (Adriano, 1986).

Agricultural and Atmospheric Inputs of Fe, Mn, Zn, Cu and Co to Soils

Small amounts of the micronutrient cations may be added to soils through their presence as incidental constituents of fertilizers and sewage sludges. In general, their concentrations in fertilizers are no greater than the average concentrations in soils, though Zn is sometimes present in relatively large amounts in phosphate fertilizers (Table 9.2). Exceptionally, in areas where a particular deficiency is widespread, there may be a deliberate policy of addition to fertilizers. For example, fertilizers in Denmark are sometimes supplemented with Cu (Dam Kofoed, 1980).

Sewage sludges vary widely in their concentrations of the micronutrient cations (see Table 2.8, p. 32), due mainly to variations in the extent of contamination from industrial effluent. Such contamination usually affects

Table 9.2. Some reported concentrations of Fe, Mn, Zn, Cu and Co in several fertilizers (mg kg⁻¹).

	Fe	Mn	Zn	Cu	Co	Reference
Ammonium nitrate	192	39	12	3.6	3.3	Verloo and
Superphosphate	1,150	9	244	23	15	Willaert, 1990
Potassium chloride	145	< 1	6.1	1.7	22.5	
Ammonium nitrate	–	0.5	–	< 0.6	< 0.07	Raven and
Urea	–	0.3	–	< 0.6	< 0.07	Loeppert, 1997
Diammonium phosphate	*c.* 2,600	34–330	*c.* 380	< 2–42	0.7–3.2	
Triple superphosphate	*c.* 17,300	300–350	*c.* 60	3.2–3.5	2.2–6.6	
Potassium chloride	*c.* 1,440	0.2–5.3	4.6	< 2–3.5	< 0.07	
Calcium ammonium nitrate	–	25	7.6	2.8	6.6	Adriano, 1986
Urea	–	0.5	4.0	0.6	–	
Diammonium phosphate	–	93–307	112–715	1.0–7.2	16	
Superphosphate	–	890	165	15	77	
Potassium chloride	–	3.5	10	3.1	22	
Ammonium nitrate	–	–	3	3	9.2	Senesi and
Urea	–	–	1	1	1.2	Polemio, 1981
Diammonium phosphate	–	–	122	13	14.2	

Zn and Cu to a greater extent than the other elements. In some countries, including the UK, the industrial contamination of sewage sludge has declined in recent decades, and excessively high concentrations of Zn and Cu are now less frequent (Smith, 1996). Large proportions of the micronutrient cations in sewage sludge are bound to organic components and, as these are decomposed by soil microorganisms, there is a slow release of the free cations. However, in most soils, the cations are quickly rendered insoluble again by precipitation or complexation, and, when sludge is applied repeatedly, the cations tend to accumulate in the surface soil. In a study involving several years of sludge application to a long-term grassland soil, there were considerable increases of Zn, Cu and Co in the top 5 cm, but there was little, if any, effect at depths below 30 cm (Hemkes et al., 1980). The accumulation of Zn and Cu in the surface soil has also been shown in other studies (e.g. Davis et al., 1988; O'Riordan et al., 1994).

The amounts of the micronutrient cations deposited from the atmosphere are generally negligible in relation to the soil content (Table 9.3), except in localized areas where there is serious atmospheric pollution. Nevertheless, the amounts of Fe, Zn, Cu and Co deposited each year from the atmosphere may be similar to the amounts removed by a silage yield of 10,000 kg DM ha^{-1}, as shown by comparing Table 9.3 with Table 3.5 (p. 55). In areas where atmospheric inputs of Zn and Cu are high, the Zn and Cu, like those from sewage sludge, tend to accumulate in the surface soil.

Recycling of Fe, Mn, Zn, Cu and Co through the Decomposition of Organic Residues

It is possible to estimate the extent to which the micronutrient cations are recycled through the decomposition of dead plant material from assumptions about the amounts of dead material and the concentrations of the cations (Table 9.4; Sposito and Page, 1984). Similarly, it is possible to estimate the amounts returned in the excreta of grazing animals (Table 9.5). The return of the micronutrient cations, except Co, is much greater in dung

Table 9.3. Amounts of Fe, Mn, Zn, Cu and Co deposited from the atmosphere to the land surface (g ha^{-1} year^{-1}) as reported in three investigations.

	Fe	Mn	Zn	Cu	Co	Reference
Europe, median	2208	68	19	536	–	Sposito and Page, 1984
North America, median	5676	236	788	441	5	Sposito and Page, 1984
UK (seven sites)	1400–7700	68–320	370–1200	98–480	2.0–8.3	Cawse, 1980
Germany	–	70	210	18	–	Kabata-Pendais and Adriano, 1995

Table 9.4. Typical amounts of Fe, Mn, Zn, Cu and Co returned to the soil through the death and decay of herbage and roots from a moderately intensive sward (g ha^{-1} year^{-1}), assuming 7000 kg dead herbage and 5000 kg dead roots ha^{-1} year^{-1}, and concentrations of the elements estimated from data on pp. 232–239.

	Fe	Mn	Zn	Cu	Co
Assumed concentration in dead herbage (mg kg^{-1})	170	115	34	8.5	0.14
Return to soil from herbage (g ha^{-1} year^{-1})	1200	800	240	60	1.0
Assumed concentration in dead roots (mg kg^{-1})	400	200	40	25	0.30
Return to soil from roots (g ha^{-1} year^{-1})	2000	1000	200	125	1.5
Total return from dead plant material (g ha^{-1} year^{-1})	3200	1800	440	185	2.5

Table 9.5. Estimated amounts of Fe, Mn, Zn and Cu returned to the soil in the excreta of dairy cattle grazing at a density of 700 cow days ha^{-1} year^{-1} (kg ha^{-1} year^{-1}), assuming 25 l urine and 3.8 kg faecal DM per cow day^{-1}, and typical concentrations derived from Tables 4.10 and 4.11 (p. 92).

	Fe	Mn	Zn	Cu
Assumed concentration in urine (mg l^{-1})	5.5	0.18	1.8	0.6
Return in urine (g ha^{-1} year^{-1})	96.3	3.2	31.5	10.5
Assumed concentration in dung (mg kg^{-1})	1800	180	160	45
Return in dung (g ha^{-1})	4790	480	424	120
Total return in excreta (g ha^{-1} year^{-1})	4890	483	456	130

than in urine (see Table 4.9, p. 91), but their release in plant-available forms is slow, due partly to complexation by organic constituents and partly to dung having an alkaline pH and thus rendering the micronutrient cations insoluble (Barrow, 1987).

Although the amounts of the micronutrient cations applied in cattle slurry are often similar, on a per hectare basis, to those returned by grazing animals, the amounts of Zn and Cu supplied in pig slurry may be much greater. This is because $CuSO_4$ is often added to pig feed to increase the efficiency of feed conversion and to control dysentery, while $ZnSO_4$ is added to prevent the Cu inducing Zn deficiency (Christie, 1990). In a survey in Denmark, the average concentration of Cu in cattle slurry was 43 mg kg^{-1} DM, while the average in pig slurry was 265 mg kg^{-1} DM (Dam Kofoed, 1980). There appears to be little information on the actual concentrations of the micronutrient cations in slurries, but, on a DM basis, they are likely to be similar to the concentrations in manures (see Table 2.7, p. 31). The rates at which the micronutrient cations are released from the organic components of slurry or manure are enhanced by soil conditions that promote microbial activity, i.e. warm, moist, well-aerated soils of approximately neutral pH.

Forms and Availability of Fe, Mn, Zn, Cu and Co in Soils

Each of the micronutrient cations can be categorized into a number of fractions in soils (Stevenson, 1986; Shuman, 1991). The most important are as follows:

1. Cations within the structure of primary and secondary minerals.
2. Cations precipitated in insoluble inorganic forms, including cations occluded by Fe and Mn oxides.
3. Cations complexed by OM, including the microbial biomass.
4. Cations specifically adsorbed to the surfaces of clay minerals, and Fe and Mn oxides.
5. Cations in exchangeable form on the negatively charged exchange sites of clay minerals and OM.
6. Water-soluble cations:
   (a) as free cations;
   (b) as complexes with organic and inorganic ligands.

Although it is difficult to distinguish some of the fractions from one another – for example, precipitated cations from those that are complexed or specifically adsorbed – it has been shown that Fe and Mn are retained mainly by precipitation and complexation, whereas Cu and Zn are retained largely by adsorption on to oxide surfaces (Harter, 1991). OM often has an important influence on the extent of retention of the micronutrient cations in soils, since it forms both insoluble and soluble complexes with them. Insoluble complexes are formed with the components of humified OM, whereas soluble complexes are formed with small organic molecules, such as some aliphatic acids, amino acids and phenolic compounds (Fig. 9.1). In general, complexes are formed more readily with ferric iron, which is trivalent, than with the divalent cations. The strength of complex formation by the divalent ions is in the order $Cu > Co > Zn > Fe^{2+} > Mn$. Ferric iron is complexed most strongly by organic groups that contain oxygen, such as carboxyl and phenolic groups, whereas Cu is complexed most strongly by groups that contain N, such as amino groups and porphyrins (Stevenson, 1991).

In most soils, the exchangeable and water-soluble fractions of any micronutrient cation (categories 5 and 6, above) account for only a small proportion of the total content; and, within the water-soluble fraction, the ions differ in the proportion that is present as the free ion, rather than in complexed form. Ferric (Fe^{3+}) and cupric (Cu^{2+}) ions, are almost entirely in complexed form in the soil solution, whereas Mn^{2+}, Zn^{2+} and Co^{2+} are complexed to a smaller extent (McBride, 1989). However, the solubility of each of the cations in soils tends to be enhanced by the microbial production of small-molecular chelating agents. In addition, microbial respiration tends to induce chemically reducing conditions in the soil, thus increasing the solubility of Fe and Mn by favouring the formation of ferrous and manganous

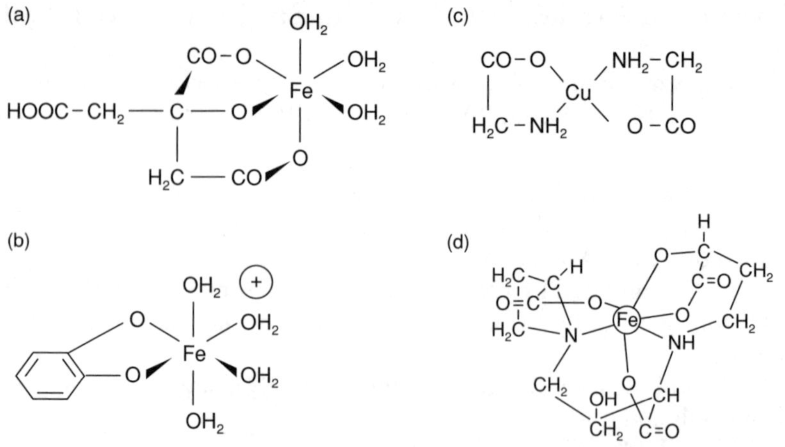

Fig. 9.1. Chemical structures of some complexes formed between micronutrient cations and organic compounds: (a) Fe^{3+} with citric acid; (b) Fe^{3+} with catechol; (c) Cu^{2+} with glycine; (d) Fe^{3+} with mugineic acid.

forms. On the other hand, when cations are immobilized in the microbial biomass, their solubility, and hence their availability, is reduced.

Much of the Fe in soils is present in hydrous ferric oxides and hydroxides, of which there are several different types, each with a different solubility. Haematite (Fe_2O_3) and goethite (FeOOH) are the least soluble, followed in order of increasing solubility by lepidocrite (also FeOOH but differing from goethite in chemical structure) and amorphous $Fe(OH)_3$. These oxides are present partly as small particles and partly as amorphous coatings on other soil constituents. Depending on conditions, they may be transformed slowly from one to another (Lindsay, 1991). The availability of Fe in non-acid well-drained soils is normally extremely low, but it tends to increase when conditions are anaerobic, due to the conversion of the ferric to the ferrous form, whose hydroxides are more soluble. Even in soils that are well drained, there may be zones that are anaerobic, due to the presence of decomposing OM, and so some conversion of ferric to ferrous iron is possible.

Manganese has three possible oxidation states in soils, Mn^{2+}, Mn^{3+} and Mn^{4+}, but only Mn^{2+} is stable in solution. The ions Mn^{3+} and Mn^{4+} form insoluble oxides and hydroxides in soils and are therefore precipitated. Although Mn^{2+} is stable in solution, it is subject to adsorption by OM, iron oxides and silicates, especially at pH > 6. In well-aerated soils at pH > 7, Mn^{2+} tends to precipitate as an oxide or hydroxide, often mixed with iron oxides as an amorphous coating on soil particles, though sometimes as MnO_2 in nodules. The proportion of the soil Mn in exchangeable and soluble forms is highest in acid conditions (Marschner, 1988), partly because acidity increases the concentration of Mn^{2+} in solution:

$$MnO_2 + 4H^+ + 2e^- \rightleftharpoons Mn^{2+} + 2H_2O$$

Conversely, the concentration of Mn^{2+} declines to about one-hundredth of its previous concentration for each unit increase in pH. The solubility of Mn is also influenced by redox conditions in the soil, and oxidation, which tends to reduce solubility, tends to occur within pores, while reduction tends to occur at anaerobic sites within aggregates (Robson, 1988). Thus, Mn may be subject to both oxidation and reduction at the same time, due to the spatial variation that occurs in the pH and redox potential of the soil. Most of the complexed Mn in soils is bound to inorganic rather than organic ligands (Laurie and Manthey, 1994).

Zinc differs from iron and manganese in that it occurs in soils in only one valence state, Zn^{2+}. Some Zn occurs as a partial replacement for Fe^{2+} and Mg^{2+} in ferromagnesian minerals, but often much of it is in the form of Zn^{2+} adsorbed by hydrous oxides, clays and soil OM. The extent of adsorption of Zn (and also of other cations) by Fe oxides depends on the types of oxide present. For example, goethite adsorbs more than haematite, due to its having more surface hydroxyl groups (Harter and Naidu, 1995). Zinc is also adsorbed on to the surface of carbonates. The Zn adsorbed by clays and OM is largely exchangeable, but, at pH > 5.5, increasing proportions are adsorbed by Fe and Mn oxides and complexed by OM, in forms that are not readily exchangeable. With an equilibrium solution concentration of 0.1 M, the adsorption of Zn by soils is about ten times greater at pH 6.5 than at pH 5.5 (Moraghan and Mascagni, 1991). Of the Zn present in the soil solution, about half exists as the free hydrated cation and the remainder is complexed by inorganic or organic ligands (Barak and Helmke, 1993).

Some soil Cu is normally present in unweathered minerals, and some is adsorbed or complexed. Copper tends to be coprecipitated or occluded by hydrous oxides of Al, Fe and Mn, and these reactions increase with increasing pH (Kabata-Pendais and Pendais, 1992). Copper is also complexed by OM more tightly than any other of the divalent micronutrient cations, and almost all of the Cu in the soil solution is complexed (McBride, 1989). Whether the Cu is adsorbed, complexed or in free inorganic form, in most soils it exists almost entirely as the divalent Cu^{2+}. However, in waterlogged soils, it is partially reduced to Cu^+, which may then form insoluble CuS.

Much of the Co in soils occurs in unweathered ferromagnesian minerals and, after being released by weathering, is adsorbed by Mn and Fe oxides or is complexed by OM. Adsorption and complexation tend to increase with increasing soil pH. The Mn oxides have a particularly high adsorption capacity for Co (Fleming, 1988; Smith and Paterson, 1995). Although adsorption and the formation of complexes reduce the solubility and availability of Co, some of the Co associated with the Mn oxides is thought to be fairly labile (Norrish, 1975), as are some complexes of Co^{2+} with OM. Availability is increased when drainage is poor (Berrow *et al.*, 1983).

However, only a small proportion of the soil Co occurs in exchangeable form.

Losses of Fe, Mn, Zn, Cu and Co from Soils

Losses of these cations by leaching are usually negligible, and there are no losses through volatilization. In most soils, reactions involving precipitation and adsorption ensure that the micronutrient cations are not leached below rooting depth (Hemkes *et al.*, 1980; Ross, 1994). However, two factors which tend to increase the mobility of the cations and may result in small losses by leaching are acidity and the production in the soil of soluble organic chelates (Ross, 1994). Losses by surface runoff are also usually negligible, though small amounts of the cations may be lost in runoff following the application of sewage sludge or slurry.

Assessment of Plant-available Fe, Mn, Zn, Cu and Co in Soils

The potentially available fraction of any micronutrient cation consists of the soluble and exchangeable forms, together with a small proportion of the insoluble complexed and precipitated forms. A reagent suitable for the assessment of the potentially available fraction should therefore: (i) displace the exchangeable fraction by another cation; (ii) be sufficiently acid or have sufficient complexing ability to dissolve those precipitated forms that contribute to plant uptake; and (iii) extract labile organic complexes. Of several extractants that have been proposed, three that are commonly used are acidic ammonium acetate and the chelating agents EDTA and DTPA, often buffered at a specific pH (Sims and Johnson, 1991). Another is a combination of 0.5 M ammonium acetate + 0.02 M EDTA at pH 4.65 (Kiekens, 1995). As the effectiveness of the various extractants varies with soil conditions, their usefulness is limited, though it can sometimes be improved by taking into account other soil factors, such as pH, cation exchange capacity and the content of OM (Sims and Johnson, 1991).

Uptake of Fe, Mn, Zn, Cu and Co by Herbage Plants

The plant : soil concentration ratio of all these elements is low (< 0.5) and those of Fe and Co are especially low (see Table 3.6, p. 56). As a consequence, the amounts removed, for example, in herbage cut for silage (see Table 3.5) are usually negligible in relation to the total amounts present in soils (Table 9.1).

In the movement of the micronutrient cations in the soil, diffusion is more important than mass flow (Moraghan and Mascagni, 1991). The movement of the micronutrient cations, especially Fe and Cu, occurs largely in the form of soluble complexes, many of which involve organic molecules that are produced by soil microorganisms or, in some instances, by plants themselves. The exudation of H^+ or HCO_3^- ions by plant roots may also influence the solubility and movement of the micronutrient cations through an effect on the pH of the rhizosphere (see p. 48). Such exudation depends largely on whether N is taken up as ammonium or nitrate: the uptake of ammonium results in the exudation of H^+, which tends to increase the solubility and uptake of the cations, especially Fe, whereas the uptake of nitrate results in the exudation of HCO_3^-, which tends to reduce the solubility and uptake of the cations. The rate of root extension appears to be important for uptake in some situations (Jarvis and Whitehead, 1981; Chen and Barak, 1982; Marschner, 1993). Evidence on the impact of mycorrhizae on the uptake of the micronutrient cations is conflicting. The inoculation of several grasses with mycorrhizal fungi was found to increase the uptake of Zn, Cu and Co (Killham, 1985), and mycorrhizae were shown to increase the uptake of Cu in white clover (Li *et al.*, 1991). However, in another investigation, there was no relationship between concentrations of Zn, Cu and Mn in several grass and legume species and the extent to which mycorrhizae had developed in the roots (Sanders and Fitter, 1992).

Plants have some capacity to secrete organic compounds, including organic acids, from their roots, thus enabling them to enhance their uptake of the micronutrient cations. With Fe, there are two distinct mechanisms of this type. One mechanism, which is used by non-graminaceous plants, involves the secretion of organic acids such as citrate and malate. These dissolve Fe from compounds such as the hydrous Fe oxides, though the rate of solubilization depends on factors such as the form of the oxide, the pH and the presence of other cations. The resulting ferric chelates are able to move readily to the root surface, where they are chemically reduced, enabling absorption to occur as Fe^{2+} (Jones *et al.*, 1996). The second mechanism, which is restricted to graminaceous species, involves the secretion of phytosiderophores, of which mugineic acid is an important example (Kochian, 1991; Manthey *et al.*, 1994; Deighton and Goodman, 1995). Mugineic acid (Fig. 9.1), which is synthesized from methionine, forms a stable chelate with Fe^{3+}, making it more soluble and therefore more mobile in the soil (Marschner, 1995; Jones *et al.*, 1996). The importance of phytosiderophores is greatest when plants are deficient in Fe – due, for example, to calcareous soil conditions – as plants react to this situation by increasing the amount secreted. Whether Fe^{3+} phytosiderophores are absorbed as intact molecules or the Fe is dissociated at the root surface is uncertain (Kochian, 1993), but, once Fe has been absorbed by the root, it appears to be transported, in the xylem, mainly as a Fe^{3+}-citrate complex (Kochian, 1991). In the leaves, some Fe is stored as a ferric phosphoprotein, phytoferritin.

As with Fe, the uptake of Mn, Zn, Cu and Co is promoted by the presence of organic acids in the soil. Phytosiderophores also form chelates with Mn, Zn and Cu, which are both stable and soluble (Treeby *et al.*, 1989), and there is evidence that the phytosiderophores are secreted in response to Zn deficiency, as well as to Fe deficiency, with similar effects on the uptake of Zn (Kochian, 1993). Soil bacteria also produce organic siderophores, which chelate Fe and probably other cations (Paul and Clark, 1996).

Although there may be some absorption of intact chelates into plant roots (Kochian, 1991), most absorption of the micronutrient cations is thought to occur in the form of the divalent ions. Thus, Mn appears to be taken up as Mn^{2+}, and to be translocated in the xylem mainly as Mn^{2+} in equilibrium with unstable organic acid complexes (Loneragan, 1988). Similarly, plants take up Zn mainly as the Zn^{2+} ion (Kochian, 1993). Strains of *Agrostis* growing on soils contaminated by Zn have evolved a tolerance to high concentrations, by retaining much of the Zn in the cell-wall material of the roots (Peterson, 1969). Copper is absorbed as the Cu^{2+} ion, and there is competition in uptake between Cu and Zn (Kochian, 1991). The transport of Cu in the xylem occurs mainly in the form of anionic complexes, though, as with Zn, roots have a strong tendency to retain Cu by adsorption to the cell walls (Kochian, 1991). There appears to be little information on the mechanisms by which Co is taken up and transported in plants (Mengel and Kirkby, 1987).

Functions of Fe, Mn, Zn, Cu and Co in Herbage Plants

The functions of the micronutrient cations in plants are generally related to their capacity to change their oxidation state (except with Zn) and/or to their ability to form complexes with organic molecules. These two properties are shown clearly by Fe, which is able to change its oxidation state between Fe^{3+} and Fe^{2+}, and to form organic complexes with both Fe^{3+} and Fe^{2+}. Iron is an essential constituent of the cytochromes, which consist of proteins linked to haem (Fig. 9.2), and are involved in various redox reactions in plants. By being susceptible to a change in oxidation state, the Fe in cytochromes enables electrons to be transferred to and from a range of substrates (Mengel and Kirkby, 1987). Iron is also involved in a second group of proteins, the Fe-S proteins, which contribute to the transfer of electrons. The Fe in these proteins is usually coordinated to the thiol group of cysteine, as in ferredoxin, which is involved in both photosynthesis and N_2 fixation (Stevenson, 1986). Iron also activates a number of enzymes, and is involved in the synthesis of ribonucleic acid (Romheld and Marschner, 1991). In legumes, Fe is a constituent of haemoglobin, which is essential for N_2 fixation, and Fe is also required for several enzymes of the nitrogenase complex (Marschner, 1995).

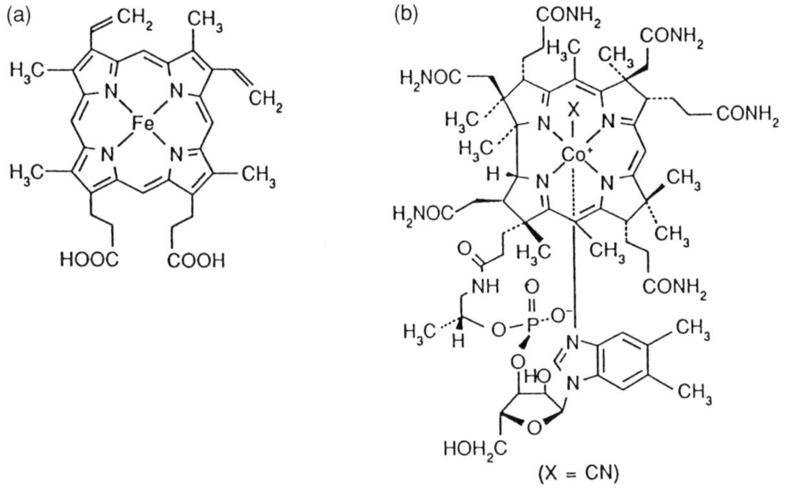

Fig. 9.2. Chemical structures of (a) haem and (b) vitamin B_{12}.

Manganese, which also has the ability to change its oxidation state, resembles Fe in being involved in redox processes. It is essential for the transfer of electrons in photosynthesis and for the detoxification of free-radical oxidants. Manganese also acts as an enzyme activator and is needed for reactions involving phosphorylation, decarboxylation, reduction and hydrolysis. In this way, it contributes to respiration, the synthesis of amino acids and the synthesis of phenolic acids and lignin (Robson, 1988; Romheld and Marschner, 1991).

Zinc, in contrast to Fe, Mn and Cu, is not subject to a change in oxidation state, and exists in plants only as Zn^{2+}. It activates a large number of enzymes, often by linking the enzyme to the corresponding substrate or by modifying the conformation of the enzyme or its substrate (Romheld and Marschner, 1991). The enzymes for which it is essential are mainly dehydrogenases and proteinases, and include carbonic anhydrase, alcohol dehydrogenase and superoxide dismutase. Zinc is thought to play a key role in stabilizing the structures of RNA and DNA and in regulating the enzymes that control their synthesis and degradation (Brown *et al.*, 1993). It is also involved in ensuring the stability and structural orientation of membrane proteins, through binding to –SH groups (Romheld and Marschner, 1991). It appears to be necessary for the production of the auxin, indole-3-acetic acid (IAA) (Marschner, 1995).

Copper is an essential component of several enzymes that catalyse oxidations involving molecular oxygen. These enzymes include cytochrome oxidase, phenol oxidase, laccase and amine oxidase (Stevenson, 1986), and their catalytic activity is thought to depend on the reversibility of oxidation between Cu^{2+} and Cu^+. Phenol oxidase and laccase are important in lignification, partly through their role in the oxidation of p-coumaric acid (Blevins, 1994). Copper also plays a role in photosynthesis, as a constituent of the Cu-protein, plastocyanin, which is involved in the transfer of electrons (Marschner, 1995). In legumes, Cu appears to be involved in nodulation and N_2 fixation (Romheld and Marschner, 1991).

Cobalt is not an essential element for plants *per se* but is essential for the N_2 fixing ability of rhizobia. It is a component of the coenzyme cobalamin, in which Co^{3+} is chelated by four N atoms at the centre of a porphyrin structure, similar to that of Fe in haem (Marschner, 1995).

Distribution of Fe, Mn, Zn, Cu and Co within Herbage Plants

In general, concentrations of Fe, Mn, Zn, Cu and Co are greater in the roots than in the shoots of grasses and legumes. This difference was shown for Fe, Mn, Zn and Cu in white clover grown in solution culture (Wilkinson and Gross, 1967), and there were also higher concentrations of Mn, Zn and Cu in the roots than in the shoots of both perennial ryegrass and white clover grown in soil (Table 9.6). Cobalt also tends to accumulate in the roots, especially in the root nodules, of legumes (Wilson and Hallsworth, 1965). Other evidence of retention in the roots has been reported for Mn (Vlamis and Williams, 1967), for Zn (e.g. Reuter *et al.*, 1982; Van Steveninck *et al.*, 1993), for Cu (Jarvis and Whitehead, 1981; Mazzolini *et al.*, 1983) and for Co (Adriano, 1986; Macklon, 1988). In contrast, in an investigation with lucerne grown in soils, concentrations of Mn, Zn and Cu, but especially Mn, were higher in the shoots than in the roots (Rominger *et al.*, 1975). The differences in concentration between roots and shoots may be influenced by the adequacy of supply, as suggested by a study with Mn. In several plant species, including subterranean clover, the development of Mn deficiency resulted in the concentration in the roots declining rapidly (Loneragan, 1988).

Within the shoots of grasses, the concentrations of all the micronutrient cations are often higher in the leaf blade than in the leaf sheath and stem (see Table 3.3, p. 53; Chiy and Phillips, 1999). Concentrations in the flowering

Table 9.6. Concentrations of Mn, Zn and Cu in the roots and shoots of perennial ryegrass and white clover (mg kg^{-1}): mean values for plants grown separately in pots of 10 soils (from Whitehead, 1987).

	Mn	Zn	Cu
Shoots of perennial ryegrass	122	21	3
Roots of perennial ryegrass	230	42	19
Shoots of white clover	80	28	6
Roots of white clover	120	43	16

spikelet are sometimes relatively high and, with Zn, the highest concentration may occur in the spikelet (see Table 3.3, p. 53; Fleming, 1965), though relatively low concentrations of Fe and Mn have also been reported (Chiy and Phillips, 1999). Higher concentrations in the leaves than the stems of grasses have also been observed for Mn (Loneragan, 1988) and for Zn and Cu (Fleming, 1965). With Co, although leaf blades sometimes contain more than leaf sheath and stem (see Table 3.3), when concentrations are low there may be no difference between leaf and stem (Fleming, 1963).

In white clover grown in solution culture, concentrations of Fe, Mn, Zn and Cu were all approximately twice as high in the leaflets as in the petioles and stolons (Wilkinson and Gross, 1967) though, in field-grown white clover, there were rather small differences between leaf and petiole (Chiy and Phillips, 1999). In lucerne, concentrations of Fe, Mn, Zn and Cu were greater in the leaves than stems, the difference being most marked with Mn, with a concentration of about 80 mg kg^{-1} in the leaves and about 16 mg kg^{-1} in the stems (Rominger *et al.*, 1975).

Ranges of Concentration and Critical Concentrations of Fe, Mn, Zn, Cu and Co in Herbage Plants

Concentrations of Fe, Mn, Zn, Cu and Co that have been reported in grassland herbage from various regions are shown in Table 9.7. In several instances, the highest concentration is more than 100 times greater than

Table 9.7. Concentrations of Fe, Mn, Zn, Cu and Co in grassland herbage: mean and range values reported from different regions (mg kg^{-1} DM).

Country	No. of samples	Fe	Mn	Zn	Cu	Co	Reference
USA	Normal range	5–150	30–100	15–60	5–30	0.05–0.3	Mayland and Wilkinson, 1996
Pennsylvania, USA	> 9000	208 (10–2600)	53 (6–1200)	27 (3–300)	13 (2–214)	–	Adams, 1975
Alberta, Canada	805	165 (22–1760)	50 (10–635)	32 (3–122)	12 (1–47)	–	Redshaw *et al.*, 1978
Manitoba, Canada	> 600	31–640	2–176	7–126	1.4–26	–	Drysdale *et al.*, 1980
UK	> 1300	70–500	99	45 (20–60)	8.3 (2–15)	0.13	MAFF, 1975; Leech and Thornton, 1987
Finland	> 2000	44 (8–800)	67 (8–510)	32 (15–119)	4.0 (0.4–20)	0.06 (0.0–0.2)	Kahari and Nissinen, 1978
France	> 700	–	9–580	5–75	2–13	–	Jarrige, 1989
New Zealand	> 3000	–	166 (25–694)	–	–	–	Smith and Edmeades, 1983

the lowest, though the data include some values that are abnormally high, possibly due to the contamination of samples with soil and dust.

Information on critical concentrations of the micronutrient cations is limited, especially for grasses. There is rather more information for Zn and Cu than for the other elements, as Zn and Cu are more likely to be deficient. Critical concentrations for Zn and Cu may be rather higher in legumes than in grasses (Table 9.8; Marschner, 1993). For Zn at least there may be variation amongst genotypes within a species in their susceptibility to deficiency, as shown for 13 genotypes of lucerne (Grewal and Williams, 1999). The critical concentration of Cu in subterranean clover appears to vary amongst cultivars; and it was found to be somewhat higher when assessed on the basis of maximum yield of seeds than when assessed on the basis of maximum vegetative growth (McFarlane, 1989).

Cobalt is not essential for higher plants, but it is required for N_2 fixation by the rhizobia in the nodules of leguminous plants (Asher, 1991). The requirement is small and a concentration in the herbage of > 0.08 mg Co kg^{-1} suggests that the supply is adequate (Hodgson et al., 1962).

Toxic Concentrations of Mn, Zn and Cu in Herbage Plants

Excessively high concentrations of Mn, Zn and Cu occasionally cause toxic effects in plants, but this problem appears not to occur with Fe or Co. Manganese toxicity is most prevalent on highly acid soils that are subject to waterlogging; legumes are more susceptible than grasses, and lucerne is particularly susceptible. Herbage concentrations of Mn that resulted in a 10% reduction in growth varied from about 340 mg kg^{-1} for lucerne to 1110 mg kg^{-1} for ryegrass, with white clover being intermediate, at about 670 mg kg^{-1} (Smith et al., 1983). Other assessments of toxic concentrations of Mn include 650 mg kg^{-1} in white clover (Andrew and Hegarty, 1969) and 500 mg kg^{-1} in ryegrass (MacNicol and Beckett, 1985). Much higher

Table 9.8. Reported critical concentrations of Fe, Mn, Zn and Cu in the herbage of grasses and legumes (mg kg^{-1}).

	Fe	Mn	Zn	Cu
Grasses	< 50[a]	20[a]	10–14[a]	4[a]
Clovers	50–70[b]	20–35[b]	12–14[c]	5–7[d]
Lucerne	20–44[e]	15–30[e]	8–20[e]	3–10[e]

[a]Mayland and Wilkinson, 1996.
[b]Dunlop and Hart, 1987.
[c]Brennan et al., 1993.
[d]Melsted et al., 1969.
[e]Kelling and Matocha, 1990.

concentrations than these can occur under certain conditions: for example, the concentration of Mn in lucerne grown in pots of a sandy soil of pH 4.7, flooded for 6 days, was > 7000 mg kg^{-1} (Graven *et al.*, 1965).

Toxic effects attributable to Zn or Cu are usually due to industrial contamination or to the repeated application of sewage sludge to the soil. Critical toxic concentrations appear to be > 200 mg kg^{-1} for Zn and > 20 mg kg^{-1} for Cu, depending to some extent on species and stage of maturity (Davis and Beckett, 1978; MacNicol and Beckett, 1985; Romheld and Marschner, 1991). In acid soils, grasses appear to be more tolerant than most dicot species of high Zn concentrations, but this difference is reversed in alkaline soils (Chaney, 1993).

Influence of Plant Species and Variety on Concentrations of Fe, Mn, Zn, Cu and Co in Herbage

Concentrations of Fe, Zn, Cu and Co are often higher in legumes than in grasses, but, with Mn, concentrations are sometimes higher in grasses than in legumes. The relative concentrations vary with the actual species of grass and legume being compared and with the adequacy of supply of the element.

With Fe, comparisons of grasses and legumes have generally shown higher concentrations in legumes than in grasses (e.g. Hemingway, 1962; Burridge, 1970; Kabata-Pendais and Pendais, 1992), but there were no appreciable differences between grass and clover from a mixed sward in New Zealand (Wheeler, 1998) or between grass and legume herbage in Canada (Miltimore *et al.*, 1970; Drysdale *et al.*, 1980). With Mn, in contrast to the other micronutrient cations, concentrations have often been found to be higher in grasses than in legumes (Burridge, 1970; Drysdale *et al.*, 1980; Grace and Clark, 1991; Kabata-Pendais and Pendais, 1992; Wheeler, 1998; Chiy and Phillips, 1999; Scott, 1999). Zinc is generally at a higher concentration in legumes than in grasses (Gladstones and Loneragan, 1967; Metson *et al.*, 1979; Drysdale *et al.*, 1980; Burridge *et al.*, 1983; Grace and Clark, 1991; Kabata-Pendais and Pendais, 1992; Wheeler, 1998). With Cu, legumes show a wider variation in concentration than do grasses, and the concentration is often but not always higher in legumes. Thus, at a number of sites in Scotland, the concentration of Cu in clovers ranged from 1.7 to 12.3 mg kg^{-1}, while in ryegrass the range was only 2.0–4.3 mg kg^{-1} (Mitchell *et al.*, 1957a). A similar result was obtained in a survey of data from USA where the median concentration was about 8 mg kg^{-1} in legumes and 4 mg kg^{-1} in grasses (Kubota, 1983a). The concentration of Cu is usually higher in clovers than in grasses when the soil is well supplied with Cu (Beck, 1962; Drysdale *et al.*, 1980; Reith *et al.*, 1984) and when the species are grown separately (e.g. Smith *et al.*, 1974; Whitehead, 1987; Grace and Clark, 1991). However, when grass and clover are grown together in a mixed sward, the concentration is sometimes higher in the grass than in the clover (Beck, 1962; Hemingway, 1962;

Burridge, 1970; Boila *et al.*, 1984). Concentrations of Co, like those of Cu, are generally higher in legumes than in grasses, except possibly on soils supplying little Co (Andrews, 1966; Forbes and Gelman, 1981; Burridge *et al.*, 1983; Reith *et al.*, 1983; Sherrell *et al.*, 1990; Kabata-Pendais and Pendais, 1992; Scott, 1999). In a study in New Zealand, there was no consistent difference between grass and clover where Co was deficient but, where the supply of Co was adequate, red clover and white clover contained more than the grasses (Andrews, 1966). However, in large areas of the USA, grasses contain insufficient Co to meet the requirements of grazing animals, whereas legumes (mainly lucerne and clovers) often do contain sufficient Co (Allaway, 1986).

   Comparisons of individual grass species, including perennial ryegrass, timothy, cocksfoot and meadow fescue, have shown cocksfoot to be relatively high in Mn and Cu, and possibly Zn, and suggest that meadow fescue is relatively high in Fe (Table 9.9). However, although cocksfoot has generally been found to have a rather high concentration of Cu in comparison with other species (Table 9.9; Forbes and Gelman, 1981), this difference is not consistent (e.g. Smith *et al.*, 1974). While the data in Table 9.9 indicate only small differences in Co between grass species, perennial ryegrass has been reported to be relatively high compared with timothy and cocksfoot (Andrews, 1966). In a comparison of several less common grass species, relatively high contents of Mn were found in *Agrostis tenuis* and crested dog's-tail (Fleming, 1973). With six cultivated grass species in USA, there was little difference in the concentration of Zn (Belesky and Jung, 1982), but substantial differences were reported between four grass species in Australia (Gladstones and Loneragan, 1967). Some plant species have the capacity to accumulate relatively large amounts of Co, and these include the couch grasses *Agropyron desertorum* and *Agropyron intermedium*, which, in the USA, sometimes contain > 8 mg Co kg^{-1} (Lambert and Blincoe, 1971).

Table 9.9. Concentrations of Fe, Mn, Zn, Cu and Co in several grass species (mg kg^{-1}): mean values reported in three investigations.

	Fe	Mn	Zn	Cu	Co	Reference
Perennial ryegrass	101	41	20	5.0	–	Fleming, 1963
Timothy	42	38	19	4.6	–	
Cocksfoot	67	105	23	7.1	–	
Meadow fescue	109	29	16	4.9	–	
Perennial ryegrass	–	22	–	8.5	0.15	Thomas *et al.*, 1952
Timothy	–	30	–	7.5	0.15	
Cocksfoot	–	46	–	10.0	0.14	
Meadow fescue	–	29	–	9.5	0.16	
Cocksfoot	174	147	36	11.8	0.23	Rinne *et al.*, 1974b
Meadow fescue	202	91	34	10.0	0.24	

Substantial differences between grass cultivars in Fe, Mn, Zn and Cu have been reported for prairie grass in New Zealand (Rumball *et al.*, 1972), but, in a comparison of four varieties of perennial ryegrass and three varieties of cocksfoot in the UK, differences in Mn, Zn, Cu and Co between the varieties were generally small and inconsistent (Forbes and Gelman, 1981).

Comparative data for the legume species are limited (Table 9.10) but suggest that white clover is relatively high in Fe, sainfoin in Mn and red clover in Cu. The relatively high concentration of Cu in red clover is supported by other data (Kubota, 1983a), and a lower concentration of Co in lucerne than in white clover and red clover has also been confirmed (Kubota, 1964). Substantial differences in Mn, Cu and Co were found to occur between cultivars of subterranean clover (Gladstones, 1962).

Although there is some evidence that various forb species have rather higher concentrations than grasses of Fe, Mn, Zn, Cu and Co (Thomas *et al.*, 1952; Baker and Reid, 1977; Forbes and Gelman, 1981), there are insufficient data to make reliable comparisons for individual species.

Influence of Stage of Maturity on Fe, Mn, Zn, Cu and Co in Herbage

Changes in herbage concentration with increasing maturity differ to some extent amongst the five elements, and are also influenced by plant species and other factors. Several investigations have shown the concentration of Fe in grasses to decrease with increasing maturity (Fig. 9.3; Thomas *et al.*, 1952; Loper and Smith, 1961; Fleming and Murphy, 1968), but, in legumes, the decline appears to be smaller (Whitehead and Jones, 1969). The change in Mn in both grasses and legumes is usually small (Fig. 9.3; Reid *et al.*, 1967b) and depends to some extent on supply. When plentiful, the concentration of Mn is generally higher in mature leaves than in young leaves and, if a

Table 9.10. Concentrations of Fe, Mn, Zn, Cu and Co in four legume species (mg kg^{-1}): mean values reported in three investigations.

	Fe	Mn	Zn	Cu	Co	Reference
White clover	117	49	25	7.3	0.19	Whitehead and Jones, 1969
Red clover	85	44	24	7.4	0.21	
Lucerne	108	42	24	7.0	0.16	
Sainfoin	93	52	28	7.0	0.17	
White clover	291	51	29	8.0	0.14	Loper and Smith, 1961
Red clover	217	50	32	10.2	0.15	
Lucerne	260	55	28	6.9	0.13	
Lucerne	115	47	20	8	–	Smith *et al.*, 1974
Sainfoin	158	62	26	7	–	

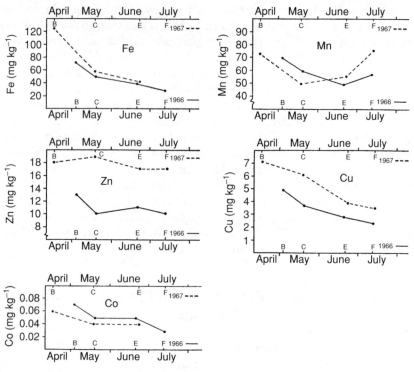

Fig. 9.3. Changes in concentrations of Fe, Mn, Zn, Cu and Co with advancing maturity in perennial ryegrass in 2 years (from Fleming and Murphy, 1968).

deficiency develops, the concentration decreases in young leaves, as well as in roots and stems, while remaining high in old leaves (Loneragan, 1988). There is little mobilization of Mn from older to younger leaves (Kochian, 1991). With Zn, there is often little change with increasing maturity in either grasses or legumes (Fig. 9.3; Reid *et al.*, 1967b; Whitehead and Jones, 1969), though a slight decline was reported in brome-grass (Loper and Smith, 1961). A decline during senescence is more likely when the supply of Zn is adequate than when it is deficient (Longnecker and Robson, 1993). In contrast to Zn, the concentration of Cu in grass herbage shows a substantial decline with advancing maturity (Fig. 9.3; Fleming, 1963, 1965; Reddy *et al.*, 1981). However, the decline in legumes is smaller and, in white clover, there appears to be little change (Fleming, 1965; Whitehead and Jones, 1969; Gladstones *et al.*, 1975). Much of the Cu in leaves appears to be bound to proteins and is not remobilized, even under Cu-deficient conditions, until the leaves senesce and the proteins are hydrolysed, when there is some translocation of Cu complexed with amino acids (Kochian, 1991). Concentrations of Co tend to decline with advancing maturity in grasses (Fig. 9.3), while in the legumes a decline was observed in red clover and sainfoin, though not in lucerne or white clover (Whitehead and Jones, 1969). In lucerne, the concentration of

Co in individual leaves increased rapidly during the early stages of development, but then declined, rather less rapidly, from the time at which the leaf began to unfold (Handreck and Riceman, 1969).

When herbage is allowed to senesce and die *in situ*, there may be further changes in the concentrations of the micronutrient cations, due to factors such as leaching and the loss of metabolic tissue relative to cell-wall material. However, Fe, Mn, Zn and Cu are not readily leached, and their concentrations may well be higher in dead than in living tissue (Waughman and Bellamy, 1981). Concentrations of Mn and Zn in dead grass litter from unfertilized, ungrazed swards were reported to be 500–800 mg Mn kg^{-1} and 100–150 mg Zn kg^{-1}, higher concentrations than are usual for living herbage material (Dickinson, 1983). However, no difference in Cu concentration was found between living and senescent/dead herbage, in an *Agrostis*–fescue sward with a normal concentration of Cu (Hunter *et al.*, 1987). Cobalt may differ from the other micronutrient cations, in that substantial amounts may be leached from dried leaves, though not from fresh leaves (Handreck and Riceman, 1969).

Influence of Season of the Year and Weather Factors on Concentrations of Fe, Mn, Zn, Cu and Co in Herbage

With regularly cut or grazed swards, there is often no consistent seasonal trend in herbage concentrations of Fe (Allaway, 1986), Mn (Metson *et al.*, 1979; Kappel *et al.*, 1985; Allaway, 1986) or Zn (Miller *et al.*, 1964; Metson *et al.*, 1979). When seasonal changes in herbage Mn do occur, they are likely to be influenced by soil water status, since herbage Mn concentration is increased by anaerobic conditions in the soil (Mitchell *et al.*, 1957b). With Zn, there is sometimes a seasonal decline in concentration from spring to late summer (Belesky and Jung, 1982; Jarvis and Austin, 1983; Allaway, 1986; White, 1993; Chowdhury *et al.*, 1997), though, in one study with Bermudagrass, there was an increase from April to July (Kappel *et al.*, 1985). Concentrations of Cu are often lowest in late summer (Reay and Marsh, 1976; Metson *et al.*, 1979; Jarvis and Austin, 1983; Allaway, 1986), though increases during the growing season have also been reported in herbage, both cut (Hemingway, 1962) and grazed (Leech and Thornton, 1987) and, in one investigation, there was no seasonal trend in Cu (Hunter *et al.*, 1987). Increasing soil temperature from 12°C to 20°C increased herbage Cu in subterranean clover (Reddy *et al.*, 1981), but differences in water-supply were found to have little, if any, effect in a study involving several species (Kilmer *et al.*, 1960). With Co, the concentration in herbage has sometimes been reported to increase in autumn and winter, and to decrease during spring and summer (Reith and Mitchell, 1964; Reith *et al.*, 1983; Sherrell *et al.*, 1990), though, following the addition of Co, there was a decrease from spring to autumn (Reith *et al.*, 1983).

Influence of Soil Type on Concentrations of Fe, Mn, Zn, Cu and Co in Herbage

Low concentrations of Fe, Mn, Zn, Cu and Co in herbage may be due to the soils having either a low total content or poor availability. Soil types that are commonly low in their total content of these elements are acid sandy soils and highly calcareous soils. In the UK, soils derived from Old Red Sandstone and Devonian rocks are particularly deficient in Co, and, in the USA, Co deficiency is associated with acid sandy soils that are strongly leached (Kubota *et al.*, 1987). Poor availability is usually due to the effects of high pH, excessive aeration and/or a high content of OM, though the impact of these three factors varies with the particular cation. Thus, a high soil pH markedly reduces the availability of Fe, Mn and Co, but has little effect on Zn and Cu. Deficiencies of Fe in grasses are exceptional, but instances have been reported in cocksfoot, and occasionally other grasses and legumes, growing on calcareous soils (Grime and Hutchinson, 1967). With Mn, an increase in soil pH by one unit in the range pH 5.0–7.5 may reduce the herbage concentration by about 50% (Stewart and McConaghy, 1963; Reith and Mitchell, 1964). Excessive soil aeration tends to reduce the availability and uptake of Fe, Mn and Co, but has a smaller effect on Cu (Adams and Honeysett, 1964) and virtually none on Zn. The effect on Fe, Mn and Co is largely due to the formation of insoluble oxides. Conversely, poor soil aeration due to poor drainage or waterlogging increases the availability of Fe, Mn and Co (Mitchell *et al.*, 1957b; Kubota *et al.*, 1963; Adams and Honeysett, 1964; Reith, 1970). The influence of soil OM on the availability of the micronutrient cations is most apparent with Cu and Co, and the availability of these elements is often low on peaty soils (Kähäri and Nissinen, 1978; Cornforth, 1984). OM may also contribute to the poor availability of Mn in some calcareous soils (Welch *et al.*, 1991). The availability of Co is reduced by a high content of Mn in the soil (Fleming, 1988), but is increased by the factors that increase the availability of Mn, including poor drainage (Berrow *et al.*, 1983).

Influence of Fertilizer N on Concentrations of Fe, Mn, Zn, Cu and Co in Herbage

The influence of fertilizer N on herbage concentrations of the micronutrient cations depends mainly on whether or not the fertilizer changes the soil pH. The soil supply of the micronutrient cation may also be important. Fertilizers, such as ammonium nitrate, which have little effect on soil pH often have little effect on the herbage concentrations of Fe, Mn, Zn, Cu or Co (Reid *et al.*, 1967b; Mudd, 1970b), though there is sometimes a decrease in Mn and small increases in Zn and Cu (Table 9.11; Mudd, 1970b). When

Table 9.11. Herbage concentrations of Mn, Zn, Cu and Co (mg kg⁻¹) as influenced by fertilizer N, applied as ammonium nitrate at 300 kg N ha⁻¹ year⁻¹: mean values for long-term and reseeded swards, 16 sites, two harvesting dates for each of 4 years (from Hopkins *et al.*, 1994).

	Mn	Zn	Cu	Co
No fertilizer N	164	28	6.9	0.10
Fertilizer N at 300 kg ha⁻¹ year⁻¹	132	31	7.2	0.09

ammonium nitrate or calcium ammonium nitrate is applied at a rate of more than about 400 kg N ha⁻¹ year⁻¹, there may be larger increases in herbage Zn (Miller *et al.*, 1964; Hodgson and Spedding, 1966; Mudd, 1970b; Sillanpää and Rinne, 1975). Conflicting results have been obtained with Cu. In several investigations, concentrations of Cu have been increased by the application of ammonium nitrate or calcium ammonium nitrate (Table 9.11; Hemingway, 1962; Havre and Dishington, 1962), whereas, in others, concentrations have been reduced. The effect may be influenced by the soil supply of Cu, as suggested by results from four sites in Scotland, where applying N as calcium ammonium nitrate (without Cu) reduced herbage Cu at three sites, whereas, at the fourth, on a soil containing adequate Cu, applying fertilizer N increased herbage Cu (Reith *et al.*, 1984). The effect of fertilizer N may also be influenced by whether or not excreta are returned during grazing, as suggested by the observation that the application of calcium ammonium nitrate increased the concentration of Cu in grazed swards but decreased the concentration in cut swards (Whitehead, 1966). Also, increasing the application of N up to 720 kg N ha⁻¹ steadily decreased the concentration of Cu in herbage that was grazed but with no return of excreta (Kreil *et al.*, 1966).

When fertilizers, such as ammonium sulphate, which lower soil pH are applied, herbage concentrations of Fe, Mn and Co tend to increase, though there is little, if any, effect on Zn and Cu. Thus, ammonium sulphate greatly increased the concentration of Mn in mixed grass herbage (Havre and Dishington, 1962). Conversely, fertilizers that tend to increase soil pH may have the opposite effect. Calcium ammonium nitrate, applied at an average rate of 245 kg N ha⁻¹ year⁻¹, reduced concentrations of Mn and Co in grass–clover herbage, but had no effect on Zn (Reith and Mitchell, 1964). Also, the same form of fertilizer N, at rates up to 600 kg N ha⁻¹ year⁻¹, decreased Fe and Mn, while slightly increasing Zn and Cu and having a negligible effect on Co (Rinne *et al.*, 1974b).

Influence of Fertilizer P and K and Lime on Concentrations of Fe, Mn, Zn, Cu and Co in Herbage

Fertilizer P tends to decrease plant concentrations of Fe, Zn and Cu and to have small or inconsistent effects on Mn (Moraghan and Mascagni, 1991).

However, the reduction in Zn may be restricted to situations where the supply of Zn is marginal (Loneragan and Webb, 1993; Chowdhury et al., 1997). Since some sources of phosphate contain appreciable amounts of Zn, the application of fertilizer P sometimes increases herbage Zn (Chowdhury et al., 1997). The effect of fertilizer P on herbage Cu is also inconsistent, with reductions occurring in some investigations (Reddy et al., 1981; Reith et al., 1984) but little, if any, effect in others (e.g. Stewart and Holmes, 1953; Hemingway, 1962; Reid and Jung, 1965; Reid et al., 1967a; Rinne et al., 1974b).

The effects of fertilizer K are often negligible or inconsistent (Hemingway, 1962; Reith and Mitchell, 1964; Reid and Jung, 1965; Reid et al., 1967a; Reith, 1970; Scottish Agricultural Colleges, 1982). However, with lucerne, fertilizer K increased herbage Mn but tended to reduce Zn and Cu (Smith, 1975).

Liming tends to decrease herbage concentrations of the micronutrient cations, especially Fe, Mn and Co, through its effect on soil pH. Liming a soil from pH 5.2 to 6.2 reduced the concentration of Mn in the herbage of a mixed sward from an average of 290 to 130 mg kg^{-1} (Reith and Mitchell, 1964); and a change in soil pH from 4.1 to 7.4 reduced the Mn content of ryegrass from 156 to 25 mg kg^{-1} (Stewart and McConaghy, 1963). Liming has a smaller effect with Zn and Cu, a change in pH from 5.4 to 6.2 reducing the Zn concentration of mixed herbage only from 32 to 28 mg kg^{-1} and the Cu concentration from 5.3 to 3.9 mg kg^{-1} (Reith and Mitchell, 1964); and a high rate of lime (16 t ha^{-1} of dolomitic limestone), which increased soil pH from 5 to 7, reduced the concentration of Zn in coastal Bermuda grass only from 33 to 27 mg kg^{-1} (Miller et al., 1964). Liming caused a marked decrease in plant uptake of both native soil Co and of applied Co (Reith and Mitchell, 1964), and also decreased Co in tall fescue (Pinkerton and Brown, 1985). Cobalt is more likely than the other micronutrient cations to become deficient for grazing animals as a result of liming.

Influence of Livestock Excreta on Concentrations of Fe, Mn, Zn, Cu and Co in Herbage

In a comparison of herbage from patches of a ryegrass sward that received urine at least 6 weeks before sampling with herbage that was not affected by urine, the urine was found to depress Mn but to have no effect on Zn and Cu (Table 9.12). The effect on Mn was probably due to the urine increasing, temporarily, the soil pH.

Influence of the Addition of Fe, Mn, Zn, Cu or Co on the Concentrations of these Elements in Herbage

When Mn, Zn, Cu or Co is deficient for either plant growth of animal nutrition, it is possible to increase the concentration in herbage by the addition of

Table 9.12. Herbage concentrations of Fe, Mn, Zn and Cu in ryegrass (mg kg⁻¹ DM) as influenced by the application of sheep urine (from Joblin and Keogh, 1979).

	Fe	Mn	Zn	Cu
Without urine	400	175	31	11
With urine	330	104	31	11

an appropriate salt, usually the sulphate. Several studies have examined the effects of such additions. For example, applications of $ZnSO_4.7H_2O$, at 1.2–4.6 kg Zn ha⁻¹, increased concentrations of Zn in both grass and clover, but the increases were relatively small, and generally less than 7% of the applied Zn was recovered in the herbage (McLaren *et al.*, 1991). The application of $CuSO_4.xH_2O$ has produced inconsistent results in terms of herbage Cu concentration (Sherrell and Rawnsley, 1982), though, when applied at three sites at a rate of 6 kg Cu ha⁻¹, it generally produced a small increase in herbage Cu (Reith *et al.*, 1984). In order to maintain herbage Co at 0.08 mg kg⁻¹ (sufficient to prevent deficiency in livestock) on soils deficient in Co, the application of $CoSO_4$ at a rate of between 0.8 and 1.6 kg Co ha⁻¹ every 3 years has been recommended (Reith and Mitchell, 1964; Reid and Jung, 1974), though smaller amounts were found to be adequate on pumice soils in New Zealand (Sherrell, 1990a, b). However, larger and/or more frequent applications of Co may be necessary on soils with relatively large amounts of exchangeable and easily reducible Mn (Evans, 1985). For non-calcareous soils, the more expensive Co-EDTA is no more effective than $CoSO_4$ in increasing herbage Co (McLaren *et al.*, 1985; Sherrell, 1990b).

The application of high rates of sewage sludge, enriched in Zn or Cu, may increase herbage concentrations of these elements for several years after the sludge application. For example, sludges applied 8 years previously resulted in the concentration of Zn in timothy being increased from 21 to 50–70 mg kg⁻¹ and the concentration of Cu from 6.7 to 7.3–10.9 mg kg⁻¹ (Berrow and Burridge, 1980). The application of sewage sludge also increased Cu and Zn in brome-grass, though the increase in Zn was small and the increase in Cu was no more than that induced by an equivalent amount of N in ammonium nitrate (Soon *et al.*, 1980). The sewage sludge also reduced the concentration of Mn.

Chemical Forms of Fe, Mn, Zn, Cu and Co in Herbage

Although the micronutrient cations function primarily as constituents of enzymes, the proportion of any individual cation present in enzyme form is thought to be less than 50%. Each of the cations also occurs in various organic complexes, some of which are soluble, and some involve links to the plant cell walls. Iron, in addition to occurring partly in various enzymes, is present in complexes with organic acids and in the storage protein phytoferritin (Butler and Jones, 1973). Much of the Fe occurring in enzymes is present in the haem complex, which forms the prosthetic group of various enzymes, including cytochrome, cytochrome oxidase, catalase and peroxidase. Manganese also occurs partly in enzymes (see p. 231), but the form of the remainder is uncertain. In a study with lucerne, almost all the Mn appeared to exist as a complex of low molecular weight (Grady *et al.*, 1978), but, in another study involving ryegrass, more than 60% of the Mn appeared to be present as a free cation (Bremner and Knight, 1970). Of the Zn in plants, about 17% is thought to be associated with two enzymes, superoxide dismutase and carbonic anhydrase (Deighton and Goodman, 1995), while much of the remainder occurs as anionic complexes of low molecular weight and a small proportion is associated with proteins (Brown *et al.*, 1993). In addition, some Zn is associated with the cell walls (Peterson, 1969; Bremner and Knight, 1970; Turner, 1970). Most of the Cu in herbage appears to exist as anionic complexes (Bremner and Knight, 1970).

Assessments of the solubility of these elements in water show considerable variation. In an investigation of several grasses and legumes, the solubility of Fe was always less than 35%, while the solubility of Zn and Cu varied between 18 and 67%: the solubility of Cu was generally greater than that of Zn (Whitehead *et al.*, 1985). However, in another study of six grass species in the USA, the water-solubility of Zn at three sampling dates between May and October was generally higher, within the range 50–99% (Belesky and Jung, 1982).

Occurrence and Metabolic Functions of Fe, Mn, Zn, Cu and Co in Ruminant Animals

Typical concentrations of Fe, Mn, Zn and Cu in the body tissue of ruminant animals are shown in Table 4.1 (p. 73); concentrations in milk in Table 4.4 (p. 78) and the plant : animal concentration ratios in Table 4.6 (p. 80). With the exception of the Fe in haemoglobin, these elements function mainly as constituents of enzymes, or as enzyme activators.

Concentrations of Fe in animals and plants are broadly similar, at about 100–150 mg kg^{-1} DM (see Table 4.6). Iron is an essential constituent of haemoglobin (Fig. 9.2), and this accounts for about 60–70% of the total Fe in the body (Underwood and Suttle, 1999). The haemoglobin is present in the erythrocytes and enables the blood to carry about 65 times more oxygen than is carried in solution in the plasma. Iron is also a constituent of myoglobin, a protein that is present in muscle and provides a storage site for oxygen, which combines reversibly with myoglobin. Iron also acts as a cofactor for a number of enzymes involved in the utilization of oxygen, enzymes that include succinate dehydrogenase, cytochromes and catalase (Underwood and Suttle, 1999). Iron is stored in the body as ferritin and haemosiderin, and these account for about 25% of the body total, with the highest concentrations occurring in liver, spleen and bone marrow. Ferritin is soluble in water and contains 20% Fe, whereas haemosiderin is insoluble and contains 35% Fe (Miller *et al.*, 1991). There is considerable recycling of the Fe within the body, since erythrocytes have a lifespan of about 2 months, and the Fe released during the breakdown of their haemoglobin is recycled via the liver and bile to the digestive tract (McDowell, 1992).

In comparison with soils and plants, animal body tissue has a low concentration of Mn with an average of about 0.3 mg Mn kg^{-1} (see Table 4.1) and an animal : plant concentration ratio as low as 0.007 : 1 (see Table 4.6). The Mn is fairly uniformly distributed, though there is some enrichment in liver and bone and in the tissues of the digestive tract (Grace and Clark, 1991; Miller *et al.*, 1991). Manganese is a constituent of a number of enzymes, including arginase and pyruvate carboxylase, and is an activator of others. The enzymes activated by Mn include some that are involved in the synthesis of chondroitin sulphate, which is required for the organic matrix of bones, teeth and cartilage (Ammermann and Goodrich, 1983), and others that are involved in the synthesis of steroid hormones and in glucose metabolism (Minson, 1990) and also in reproductive processes (Hurley and Doane, 1989).

The average concentration of Zn in the animal body is much higher than that of Mn, at about 20–30 mg kg^{-1} fresh weight, and the animal : plant concentration ratio is also much higher, at about 2.2 : 1. Although Zn is fairly uniformly distributed, there are relatively high concentrations in wool (see Table 4.5, p. 79) and, to a lesser extent, in bone; it probably has a role in maintaining the structural integrity of wool (Lee *et al.*, 1999). Zinc is important in stabilizing the structures of RNA, DNA and ribosomes and in the structure and function of membranes (Hambidge *et al.*, 1986). It is also a component of various enzymes involved in carbohydrate, protein and nucleic acid metabolism (Hambidge *et al.*, 1986), including carbonic anhydrase and alcohol dehydrogenase (Underwood and Suttle, 1999). It is involved in the regulation of appetite, though the mechanisms are obscure (Underwood and Suttle, 1999), and in reproductive processes (Hurley and Doane, 1989). Cows' milk usually contains about 3–5 mg Zn l^{-1} (see Table 4.4), though the concentration may be as low as about 1 mg l^{-1} when the diet is low in Zn (Miller, 1970). Colostrum has a higher concentration of about 7.2 mg l^{-1} (Miller *et al.*, 1991). Considerable mobilization from bone (though not from wool) can occur during lactation (Lee *et al.*, 1999).

Most of the Cu in the body of ruminant animals is stored in the liver, which has a much higher concentration than other body tissues (Grace, 1983a). With sheep, the proportion in the liver may be as high as 72–79%, compared with 8–12% in muscle, 9% in skin and wool and 2% in the skeleton (Davis and Mertz, 1987). Copper is a constituent of more than 30 enzymes, including superoxide dismutase, which contributes to the prevention of oxidative damage to cells, and cytochrome oxidase, which is involved in the synthesis of ATP (Underwood and Suttle, 1999). In milk, the concentration of Cu is much less than that of Zn (see Table 4.4). The concentration of Cu in whole blood or plasma is normally in the range 0.6–1.5 mg l^{-1} and, in general, reflects dietary intake (McDowell, 1992). There is evidence that animals deficient in Cu become more susceptible to infection (Suttle, 1988; Lee et al., 1999).

Although about 43% of the Co in the body is stored in muscle and about 14% in bone, the highest concentrations of Co occur in the kidneys and liver (National Research Council, Subcommittee, 1989). Only small amounts of Co occur in blood and milk. With cattle, the blood has been reported to · contain 0.06–0.09 mg Co l^{-1} (McAdam and O'Dell, 1982), while in milk, concentrations are usually in the range 0.4–1.0 µg Co l^{-1} (see Table 4.4). Cobalt in the form of vitamin B_{12} (Fig. 9.2) is an essential cofactor for several enzyme systems, including those involved in the synthesis of methionine and the metabolism of propionic acid (McDowell, 1992). A deficiency of Co impairs reproduction (Hurley and Doane, 1989) and may also increase susceptibility to infection (Suttle, 1988; Lee et al., 1999).

Absorption of Dietary Fe, Mn, Zn, Cu and Co by Ruminant Animals

The micronutrient cations are absorbed mainly from the small intestine and, sometimes, to a small extent, from the large intestine (Davis and Mertz, 1987). In general, the extent of absorption depends on the balance between supply in the diet and metabolic demand, with excessive absorption being prevented by homeostatic mechanisms. The homeostatic control of absorption is particularly important for iron, as animals have only a limited capacity for excretion. Absorption is also influenced by the physiological status of the animal, by the chemical form of the element ingested and by other components of the diet. During growth, pregnancy and lactation, there is an increased demand, particularly for Fe, Mn and Zn, and therefore an increase in percentage absorption. In relation to the effects of other components of the diet, a high intake of any one of the micronutrient cations is likely to interfere with the absorption of others, though not necessarily all. For example, Mn, Zn and Cu interfere with the absorption of Fe, presumably through competition for binding sites in the digestive tract (Morris, 1987; Miller et al., 1991), and a high intake of Fe depresses the absorption of Cu (Underwood and Suttle, 1999).

Most studies of the absorption of the micronutrient cations have been carried out with diets based on concentrate feeds and, apart from some data for Cu, there is a lack of information on the extent of absorption from grassland herbage. As indicated above, Fe shows a wide variation in absorption. With Mn, estimates include 1–4% (Hurley and Keen, 1987) and 10–20% (Underwood and Suttle, 1999). However, Mn is secreted into the intestine, mainly in chelated form in the bile, and some of this Mn is reabsorbed. With

Zn, the usual extent of absorption is probably around 60–70% (Underwood and Suttle, 1999), but absorption may approach 100% if the diet is deficient in Zn (Payne, 1989). However, the absorption of Zn tends to be reduced by high concentrations of P, Ca, Fe, Cu, fibre and vitamin D (Miller *et al.*, 1991; Payne, 1989). There is normally a large secretion of Zn in the duodenum and a progressive absorption in the rest of the small intestine (Larvor, 1983). The absorbed Zn is bound to albumin in the bloodstream (Hambidge *et al.*, 1986).

The usual extent of absorption of Cu is probably in the range 1–10%, with about 3% being typical for grazed herbage (Underwood and Suttle, 1999). As well as being influenced by the Cu status of the animal, absorption is influenced by the forms of Cu in the diet and by interactions with other constituents (Miller *et al.*, 1991). In addition, animals show considerable genetic variation in their ability to absorb Cu. For example, Welsh mountain sheep are 50% more efficient in absorbing Cu than Scottish Blackface; and the Welsh mountain sheep also have a higher concentration of plasma Cu and a greater retention of Cu in the liver, where any excess is stored (Grace and Clark, 1991). The absorption and utilization of Cu are strongly influenced by dietary concentrations of Mo and S (Mills and Davis, 1987; Underwood and Suttle, 1999), and the three elements are thought to be involved in a sequence of reactions as follows:

1. Sulphide (S^{2-}) or hydrosulphide (HS^-) ions are produced in the rumen by the degradation of S amino acids and/or by the microbial reduction of sulphate.
2. The sulphide progressively displaces oxygen from molybdate ions to yield oxythio- or tetrathiomolybdate ions:

$$MoO_4^{2-} \rightarrow MoOS^{2-} \rightarrow MoO_2S_2^{2-} \rightarrow MoOS_3^{2-} \rightarrow MoS_4^{2-}$$

3. The thio- or oxythiomolybdates react with Cu to form compounds that render the Cu physiologically unavailable.

The relationships between the availability of Cu and herbage concentrations of Mo and S contents are illustrated in Fig. 9.4, which shows that increases in the herbage concentrations of Mo and S within the normal ranges, i.e. up to 5 mg Mo kg^{-1}, and from 0.1 to 0.4% S, may reduce the availability of Cu from 6% to 2%. Clearly, this trebles the animal's requirement for Cu, e.g. from a dietary concentration of 4 mg kg^{-1} to 12 mg kg^{-1}. Recent work suggests that concentrations of Mo as low as 0.5 mg kg^{-1} may significantly reduce the absorption of Cu (Lee *et al.*, 1999). However, the effects of Mo and S on the absorption of Cu are less in hay than in fresh herbage (Suttle, 1983).

There is also antagonism between Zn and Cu, though the dietary concentration of Zn at which Cu deficiency may be induced is uncertain. Evidence from seven farms in The Netherlands suggested that herbage Zn concentrations of more than about 100 mg kg^{-1} might induce Cu deficiency (Van Koetsveld and Lehr, 1961) but, subsequently, in the USA it was concluded

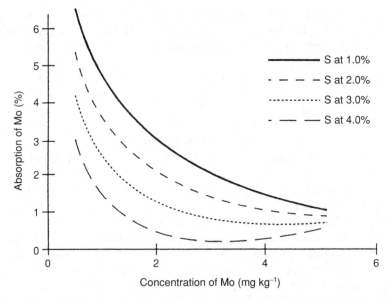

Fig. 9.4. The availability, expressed as % absorption, of herbage Cu as influenced by concentrations of S and Mo in the herbage (from Underwood and Suttle, 1999).

that concentrations of more than 300 mg Zn kg^{-1} were required (National Research Council, Subcommittee, 1989). Clearly the supply of Cu itself will have a substantial effect. The ingestion of soil may also influence the availability of Cu and, at about 10% of DM intake, this has been found to impair the availability of Cu under some, though not all, experimental conditions (Suttle, 1988). The mechanism by which soil reduces the absorption of Cu is not established, but possibilities include the adsorption of Cu by cation exchange sites or by oxides of Fe, Al and Mn (Grace *et al.*, 1996a). Certainly, there is antagonism between Cu and Fe (Miller *et al.*, 1991).

Animals require Co in the form of vitamin B$_{12}$ (Fig. 9.2), and ruminant animals have the advantage that the rumen bacteria synthesize vitamin B$_{12}$ from other sources of Co (Welch *et al.*, 1991). However, only a small proportion of the Co in the diet is converted to vitamin B$_{12}$ in the rumen and, of this, only a small proportion is absorbed (Smith, 1987; National Research Council, Subcommittee, 1989). Vitamin B$_{12}$ may be stored in the liver when the dietary supply is plentiful, but storage is limited and there is a need for a continued supply of Co in the diet (National Research Council, Subcommittee, 1989). There may be some absorption of Co that is not in the form of vitamin B$_{12}$, and, although this may be stored in body tissues, it is not readily returned to the rumen for conversion to vitamin B$_{12}$ (National Research Council, Subcommittee, 1989).

Although soil may impair the absorption of Cu, there is some evidence that it can be a direct source of Mn, Zn and Co for grazing animals (Healy *et al.*, 1970). However, in a study with lambs, soil ingestion had no effect on the concentrations in the liver of Fe, Mn, Zn or Cu, though it did increase the concentration of vitamin B_{12} (Grace *et al.*, 1996a).

Nutritional Requirements for Fe, Mn, Zn, Cu and Co in Ruminant Animals

Although the dietary concentrations of the micronutrient cations recommended for cattle and sheep are broadly similar in various countries (Table 9.13), actual requirements are not well established and the recommended concentrations probably include an appreciable safety margin.

With Fe, one reason why the requirement is not well established is that the concentrations in herbage and other feeds are normally more than adequate. Also, when animals are grazing, they consume an additional amount of Fe through the ingestion of soil. However, it is clear that the requirement is much greater for growth than for maintenance (Suttle, 1979). With Mn, several studies have indicated that 20–25 mg Mn kg^{-1} is adequate for growth and reproduction (Agricultural Research Council, 1980; Hurley and Keen, 1987; Underwood and Suttle, 1999) and, in at least one investigation, as little as 10 mg Mn kg^{-1} appeared to be adequate for the growth of heifers (Bentley and Phillips, 1951). Requirements for Zn, as for Fe, are greater for growth and reproduction than for maintenance (Underwood and Suttle, 1999), but, during a temporary shortage, there may be some mobilization of Zn from bone (Lee *et al.*, 1999).

Although the dietary recommendations for Cu are generally in the range 7–11 mg kg^{-1} (Table 9.13), the actual requirement under some conditions may be as low as 4 mg kg^{-1} (National Research Council, Subcommittee, 1989). As noted above, requirements increase over the range 4–12 mg Cu kg^{-1} with increasing concentrations of Mo and S in the diet. Other antagonists of Cu, such as Fe, may also have a significant effect on the

Table 9.13. Dietary concentrations of Fe, Mn, Zn, Cu and Co (mg kg^{-1} DM) recommended for ruminant livestock.

	Fe	Mn	Zn	Cu	Co	Reference
Dairy cattle, USA	50	40	40	10	0.1	National Research Council, Subcommittee, 1989
Dairy cattle, UK	30	–	30	10	0.1	Agricultural Research Council, 1980
Beef cattle, USA	50	40	30	8	0.1	National Research Council, 1984
Cattle, Australia	30–40	20–30	20	6–9	0.11	Australian Agricultural Council, 1990
Sheep, USA	30–50	20–40	20–33	7–11	0.1–0.2	National Research Council, 1985
Sheep, Australia	30–40	20–30	20	4–6	0.11	Australian Agricultural Council, 1990
Ruminants, France	–	50	50	10	0.1	Gueguen *et al.*, 1989

requirement. However, when there is a change from a diet that is adequate in Cu to one that is inadequate, the appearance of deficiency symptoms may be delayed by the mobilization of reserve Cu from the liver (Lee *et al.*, 1999). Copper sometimes occurs in concentrations that have toxic effects on livestock, with sheep being more sensitive than cattle, apparently due to their inability to excrete Cu from the body (Gooneratne *et al.*, 1989). The risk is greatest where pig slurries with high concentrations of Cu are applied to grassland (Welch *et al.*, 1991). Herbage concentrations of more than 15–20 mg Cu kg^{-1} may have toxic effects if such herbage is consumed for periods of weeks or months (Dam Kofoed, 1980). With Co, although the dietary concentration usually recommended is about 0.1 mg kg^{-1}, there is evidence that minimum requirements are 0.05–0.08 mg Co kg^{-1} for sheep and 0.04–0.06 mg kg^{-1} for cattle (Underwood and Suttle, 1999).

Effects of Deficiencies of Fe, Mn, Zn, Cu and Co on Ruminant Animals

Iron deficiency in grazing animals occurs only when an animal suffers a loss of blood or has a serious parasitic infection or disease (Morris, 1987). However, housed animals that are fed for long periods on low-quality forages, such as straw, occasionally become deficient (McDowell, 1992). Calves that are fed entirely on milk may also become deficient and thus anaemic, since milk is rather low in Fe.

Manganese deficiency is also uncommon, but, when it does occur, it often results in lameness, presumably due to Mn being involved in the formation of the matrix of bone and cartilage (Payne, 1989). Although several claims were made during the period 1950–1970 that supplementation with Mn improved the fertility of dairy cows, other investigations showed no effects, but the reasons for these inconsistencies are uncertain (Hurley and Keen, 1987). There are no biochemical analyses that enable a marginal Mn deficiency to be diagnosed (National Research Council, Subcommittee, 1989).

When Zn is marginally deficient in cattle and sheep, the effects may include loss of appetite, poor growth, reduced fertility and mild skin disorders. These symptoms may occur in animals grazing herbage with less than about 20 mg Zn kg^{-1} (White, 1993). Excessive salivation is also an early symptom (Underwood and Suttle, 1999) and the concentration of Zn in milk is likely to be low (Miller *et al.*, 1991). With a more severe deficiency, resulting from the diet containing less than 5 mg Zn kg^{-1}, the symptoms include loss of hair and wool, reduced resistance to disease and thickening, hardening and fissuring of the skin (Underwood and Suttle, 1999). As with Mn, there is no effective means of diagnosing mild Zn deficiency in ruminants (Spears, 1991).

The symptoms of mild Cu deficiency are, like those of Zn, non-specific and include poor growth and, in cattle, roughness of the coat and loss of hair pigmentation. Copper deficiency may also increase the susceptibility of animals to disease (Suttle, 1988; MacPherson, 1989; Underwood and Suttle, 1999). With more serious or prolonged deficiency, diarrhoea, lameness and anaemia may develop (Payne, 1989). In sheep, the most common symptom of Cu deficiency is the occurrence of swayback in young lambs. In cattle, the most serious problem in economic terms is the reduced growth rate in young animals. On the basis of a survey carried out in the early 1980s, it was estimated that clinical symptoms of Cu deficiency occurred each year in nearly 1% of the cattle population of the UK, and subclinical deficiency in about twice this proportion (Price, 1989). Although Cu deficiency may develop in livestock grazing herbage low in Cu, it often reflects a poor utilization of dietary Cu, due to high intakes of Mo and/or S (see p. 247). It may also be induced by a high concentration of Zn in the diet, in the range 300–1000 mg Zn kg^{-1} or more (National Research Council, Subcommittee, 1989). A deficiency of Cu can be overcome by providing supplementary Cu, for example as $CuSO_4$, which appears to be just as effective as Cu in chelated form (Suttle, 1994). Dosing with cupric oxide needles is also effective, possibly more so than $CuSO_4$ (Sinclair *et al.*, 1993).

The first symptom of Co deficiency in ruminants is a loss of appetite, and the reduced food intake is then liable to result in other symptoms, such as poor growth rate, anaemia and abnormalities in wool or hair. A severe deficiency results in a loss in body weight, followed by emaciation, listlessness and eventually death (Miller *et al.*, 1991). There is also evidence that Co deficiency results in increased susceptibility to disease (MacPherson, 1989). Severe deficiencies occur mainly in fairly well-defined areas, but subclinical deficiencies are more widespread. In contrast to the situation with Cu, Co deficiency is more common in sheep than in cattle, due to their slightly higher requirement (see p. 250). Cobalt deficiency in sheep occurs in many countries, and the symptoms are known as 'pining' in the UK, 'bush sickness' in New Zealand and 'coast disease' in Australia (Underwood and Suttle, 1999). It is most frequent in areas where the soils either are inherently low in Co (less than about 5 mg Co kg^{-1}) or have high contents of iron and manganese oxides, which impair the availability of Co to plants. Deficiency symptoms are liable to occur when the concentration of vitamin B_{12} in the blood plasma is below about 0.2 ng ml^{-1} (Smith, 1987). Cobalt can be provided by adding it to supplementary feeds, in salt-licks, by a slow-release Co bullet (e.g. 90% CoO + china clay) or by the intramuscular injection of vitamin B_{12}.

Excretion of Fe, Mn, Zn, Cu and Co by Ruminant Animals

All the micronutrient cations except Co are excreted to a greater extent in faeces than in urine (see Table 4.9). The amount in the faeces includes both the proportion not absorbed from the diet and that excreted endogenously into the digestive system. Much of the endogenous loss of Fe, Mn, Zn and Cu results from secretion via the saliva, bile and pancreatic juice (Grace and Clark, 1991), bile being particularly important for the excretion of Cu (Gooneratne *et al.*, 1989). Similarly, the endogenous excretion of Zn in the faeces appears to be an important homeostatic mechanism, while urinary Zn shows little change with increased intake (Miller *et al.*, 1991). In contrast to the other micronutrient cations, the endogenous excretion of Co occurs mainly in the urine, and the urinary concentration increases with intake (Underwood and Suttle, 1999). There is little information on the actual concentrations of Fe, Mn, Zn, Cu and Co in faeces and urine, apart from the data in Tables 4.10 and 4.11. However, it has been suggested that the concentration of Zn in urine is typically about 0.25 mg Zn l^{-1} (Hambidge *et al.*, 1986), which contrasts with the 1.8 mg Zn l^{-1} in Table 4.11.

Transformations of Fe, Mn, Zn, Cu and Co in Grassland Systems

Although Fe, Mn, Zn, Cu and Co are broadly similar in their transformations in soil, plants and animals (Fig. 9.5), there are marked differences between them in some of their functions in both plants and animals. An important characteristic for each of them, associated with having a valency of 2 or 3, is their susceptibility to forming organic complexes. This may either increase or decrease their solubility in the soil or animal digestive system, depending on the size and solubility of the molecular complex. Within the tissues of both plants and animals, complexation by organic compounds is essential for each of these ions to fulfil many of its metabolic functions.

Quantitative Balances of Fe, Mn, Zn, Cu and Co in Grassland Systems

The estimated balances of Fe, Mn, Zn, Cu and Co for grazed grassland (Table 9.14) show that outputs as milk and/or live-weight gain are extremely small in relation to the amounts of these elements in circulation. They also show that substantial inputs, relative to the outputs, may arise from atmospheric deposition and that large proportions of these inputs are accounted for by gains to the soil.

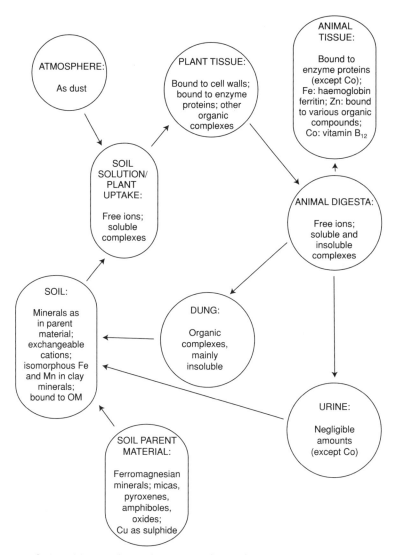

Fig. 9.5. Outline of the main forms of Fe, Mn, Zn, Cu and Co involved in the cycling of these elements in grassland.

Table 9.14. Estimated balances of Fe, Mn, Zn, Cu and Co (g ha^{-1} year^{-1}) typical of two systems of grassland management, intensively managed grass and extensively managed grass–clover, both grazed by cattle.

	Intensive grass (dairy cows)					Extensive grass–clover (beef cattle)				
	Fe	Mn	Zn	Cu	Co	Fe	Mn	Zn	Cu	Co
Inputs										
Fertilizer	300	15	7	2	2	0	0	0	0	0
Atmosphere	3500	100	700	210	4	3500	100	700	210	4
Aspects of recycling										
Uptake into herbage	3000	1000	600	150	2.5	150	60	45	15	0.15
Consumption of herbage by animals	2400	800	480	120	2.0	75	30	23	8	0.08
Dead stubble + herbage to soil	1200	400	240	60	1.0	225	90	66	21	0.21
Dead roots to soil	2000	1000	200	125	1.5	150	60	44	14	0.14
Excreta to soil of grazed area[a]	4540	540	405	110	1.3	1175	30	21	8	0.08
Outputs										
Milk/live-weight gain	2.5	0.25	42	0.7	0.007	0.13	0.013	0.12	0.04	0.0004
Leaching/runoff	25	2.5	2.3	0.7	<0.01	0	0	0	0	0
Loss through excreta off sward	360	95	70	20	0.2	0	0	0	0	0
Gain/loss to soil	3000	17	580	190	6	3500	100	700	210	4

[a]Any excess over consumption of herbage due to soil ingestion.

Chapter 10

The Nutrient Halogens: Chlorine and Iodine

Sources and Concentrations of Cl and I in Soils

Only a few of the minerals in igneous rocks contain a halogen as an essential constituent, and none of those that contain Cl or I is widespread. A small amount of Cl is sometimes present in the place of hydroxyl groups in micas and amphiboles, and a small amount is sometimes present in apatites. In sedimentary rocks, Cl may occur in clays and marine sediments, partly as a substituent for hydroxyl groups and partly in soluble forms (Fuge, 1988). Concentrations of Cl are < 300 mg kg^{-1} in most igneous, metamorphic and sedimentary rocks: they are generally higher in basaltic rocks than in granitic rocks, and are higher in shales and limestones than in sandstones (see Table 2.2, p. 20). Iodine does not fit readily into the lattice structures of common minerals, and concentrations in igneous rocks are consistently less than 1 mg I kg^{-1}, though they may be higher than the < 0.03 mg kg^{-1} indicated as normal in Table 2.2 (Fuge and Johnson, 1986). However, I does occur in sediments, particularly marine sediments, largely associated with OM. Some recent marine sediments are particularly rich in I, with up to 200 mg kg^{-1} (Fuge, 1996), while many sandstones and limestones contain less than 5 mg kg^{-1} (Fuge and Johnson, 1986).

The Cl in soils is derived partly from the parent material and partly from atmospheric and agricultural inputs. In humid regions, there is normally little retention of Cl$^-$ by soils, though, under acid conditions, there may be a slight adsorption by kaolinitic clays and by hydrous oxides of Al and Fe (Fuge, 1988). However, in the saline soils of more arid regions, the absence of leaching may result in an accumulation of Cl. Sources of I in soils also include the parent material and atmospheric and agricultural inputs. In general, relatively little I is released by the weathering of rocks, and the main input is from the atmosphere. As a result, the concentration of I in soil is often much higher than the concentration in the rocks from which it is derived (Fuge and

Johnson, 1986). However, the actual concentration in a particular soil is influenced by the soil's ability to retain I against leaching and volatilization, as well as by the magnitude of the inputs. Both elements are at relatively high concentrations in the oceans, and transfers from the sea via the atmosphere are important for many soils, especially those near to the coast.

Both Cl and I show a wide range of concentration in soils, Cl ranging from < 10 to > 1000 mg kg^{-1} and I ranging from < 0.5 to > 100 mg kg^{-1} (Table 10.1). However, concentrations of Cl are generally less than 200 mg kg^{-1} in soils that are subject to leaching. The average concentration of I, for surface soils on a worldwide basis, appears to be between 1 and 5 mg kg^{-1} (Fuge, 1996), with generally higher concentrations in soils near the coast than in soils from the interior of continents. In a study involving 132 surface soils from the UK, there were marked differences in the concentration of I between soils developed on different categories of parent material, with the highest concentrations in soils derived from peat and from marine and estuarine alluvium, and the lowest in soils derived from sand or sandstone (Table 10.2). In another study, in Ireland, soil I was higher in clay loam than in sandy loam soils, and, within the area studied, the effect of soil composition appeared to be greater than the effect of proximity to the coast, as the concentration of I in sandy loam soils only 4 km from the coast was, on

Table 10.1. Reported concentrations of Cl and I in soils (mg kg^{-1} dry soil).

	No. of soils	Cl	I	Reference
Worldwide	–	50–1800	c. 5 (0.1–150)	Kabata-Pendais and Pendais, 1992
USSR	–	26 (3.5–40)	0.4–42	Fuge, 1988; Fuge and Johnson, 1986
USSR	–	–	0.4–13	Zyrin and Zborischuk, 1975
UK	132	–	9.2 (0.5–98.2)	Whitehead, 1979
UK	–	< 10–140	1.3–149	Fuge, 1988
Ireland	42	–	2.9–30.2	McGrath and Fleming, 1988

Table 10.2. Concentration of I (mg kg^{-1} dry soil) in the top 15 cm of soils derived from nine categories of parent material in the UK (from Whitehead, 1979).

Category of parent material	No. of samples	Range	Mean
Acid igneous rocks and associated till	6	4.4–15.7	10.4
Till associated with basic igneous rocks	4	3.4–16.3	10.9
Slate, shale and associated till	18	4.4–27.6	9.8
Sand and sandstone	7	1.7–5.4	3.7
Chalk and limestone	7	7.9–21.8	12.3
Clay	13	2.1–8.9	5.2
River and river terrace alluvium	10	0.5–7.1	3.8
Marine and estuarine alluvium	7	8.8–36.9	19.6
Peat	4	18.7–98.2	46.8

average, one-third that in clay loam soils 27 km from the coast (Fleming, 1988). The concentration of I in soils has been found to be positively correlated with extractable Al and Fe and with OM (Whitehead, 1973a; McGrath and Fleming, 1988), and these soil constituents tend to adsorb I⁻ from solution. Retention by these constituents provides an explanation for the relatively high concentrations in clay soils and in peats.

Atmospheric and Agricultural Inputs of Cl and I to Soils

Both Cl and I are unusual amongst the nutrient elements in the quantitative importance of transfers to and from the atmosphere. Both are released in substantial amounts from the surface of the sea to the atmosphere, from which they are subsequently deposited on to the land or sea surface. The transfer of Cl from the sea to the atmosphere is probably due mainly to the evaporation of spray, resulting in the formation of aerosol particles containing Cl. Although some transfer of I occurs in this way, there is also some volatilization of gaseous forms (Fuge, 1988), with methyl iodide, produced by marine algae and microorganisms, thought to be particularly important (Campos *et al.*, 1996; Saini *et al.*, 1995). The I released in the form of methyl iodide is slowly converted to free I⁻ by photolytic dissociation. In addition to transfer from the sea, Cl is also released to the atmosphere, as HCl, through volcanic activity and through the combustion of fossil fuels (Goulding *et al.*, 1986; Fuge, 1988). Small amounts of I, as I_2 or I⁻, are also released to the atmosphere from the combustion of fossil fuels (Fuge and Johnson, 1986), and through volatilization from the soil surface (Whitehead, 1981; Fuge, 1996). As a result of the transfers to the atmosphere from the sea and other sources, the amounts of both Cl and I deposited on the land surface vary widely with geographical location. Proximity to the coast, the direction of the prevailing wind and the amount of rainfall are important influences on the actual amounts.

Near the coast, the input of Cl in precipitation may be as much as 100 kg Cl ha⁻¹ year⁻¹ but, at 200 km inland, it may be only 20 kg ha⁻¹, and further inland, for example in the midwestern USA, the atmospheric input has been reported to be only 0.6–1.2 kg Cl ha⁻¹ year⁻¹ (Fixen, 1993). These differences reflect the concentration of Cl in rainfall, which ranges from about 2 mg l⁻¹ near the Pacific coast of the USA to < 0.2 mg l⁻¹ in central USA (Mikkelsen and Camberato, 1995). The influence of distance from the coast on the deposition of Cl has also been observed in Australia, with amounts of 60–70 kg Cl ha⁻¹ year⁻¹ at the coast and < 1 kg ha⁻¹ year⁻¹ in the interior of the continent (Keywood *et al.*, 1997). In the UK, a total deposition of about 40 kg Cl ha⁻¹ year⁻¹ was recorded at Rothamsted, about 100 km from the coast, and 50–60 kg ha⁻¹ year⁻¹ at Saxmundham, about 10 km from the coast (Goulding *et al.*, 1986). Atmospheric inputs at three rural locations in UK were 42–74 kg Cl ha⁻¹ year⁻¹ (Cawse, 1987) and, in a wet upland area of north-west England, deposition was about 44 kg Cl ha^{-1} year^{-1}, 60 km from the coast (Taylor *et al.*, 1999).

Although the concentration of I in rainfall is also generally higher in coastal areas than inland, the gaseous component of atmospheric I can be carried over large distances. The concentration of I in rainfall over land is generally in the range 0–5 µg l^{-1} (Whitehead, 1984; Fuge, 1988) and is usually between 1.5 and 2.5 µg l^{-1} for locations more than a few kilometres from the coast. There is some evidence that the I concentration in rainfall is higher in the first rain that falls after a period of dry weather (Fuge and Johnson, 1986). If the annual rainfall were 800 mm and the average I concentration were 2.0 µg l^{-1}, wet deposition would supply 16 g I ha^{-1} year^{-1}. Expressed in another way, soil I content to a depth of 15 cm would be increased by about 0.7 mg kg^{-1} in 100 years, if it were all retained within this depth. Atmospheric I may also be transferred to the soil through dry deposition, including its absorption by plant leaves and the subsequent decomposition of the plant material. The amount of dry deposition varies widely with factors such as weather conditions and the type of vegetative cover, but an average value might be about 10 g ha^{-1} year^{-1} (Whitehead, 1984).

In addition to the atmospheric inputs, substantial amounts of Cl may be added to agricultural soils in fertilizers, and smaller amounts may be added in sewage sludge and slurries. Fertilizers are often a major source, since the widely used KCl contains 47% Cl, though other fertilizers contain negligible amounts. Sewage sludges vary widely in Cl concentration, the range for 16 samples from cities in the USA being 0.05–1.02%, with a mean of 0.38% (see Table 2.8, p. 32).

Agricultural inputs of I generally have little influence on the I content of soil, with the possible exception of the repeated application of seaweed-based fertilizers. Fertilizers generally contain little I, < 30 mg kg^{-1}, apart from Chilean nitrate, which contains around 80 mg I kg^{-1} (Table 10.3), and fertilizers based on seaweed, which may contain up to 1000 mg kg^{-1} or more (Whitehead, 1979). The I content of sewage sludges varies considerably, but is usually in the range 1–20 mg kg^{-1} (see Table 2.8), broadly similar to the

Table 10.3. Concentrations of I in several fertilizers (from Whitehead, 1973a).

	I (mg kg^{-1})
Ammonium nitrate	< 0.1
Calcium ammonium nitrate	< 0.1
Chilean nitrate	80
Ammonium sulphate	< 0.1
Superphosphate	26
Triple superphosphate	19
Potassium chloride	0.3
Compound fertilizers (eight)	< 0.1–20.6

range found in soils (Table 10.1). An additional agricultural input of I for grassland soils may arise through the excreta of animals that are provided, intentionally or unintentionally, with supplementary I.

Recycling of Cl and I through the Decomposition of Organic Residues

Although there are some data on the concentrations of Cl and I in living grassland herbage, there appear to be none for dead herbage. Chlorine, and possibly I, may well be partially leached by rainfall from senescent tissue, and it is likely that both elements would be released readily in soluble forms during decomposition in the soil. Assuming the decomposition of 7000 kg herbage containing 0.6% Cl and 5000 kg roots containing 0.3% Cl, the amount of Cl recycled would be 57 kg ha^{-1} year^{-1}. For I, assuming concentrations of 0.25 mg kg^{-1} in herbage and 0.75 mg kg^{-1} in roots, the amount recycled would be 5.5 g ha^{-1} year^{-1}.

In animal excreta, both Cl and I usually occur to a greater extent in urine than in faeces, though the difference between urine and faeces is much greater with Cl than with I. Assuming concentrations in urine of 2.3 g Cl l^{-1} (see Table 4.11, p. 92) and 0.15 mg I l^{-1} (Whitehead, 1973a), a grazing intensity of 700 cow-days ha^{-1} year^{-1} and 25 l urine per cow day^{-1}, then the amounts recycled in urine would be about 35 kg Cl and 2.6 g I ha^{-1} year^{-1}. With animals supplied with large amounts of supplementary I – for example, in concentrate feeds or as mineral licks – the concentration in urine would be increased (Whitehead, 1973a). Assuming concentrations in faeces of 0.6% Cl (see Table 4.10, p. 92) and 0.2 mg I kg^{-1} (Whitehead, 1973a) and a grazing intensity as above, with 3.8 kg faecal DM cow^{-1} day^{-1}, the amounts recycled in the faeces would be about 15 kg Cl and 0.5 g I ha^{-1} year^{-1}.

Forms and Availability of Cl and I in Soils

Since soil minerals contain little Cl and the commonly occurring Cl salts are soluble, most of the Cl in soils is derived from recent inputs, and much is present in the soil solution. However, acid soils with kaolinitic clays and/or oxides of Fe and Al may show some adsorption of Cl$^-$ (Fixen, 1993), and in soils of arid regions there is often an accumulation of chloride salts. Most of the Cl in soils is readily available for plant uptake.

The I in soils is partly inorganic and partly organic. Iodide and iodate are the major inorganic forms and, of these, iodide (as I$^-$ or I$_3^-$) is likely to dominate in the soils of humid temperate areas, though iodate may dominate in the alkaline soils of arid areas (Fleming, 1980). Both iodide and iodate are retained strongly by soil OM and by oxides of Fe and Al, and these soil components are thought to be largely responsible for the retention of I in soils

(Whitehead, 1984; McGrath and Fleming, 1988; Sheppard *et al.*, 1996). Some of the retention of I by OM in soils appears to be due to immobilization by the soil microorganisms (Koch *et al.*, 1989). Soils vary in the extent to which the concentration of I changes with depth: sometimes there is a decrease with increasing depth, sometimes little change and sometimes an accumulation at intermediate depth (Whitehead, 1978). These differences appear to be due partly to differences in the distribution, within the soil profile, of Fe and Al oxides and OM and partly to differences in the length of time during which the soil material has been exposed to atmospheric inputs.

Losses of Cl and I from Soils

Losses from soils by leaching occur readily with Cl, but much less readily with I. On the other hand, losses by volatilization are potentially important with I, but not with Cl.

In intensively managed grassland, leaching may well account for a large proportion of the Cl added in the form of fertilizer and as deposition from the atmosphere. Thus, grassland receiving 94 kg Cl ha^{-1} year^{-1} as KCl and grazed by sheep lost 72–118 kg Cl ha^{-1} year^{-1} by leaching (Heng *et al.*, 1991); and grassland receiving 152 kg Cl ha^{-1} year^{-1} as KCl, and grazed by cattle during the summer, lost about 116 kg Cl by leaching and 12 kg ha^{-1} by surface runoff (Owens *et al.*, 1998). The Cl returned in urine patches is particularly susceptible to leaching (Hogg, 1981).

The leaching of I has been studied mainly in the context of the movement through soil profiles of radioactive isotopes, e.g. ^{129}I derived from the reprocessing of nuclear fuels. These studies have indicated that there is some leaching when water moves rapidly through macropores, but that there is often considerable retention of I by soil constituents (Whitehead, 1984). Nevertheless, appreciable concentrations of I (2.5–6.5 µg l^{-1}) were detected in leachates from lysimeters containing a sandy loam soil sown to ryegrass, suggesting that, with this soil, there was no appreciable adsorption of the atmospheric input (Whitehead, 1979).

The volatilization of I from soil to the atmosphere may occur as a result of both purely chemical and microbiological processes. In general, the chemical processes yield elemental iodine (I_2) or hydrogen iodide, and the microbiological processes yield organic compounds, such as methyl iodide (Whitehead, 1984). Volatilization by chemical processes is most likely to occur from soils which have low contents of the constituents responsible for the retention of I, namely Fe and Al oxides and OM (Whitehead, 1981). It has been suggested that I deposited from the atmosphere in coastal regions may be partially volatilized and redeposited further inland, and that this process might be repeated a number of times across a land mass, depending on the ability of successive soil types to retain I (Fuge, 1996).

Uptake of Cl and I by Herbage Plants

In general, plants take up Cl readily, and the amount taken up reflects the available supply. However, there are competitive effects on chloride uptake from sulphate and nitrate, though not from fluoride or iodide (Fixen, 1993). Typically, the plant : soil concentration ratio for Cl is high, at about 7 : 1 (see Table 3.6, p. 56), reflecting not only the ease with which it is absorbed but also the lack of retention by soil constituents.

Iodine is also taken up readily from solution, despite the fact that I appears not to be essential for plants, but the amounts involved are much smaller than those of Cl. Roots absorb both iodide and iodate ions, though iodide is absorbed preferentially, and the amounts increase with increasing supply. In soils, adsorption by hydrous Fe and Al oxides and OM restricts availability and uptake, and the presence of large amounts of $CaCO_3$ and/or high pH has a similar effect (Whitehead, 1984). As well as uptake through the roots, plants can absorb I through their leaves, with gaseous forms of I being able to enter the stomata of plant leaves and becoming incorporated into the leaf tissue (Whitehead, 1984). While both soil and atmospheric sources contribute to the I content of plants grown in the field, the relative importance of the two sources is uncertain. There is some evidence that, if the I deposited superficially on leaves is included as plant I, the atmosphere may often be the dominant source. For example, in the vicinity of a reprocessing plant for nuclear fuel, the leaves of plants, including grasses, were found to have much higher ratios of $^{129}I : ^{127}I$ than samples of surface soils from the same sites, suggesting an appreciable direct uptake of atmospheric ^{129}I (Brauer and Ballou, 1975). On the other hand, the fact that goitre in animals caused by I deficiency occurs mainly in localized, well-defined areas tends to support the view that the soil is the major source for plants.

The uptake of soil I by plants is proportionately much less than the uptake of Cl, as illustrated by the typical plant : soil concentration ratio for I of about 0.04 : 1 (see Table 3.6; Sheppard *et al.*, 1993). This is probably due mainly to the much greater retention of I by soils. If some of the plant I were due to direct absorption from the atmosphere, the difference between Cl and I in uptake from the soil would be even greater. However, as well as absorbing I from the soil and the atmosphere, plants can also volatilize small amounts of I (Amiro and Johnston, 1989), probably as methyl iodide (Saini *et al.*, 1995), but the quantitative importance of this process is unknown.

Functions of Cl and I in Herbage Plants

An important function of Cl in plants is, as chloride, to balance the positive charges of the major cations. It thus contributes to maintaining the stability of osmotic pressure and pH (Romheld and Marschner, 1991; Fixen, 1993). As it is readily translocated, Cl⁻ is able to balance the cations during transport in the vascular system. It is also important, with K^+ and malate, in maintaining the turgor of guard cells (Clarkson and Hansen, 1980; Fixen, 1993). Chloride may also be involved in the functioning of enzymes, such as ATPase (Mengel and Kirkby, 1987), though it appears not to be incorporated into any plant metabolite. In most regions, atmospheric inputs ensure that there is little possibility of plants suffering from a lack of Cl, but deficiencies have been reported at a few locations far from the sea, e.g. in Oregon and Dakota, USA (Fixen, 1993).

Iodine appears to have no metabolic function in plants, though there are a few reports that, when added to a purified nutrient solution, it stimulated plant growth (e.g. Umaly and Poel, 1970). However, despite its apparent lack of function, a small proportion of the plant I is synthesized into iodo-amino acids, such as 3-iodotyrosine, 3,5-diiodotyrosine and 3,3,5-triiodothyronine (Mengel and Kirkby, 1987).

Distribution of Cl and I in Herbage Plants

There is little information on the relative concentrations of Cl in the roots and shoots of grassland plants. In tall fescue grown in culture solution, the concentration of Cl was about twice as great in the shoots as in the roots (Wu et al., 1988) and, in lucerne grown in soil, the concentration in the shoots was also greater than that in the roots (Rominger et al., 1975). Within the shoots of grasses, Cl was found to be higher in stems than in leaves (Fleming, 1965; Smith and Rominger, 1974) and higher in leaves than in flowering heads (Fleming, 1965). However, in another study, Cl tended to be high in the inflorescence of ryegrass relative to leaves and stems (Chiy and Phillips, 1996a).

The concentration of I is generally greater in roots than in shoots, and much of the I absorbed by the roots is retained there. However, in solution culture, the proportion transported to the shoots of two grasses and two clovers increased with increasing concentration of iodide in the nutrient solution (Whitehead, 1973b). Within the shoots of plants grown with an enhanced supply of iodide, dead and senescent material had a concentration more than twice that of living leaves and stems (Whitehead, 1973b).

Ranges of Concentration of Cl and I and Critical Concentrations of Cl in Herbage Plants

The concentration of Cl in herbage varies widely, though it is usually within the range 0.2–2.5% (Table 10.4; Cunningham, 1964; Kemp and Geurink, 1978; Whitehead et al., 1983; Roche et al., 2000). Much of the variation is due to differences in current inputs, particularly from fertilizer KCl, and to some extent from rainfall. There are also differences between soils in the susceptibility with which Cl is lost by leaching.

There is also a wide variation in reported concentrations of I in herbage, though some of this variation is probably due to shortcomings in analytical procedures. Most recent studies have reported herbage concentrations in the range 0.04–0.8 mg I kg^{-1} (Table 10.4; McCoy *et al.*, 1997), but, in a number of studies, mainly carried out before 1960, concentrations up to 5 mg I kg^{-1} were reported. Such high concentrations are suspect, and probably reflect inaccuracies in analysis or contamination of the herbage by soil, a factor that can greatly increase the apparent concentration of I (McGrath and Fleming, 1988).

The concept of critical concentration applies only to Cl. For its specific biochemical functions, the critical concentration appears to be about 150–300 mg Cl kg^{-1} (Romheld and Marschner, 1991), but, if the role of chloride in the regulation of osmotic pressure and pH is taken into account, the critical concentration would be much higher, though not well defined (Fixen, 1993). Concentrations of more than about 3–4% Cl may be toxic (Jones, 1991).

Influence of Plant Species and Variety on Concentrations of Cl and I in Herbage

In a comparison of grass and clover grown together in mixed swards, the concentration of Cl was almost twice as high in the grass as in the clover (Whitehead *et al.*, 1983), but, in a comparison of Cl in more than 8000 samples of hay and silage, of either grasses or legumes, there was little difference between the grasses and legumes in concentration (Mayland and Wilkinson, 1996). Differences in Cl concentration between the main cultivated grass species appear to be small and inconsistent (Table 10.5), but some species, such as Yorkshire fog and rough-stalked meadow-grass, have been reported to have higher concentrations than ryegrass (Fleming, 1965; Hartmans, 1974). Amongst the legume species, white clover has a relatively high concentration of Cl (Whitehead and Jones, 1969).

Table 10.4. Concentrations of Cl and I in grassland herbage reported from various regions.

	Cl (%)	I (mg kg^{-1})	Reference
UK	0.4–3.0	0.08–0.5	MAFF, 1975; Whitehead *et al.*, 1983
USA	0.07–1.85	0.07–0.49	National Research Council, 1985
USA	0.1–0.5	0.04–0.8	Mayland and Wilkinson, 1996
Ireland	–	0.08–0.23	McGrath and Fleming, 1988
New Zealand	–	0.17–0.26	Grace *et al.*, 1996b
Tasmania	–	0.05–0.4	Statham and Bray, 1975

Table 10.5. Concentrations of Cl and I in four grass species: mean values.

	Cl (%)		I (mg kg^{-1})	
	Whitehead, 1966	Thomas *et al.*, 1952	Alderman and Jones, 1967	Hartmans, 1974
Perennial ryegrass	0.94	0.51	0.22	0.07
Timothy	0.81	0.56	0.31	0.09
Cocksfoot	0.99	0.31	0.32	0.12
Meadow fescue	0.85	0.66	0.22	0.08

With I, some studies carried out in the 1950s and 1960s suggested that grassland species and strains showed marked differences in their ability to accumulate I (Butler and Jones, 1973), though some later studies failed to confirm the earlier emphasis on genetic differences. Nevertheless, white clover appears to have a rather higher concentration of I than do grass species (Alderman and Jones, 1967; Hartmans, 1974), although there was no marked difference between subterranean clover and ryegrass (Statham and Bray, 1975). Amongst the grasses, cocksfoot has been found to have a rather higher concentration than several other species (Table 10.5; Hartmans, 1974). There appear to be considerable differences between varieties of white clover as the concentration of I in ten varieties ranged from 0.14 to 0.44 mg kg^{-1} (Alderman and Jones, 1967), and a survey of 51 cultivars of white clover in New Zealand showed I concentration to range from 0.08 to 0.21 mg kg^{-1} DM (Crush and Caradus, 1995).

Influence of Stage of Maturity on Concentrations of Cl and I in Herbage

The concentration of Cl in grass herbage shows little change with advancing maturity, though, in one study, there was a small reduction after inflorescence emergence (Whitehead, 1966) and, in another, an increase with increasing age of leaf (Chiy and Phillips, 1997). In legumes, there appears to be no consistent change (Whitehead and Jones, 1969; Chiy and Phillips, 1997).

There is little information for I, apart from one finding that herbage concentrations tended to decline with age (Alderman and Jones, 1967).

Influence of Season of the Year and Weather Factors on Concentrations of Cl and I in Herbage

With swards that are cut or grazed on several occasions during the season, variation in herbage Cl often depends on the pattern of application of fertilizer KCl. When this fertilizer is applied only at the beginning of the season, herbage concentrations of Cl tend to decline during the season, e.g. from about 1.2% in early June to about 0.7% in October, reflecting the decline in supply (Whitehead, 1966). However, when fertilizer KCl is applied to cut swards after each cut, the concentration of Cl may well increase during the season (Whitehead *et al.*, 1983).

In general, concentrations of I appear to be highest at the first harvest in spring, then decline to a minimum during summer and increase again in autumn (Alderman and Jones, 1967; Mudd, 1970b; Hartmans, 1974; Statham and Bray, 1975). This pattern suggests that concentrations of I are highest after periods of slow herbage growth, and this may be due to the greater length of time during which the herbage is exposed to deposition from the atmosphere.

Influence of Soil Type on Concentrations of Cl and I in Herbage

Soil type has little influence on herbage concentrations of Cl as, even in clay soils, there is little retention by soil constituents, though the rate of leaching may be slower than in sandy soils. The supply of Cl depends mainly on current inputs from the atmosphere and fertilizers.

There are few data relating herbage I concentrations directly to differences in soil type though, with mixed swards of perennial ryegrass and subterranean clover, higher concentrations were found in both species on clay than on sandy soils (Table 10.6). The geographical distribution of areas in which goitre is (or was in the past) prevalent in livestock and/or human populations (Schutte, 1964) shows some relationship to soil type on a continental scale, but the relationship is less close in more restricted areas. Many of the areas of goitre were glaciated during relatively recent geological times, and the glaciation is thought to have removed the previously existing surface soil, leaving new soil parent material low in I. Some of the other main areas of goitre are in regions where atmospheric inputs of I are low, due to distance from the sea or to low rainfall, while others show a combination of high altitude and high rainfall, which is likely to result in intense leaching (Welch *et al.*, 1991). However, these factors of glaciation, distance from the coast and leaching cannot account for all the known areas of goitre, and they do not

Table 10.6. Concentrations of I (mg kg^{-1}) in perennial ryegrass and subterranean clover grown on two soil types (from Statham and Bray, 1975).

	Sand	Clay
Perennial ryegrass	0.05–0.39	0.15–0.62
Subterranean clover	0.08–0.39	0.08–0.59

provide a satisfactory explanation for some of the areas in the UK, such as those in Devon. Other possible factors that may contribute to the geographical distribution of goitre are differences (due to unknown factors) in the availability of soil I and the localized occurrence of goitrogens. It has been suggested that a goitrogenic compound that interferes with the assimilation of I in animals may be present in various plants and be related to a geochemical characteristic with a localized distribution (Stewart and Pharoah, 1996).

Influence of Fertilizer N on Concentrations of Cl and I in Herbage

The effect of fertilizer N on herbage Cl is inconsistent. Thus, in two investigations, fertilizer N as ammonium nitrate caused a substantial reduction in the Cl concentration of ryegrass, particularly when Cl was also added (Dijkshoorn, 1958; Rahman et al., 1960), whereas, in a third investigation, fertilizer N, also as ammonium nitrate, had no consistent effect on Cl in ryegrass or white clover (Whitehead et al., 1983). The application of ammonium phosphate also had no significant effect on the Cl content of herbage (Kemp, 1960).

The application of fertilizer N has sometimes been found to decrease the concentration of I in herbage (Alderman and Jones, 1967; Mudd, 1970b; Hartmans, 1974), probably by a dilution effect due to increased growth. However, no effect of fertilizer N on herbage I was found in other investigations (Horn et al., 1974; McGrath and Fleming, 1988).

Influence of Fertilizer P and K, of Lime and of KI on Concentrations of Cl and I in Herbage

The application of fertilizer K in the commonly used form of KCl is an important factor influencing the concentration of Cl in herbage (Table 10.7). Other fertilizers contain negligible amounts of Cl and have little effect.

The application of fertilizer P and K was found to have little effect on herbage I (Hartmans, 1974), and liming also had no appreciable effect (McGrath and Fleming, 1988).

Table 10.7. Influence of fertilizer KCl, applied at four rates, on the herbage concentration of Cl in ryegrass (from Whitehead, 1966).

	Cl applied in KCl (kg Cl ha^{-1})			
	0	58	117	233
Cl in herbage (% in DM)	0.39	0.75	1.04	1.30

The application of KI at 2 kg I ha^{-1} increased the average concentration of I in perennial ryegrass from 0.08 to 1.41 mg kg^{-1} at the first cut, but the applied I was rapidly immobilized or lost from the soil, and the second cut contained only 0.21 mg kg^{-1} (Hartmans, 1974). In a study in New Zealand, in which KI and KIO$_3$ were applied separately to pastures at rates up to 4 kg I ha^{-1}, iodide generally had a slightly greater effect than iodate on herbage I concentration. The magnitude of the effects differed between the three sites examined though, at each site, it declined markedly during the period between 6 and 20 weeks after the application (Smith et al., 1999). In general, the recovery of I applied as KI or KIO$_3$ in the annual production of herbage appears to be less than 5% (Hartmans, 1974; Smith et al., 1999).

Influence of Livestock Excreta on Concentrations of Cl in Herbage

The concentration of Cl in the herbage of a long-term upland sward was about 0.9% in plots that had been grazed for 25 years, compared with about 0.4% in plots that had been cut (Davies et al., 1959).

Chemical Forms of Cl and I in Herbage

Herbage Cl occurs entirely, or almost so, in the form of the chloride ion and is therefore completely soluble in water (Makela, 1967).

Iodine is present mainly as inorganic iodide (Butler and Jones, 1973) but, in some plants at least, small amounts are converted to iodo-amino acids (Mengel and Kirkby, 1987).

Occurrence and Metabolic Functions of Cl and I in Ruminant Animals

The Cl in the animal body is present almost entirely in the soft tissues, with little in the skeleton, and its concentration in the body is maintained at about 0.10–0.11% Cl (see Table 4.1, p. 73). It is present mainly as the chloride ion, which is the major anion in the extracellular fluids, and it therefore contributes to the regulation of pH and osmotic pressure. Chloride is required for the formation of gastric HCl and for the activation of amylase in the intestines, and is therefore important in protein digestion (Ammermann and Goodrich, 1983). It occurs in enhanced concentrations in bile, pancreatic juice and other intestinal secretions (Fontenot and Church, 1979; Coppock, 1986). Chloride is also involved in the transport of oxygen and carbon dioxide (National Research Council, Subcommittee, 1989). The influence on blood pH occurs particularly through the exchange between Cl in the colon and bicarbonate in the blood (Kincaid, 1988) and, if the diet is deficient in Cl, the blood may become abnormally alkaline.

Normal concentrations of Cl in the whole blood of ruminant animals are 2.1–2.6 g l^{-1}, while the plasma has a higher concentration of 3.5–3.7 g Cl l^{-1}, due to the red cells being low in Cl (Kincaid, 1988). Milk contains about 1.0–1.1 g Cl l^{-1} (see Table 4.4, p. 78; Kincaid, 1988).

The metabolic need for I is due to its being a constituent of the hormones tetraiodothyronine (thyroxine) and triiodothyronine (Fig. 10.1). These hormones, which are produced in the thyroid gland, influence a variety of physiological processes, and are especially important in controlling the rate of energy metabolism, through their effect on the rate of oxidation in cells. They also influence protein synthesis, the growth and differentiation of tissues and the activity of other endocrine glands (Underwood and Suttle, 1999). The thyroid gland contains about 70–80% of the total I in the body (Miller *et al.*, 1991). A small proportion of the I not in the thyroid hormones occurs as free I_2 or I^-, which permeates all tissues (Underwood and Suttle, 1999). The normal concentration of I in the blood serum of cattle and sheep is about 0.05–0.15 mg l^{-1} (Miller *et al.*, 1991). It is important for newborn animals to be well supplied with I, and the concentration in colostrum is relatively high. With cattle, the concentration is around 0.25 mg I l^{-1} in colostrum, compared with about 0.10 mg l^{-1} in normal milk (Hetzel and Maberly, 1986). However, the concentration of I in normal milk is related to I intake (Alderman and Stranks, 1967; Hemken, 1979; McCoy *et al.*, 1997), and may be greatly increased by providing supplementary I in the diet, or by using an iodophor compound as an udder wash (Miller *et al.*, 1991). The concentration is often much higher in winter than in summer, due to the provision of supplementary concentrate feeds. The I in milk is present entirely in the form of iodide (Hetzel and Maberly, 1986).

thyroxine
(3,5,3',5'-tetraiodothyronine, T_4)

3,5,3'-triiodothyronine, T_3

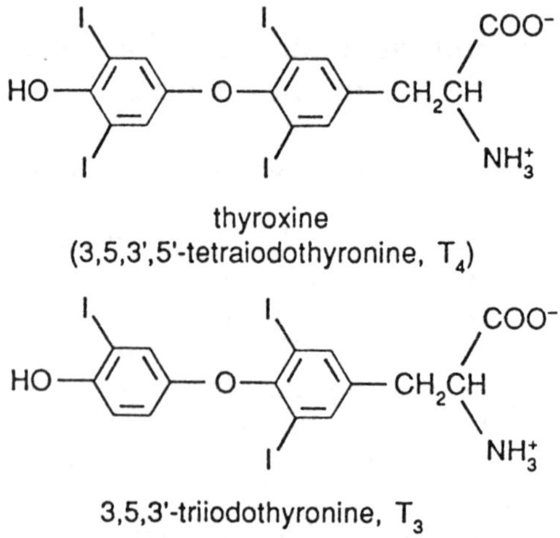

Fig. 10.1.   Chemical structure of the I-containing hormones, thyroxine and triiodothyronine.

Iodine is at a much higher concentration in animal body tissue than in grass herbage, while Cl shows little difference in concentration (see Table 4.6, p. 80).

Absorption of Dietary Cl and I by Ruminant Animals

Chlorine can be absorbed throughout the digestive tract (Coppock, 1986) but most absorption occurs from the intestines (Fontenot and Church, 1979). Absorption is almost complete, and homeostatic control of the concentration of Cl in blood is achieved almost entirely through the extent of excretion in the urine (Burkhalter *et al.*, 1979).

The initial absorption of I in ruminants occurs mainly from the rumen in the form of iodide (Miller *et al.*, 1991). However, there is considerable recycling within the animal, with iodide being secreted into the abomasum and subsequently reabsorbed from the small and large intestines (Miller *et al.*, 1975; Hetzel and Maberly, 1986). This recycling system contributes to homeostatic control and reduces the risk of deficiency. After absorption into the blood, I is bound loosely by proteins in the plasma and is transported to the thyroid gland, where the I is converted to the I-containing hormones by reaction with the amino acid tyrosine (Underwood and Suttle, 1999).

In addition to consuming I in grassland herbage, which usually contains between 0.08 and 0.50 mg I kg^{-1} (Table 10.4), many ruminant livestock, particularly dairy cattle, receive I from other sources, which are not always recognized by farmers as supplying I. For example, salt-licks or blocks often contain supplementary I, typically at a concentration between 40 and 850 mg I kg^{-1}. In the UK, the usual concentration is about 200 mg kg^{-1} (Thompson, 1980). Many commercially produced concentrate feeds, particularly those for dairy cattle, are supplemented with I and contain up to 10 mg I kg^{-1}, and these represent a major source for many dairy cattle. A further source of I for livestock is the ingestion of soil. The concentration of I in soils is, on average, about 25 times greater than that in herbage, and the consumption of soil by grazing animals often amounts to 5–20% of their DM intake. There is evidence that the incidence of goitre in lambs is less when ewes consume relatively large amounts of soil, especially clay soil, and this observation suggests that soil I is at least partly available for absorption by animals (Healy *et al.*, 1972; Statham and Bray, 1975). A source of I that applies only to dairy cattle arises from the iodophor detergents and sterilants that are often used as udder washes and for cleaning milking equipment: appreciable amounts of I from these materials can be absorbed through the skin. The veterinary use of I compounds, such as the addition to the diet of ethylenediaminedihydra-iodide (EDDI) in order to control foot-rot, may also augment the nutritional sources of I (Whitehead, 1984). Finally, animals may be able to absorb gaseous I from the atmosphere. Iodine concentrations in the atmosphere (normally 10–25 ng m^{-3}) are increased in buildings in which iodophor compounds are used (Vought *et al.*, 1964), and may result in an increased intake by housed animals.

Nutritional Requirements for Cl and I in Ruminant Animals

The requirements of ruminant animals for Cl are not well defined, and a rather wide range of recommended concentrations has been proposed. In the USA, a dietary concentration of 0.25% is recommended for lactating cows and a concentration of 0.20% for non-lactating dairy cattle (National Research Council, Subcommittee, 1989). Higher concentrations are recommended in the UK (Agricultural Research Council, 1980) and The Netherlands (Kemp and Geurink, 1978), and lower concentrations in Australia (Australian Agricultural Council, 1990). Actual minimum requirements have been estimated at about 0.20% for lactating cows and 0.04% for non-lactating animals (Coppock, 1986). However, it is unusual for herbage to contain less than 0.5% Cl, and there appear to be no reports of symptoms in ruminants that can be attributed simply to a deficiency of Cl (Fontenot and Church, 1979). Nevertheless, as plants tend to be low in Na, in comparison with animal needs, many ruminant livestock are provided with salt (NaCl) either in supplementary feed or on a 'free access' basis (McDonald *et al.*, 1995).

The requirement for I is strongly influenced by the physiological condition of the animal. A dietary concentration of 0.12 mg I kg^{-1} is normally adequate for maintenance and growth, but a concentration of 0.5–0.6 mg kg^{-1} is recommended for pregnant and lactating animals (Agricultural Research Council, 1980; National Research Council, Subcommittee, 1989). If goitrogens, which interfere with the synthesis of thyroid hormones, are present in the diet, it is advisable for the concentration of I to be increased to 2 mg kg^{-1}. Goitrogenic substances are of two main types. One type, the cyanogenetic, are the thiocyanates derived from the cyanide in white clover and subterranean clover, which reduce the uptake of I by the thyroid, and whose effect can be overcome by the provision of supplementary I. The second type, the thiouracil-type goitrogens found in brassicas, inhibit the iodination of tyrosine, and their effects are less easily overcome (Australian Agricultural Council, 1990). Symptoms of I deficiency may appear in dairy cattle consuming a diet containing 0.60 mg I kg^{-1} if as much as one-quarter of the feed is from goitrogenic brassicas (National Research Council, Subcommittee, 1989). In addition to these compounds, which are well-established as goitrogens, antithyroid activity has been demonstrated for a number of cations and anions when these are at excessive concentrations in the diet. In particular, the cations Co^{2+} and Na^+ and the anions chloride and perchlorate appear to have this effect, especially when the intake of I is low (Underwood and Suttle, 1999). Conversely, a deficiency of Se impairs the formation of the thyroid hormones and may induce an apparent deficiency of I (Underwood and Suttle, 1999).

Lactating animals secrete large amounts of I in the milk and, for dairy cows, milk analysis is a useful method of assessing I status. The concentration of I in milk usually varies linearly with intake in the normal dietary range (Miller *et al.*, 1975). However, the concentration is also influenced by the stage of lactation, season of the year, goitrogens and thyroid status. There is evidence that, when the concentration of I in bulk herd milk is less than about 20–25 μg l^{-1}, the amount of I in the diet is too low (Langlands, 1987; National Research Council, Subcommittee, 1989), though this was not confirmed by work in New Zealand (Lee *et al.*, 1999). In sheep, a concentration of I in the milk of < 80 μg l^{-1} is associated with a likelihood of enlarged thyroid in lambs (Mason, 1976; Scottish Agricultural Colleges, 1982; Australian Agricultural Council, 1990). When there is a change in the dietary supply of I, a new equilibrium concentration in milk is usually established within 1–2 weeks (Hartmans, 1974).

Effects of Deficiencies of Cl and I, and Toxicity of I, on Ruminant Animals

Despite the importance of Cl in regulating osmotic pressure and acid–base equilibrium in body fluids, deficiencies are almost unknown, as the concentration of Cl in normal diets is sufficient to meet requirements.

Deficiencies of I, however, are quite widespread. The most common sign of I deficiency in ruminant animals is the birth of dead lambs or calves with symptoms of goitre, i.e. a swelling in the upper neck due to enlargement of the thyroid gland. The thyroid is regarded as goitrous when the fresh weight exceeds 2.8 g for lambs and 13 g for calves (Scottish Agricultural Colleges, 1982) or, for sheep, when the ratio of thyroid weight (in g) to live weight (in kg) is greater than 0.4 (Lee *et al.*, 1999). Other symptoms of I deficiency include the impairment of brain and lung development, and these effects contribute to the mortality of animals with goitre (Hetzel and Maberly, 1986). Despite the occurrence of goitre and other symptoms in aborted or newly born lambs or calves, the adult ewes or cows may appear to be normal (Miller *et al.*, 1991). However, when the deficiency of I is sufficiently severe to affect older animals, there is a general decline in basal metabolism, leading to a variety of non-specific clinical signs, including failure to thrive, low milk yield and general debility (Payne, 1989). Biochemical criteria for assessing I status are poorly defined, due partly to the difficulties of analysis and partly to the involvement of other factors, such as Se deficiency (Lee *et al.*, 1999). As indicated above, goitre in animals is not always due to an inherent lack of I, but is sometimes due to the presence of goitrogenic substances or a deficiency of Se in the diet; and, although goitre occurs predominantly in localized well-defined areas, the distribution of these areas is not fully explained. An overall lack of I in soil and water-supplies is a major factor, and areas where goitre is prevalent correspond to many of the areas where I in soil and water-supplies is known to be low. Examples include the area surrounding the Great Lakes of North America and parts of the Himalayas. Both of these areas are far from the sea and atmospheric inputs of I are undoubtedly small. However, other areas where goitre occurs are much closer to the sea, and sometimes have relatively high concentrations of soil I (Fuge, 1996). Although the known goitrogenic compounds may be involved in some instances, they do not explain completely the occurrence of goitre in areas where soil I is relatively high. Possible contributory factors include the presence of excessive amounts of Co in the soil and a deficiency of Se, which impairs the metabolism of the thyroid hormones (Underwood and Suttle, 1999). However, in many of the areas where goitre was prevalent a few decades ago, its incidence is now much less. This is mainly because many ruminant animals, particularly dairy cattle, receive I from other sources, such as concentrate feeds, salt-licks and blocks and iodophor detergents and sterilants. Where deficiency does occur in cattle, it is most likely in calves born in autumn, when their mothers have been grazed during spring and summer with a minimum of supplementary feed. Sheep generally receive much less I from incidental sources than do cattle, and their need for deliberate supplementation may therefore be greater. Possible ways of providing supplementary I include the addition to the diet of slow-release forms, which persist in the rumen, and the intramuscular injection of iodized poppy-seed oil (Australian Agricultural Council, 1990). The application of KI to a sward scheduled for grazing may also be effective for one season (Smith *et al.*, 1999).

Iodine toxicity can occur in animals that are fed diets with excessively high concentrations, or that receive prolonged treatment with organic I compounds in order to overcome problems such as foot-rot. The maximum dietary concentration that can be tolerated by cattle and sheep is about 50 mg I kg^{-1}, and higher concentrations consumed for a period of weeks are liable to result in symptoms of toxicity, such as excessive salivation, nasal discharge and coughing (Miller *et al.*, 1991).

Excretion of Cl and I by Ruminant Animals

The excretion of Cl usually occurs to the extent of 95–98% in the urine (Fontenot and Church, 1979), with only a small amount in the faeces (Ammermann and Goodrich, 1983). However, excretion in the urine is a major homeostatic mechanism for Cl and therefore, when the animal needs to conserve Cl, there may be little excretion (Agricultural Research Council, 1980). Sometimes, there is substantial excretion in the faeces, suggesting that, under some conditions, Cl may be secreted into the large intestine or colon (Coppock, 1986). Perspiration is another route for the excretion of Cl: the loss from a 500 kg cow may be 0.4 g Cl day^{-1} in temperate conditions and increase to 1.6 g Cl day^{-1} in tropical conditions (Agricultural Research Council, 1980). Substantial quantities of Cl are secreted in milk (see Table 4.4), with rather more in cows' milk than in ewes' milk (Australian Agricultural Council, 1990).

Iodine, like Cl, is usually excreted mainly in the urine (Hetzel and Maberley, 1986), with a smaller amount in faeces and possibly a little in perspiration (Miller *et al.*, 1975). Excretion in the faeces is limited by reabsorption. It has been estimated that, at a normal I intake for dairy cows, about 40% is excreted in urine and about 30% in faeces, and about 8% is secreted in milk (Miller *et al.*, 1975). The I in urine is present mainly as iodide (Underwood and Suttle, 1999). On the basis of limited information, it appears that the faeces of ruminant animals may contain up to 10 mg I kg^{-1} DM and that urine may contain up to 4 mg I l^{-1} (Whitehead, 1973a; 1984).

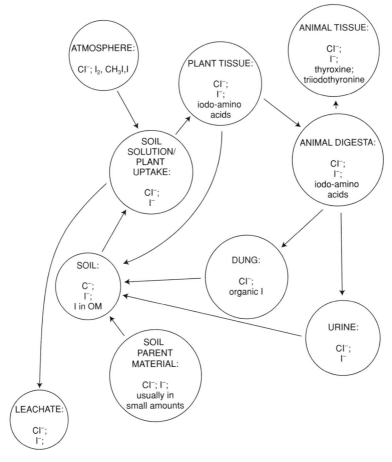

Fig. 10.2. Outline of the main forms of Cl and I involved in the cycling of these elements in grassland.

Transformations of Cl and I in Grassland Systems

A large proportion of the Cl moving from the soil to the plant to the animal and back to the soil remains in the form of chloride and undergoes no chemical transformation during this sequence (Fig. 10.2). A small proportion of the Cl in soil may be temporarily adsorbed or incorporated into the microbial biomass, and a small proportion of the Cl in plants and animals may be incorporated into organic molecules, but most remains in the form of chloride.

Transformations of I are much more complex. Several reactions lead to the retention of I by both inorganic and organic constituents of soils and, although iodide taken up by plants undergoes little change in the plant tissue, once consumed by animals it is largely incorporated into organic forms.

Quantitative Balances of Cl and I in Grassland Systems

Despite both elements being halogens, there are large differences in the amounts of Cl and I involved in grassland systems (Table 10.8). In comparison with other nutrient elements, there are few data from which quantitative balances for Cl and I can be prepared, and the values (particularly for the 'outputs') in Table 10.8 are therefore rather uncertain.

Table 10.8. Estimated balances of Cl and I (g ha^{-1} year^{-1}) typical of two systems of grassland management, intensively managed grass and extensively managed grass–clover, both grazed by cattle (except during the winter) in UK conditions.

	Intensive grass (dairy cows)		Extensive grass–clover (beef cattle)	
	Cl	I	Cl	I
Inputs				
Fertilizer	60,000	0	0	0
Atmosphere	40,000	25	40,000	25
Aspects of recycling				
Uptake into herbage	80,000	5.0	6,000	0.4
Consumption of herbage by animals	64,000	4.0	3,000	0.2
Dead stubble + herbage to soil	32,000	2.0	9,000	0.45
Dead roots to soil	11,000	3.0	3,000	0.68
Excreta to soil of grazed area	44,000	2.7	2,700	0.17
Outputs				
Milk/live-weight gain	10,000	0.8	300	0.03
Leaching/runoff/volatilization	80,000	2.0	39,400	1.0
Loss through excreta off sward	10,000	0.5	0	0
Gain/loss to soil	0	22	0	24

Chapter 11

Boron, Molybdenum and Selenium

Sources and Concentrations of B, Mo and Se in Soils

The B in igneous rocks occurs partly in the mineral tourmaline ($Na(Fe,Mg)_3$ $Al_6(OH)_4(BO_3)_3Si_6O_8$), which contains 3–4% B, and partly as a substituent for Si in various silicate minerals. Although tourmaline is a resistant mineral and weathers slowly, there is a slow release of B from this and other minerals, mainly as boric acid, H_3BO_3 (Gupta, 1979). At pH > 7, the boric acid tends increasingly to dissociate to the borate ion, $H_2BO_3^-$, which is less susceptible than boric acid to leaching. The borate ion is retained by hydrous Fe and Al oxides and, to a lesser extent, by clay minerals, with maximum adsorption occurring in the pH range 7–9 (Harter, 1991). Some of the adsorbed borate is easily exchanged, but that adsorbed by the Fe and Al oxides is bound rather strongly and tends to become less exchangeable with time. The total B content of soils depends largely on the nature of the soil parent material and on whether or not weather conditions induce leaching. In humid regions, concentrations are generally in the range 2–100 mg kg^{-1} (Table 11.1; Gupta, 1979), with sandy soils often containing 2–6 mg kg^{-1} and clay soils 30–60 mg kg^{-1} (Adriano, 1986). In the UK, half of almost 400 soils had concentrations of B between 20 and 40 mg kg^{-1} (Archer and Hodgson, 1987). The highest

Table 11.1. Reported concentrations of B, Mo and Se in soils (mg kg^{-1} dry soil).

	B	Mo	Se	Reference
Temperate (range)	1–450	0.1–18	0.01–4.0	Kabata-Pendais and Pendais, 1992
Temperate (usual range)	2–100	0.2–5.0	0.1–2.0	Berrow and Burridge, 1980
UK (> 200 soils)	33 (7–119)	0.8 (0.03–13.0)	0.5 (0.02–2.0)	Archer and Hodgson, 1987
USA	33 (< 20–300)	1.2 (0.1–40.0)	0.4 (< 0.1–6.0)	Adriano, 1986

concentrations of B occur in the saline soils of arid regions, where concentrations may be up to 200 mg kg^{-1}, and occasionally more.

The Mo in igneous rocks occurs mainly in the ferromagnesian minerals, such as olivine. However, the only important mineral with a high concentration of Mo is molybdenite (MoS_2), which occurs, in small amounts, in some granites (Chesworth, 1991). Concentrations of Mo in rocks are usually less than 10 mg kg^{-1} (see Table 2.2, p. 20), though, exceptionally, they may be as high as 3000 mg kg^{-1} (Mills and Davis, 1987). Weathering releases the anion molybdate, which is partially retained in sedimentary rocks by the hydrous oxides of Fe, Mn and Al, and also by OM and by carbonates. The retention of Mo by OM results in relatively high concentrations in shales. In general, shales and granites have higher concentrations of Mo than do basalts, limestones and sandstones (see Table 2.2; Gupta and Lipsett, 1981). Concentrations of Mo in soils are normally in the range 0.2–5 mg kg^{-1} (Table 11.1; Stevenson, 1986), though soils formed from shale are often relatively rich in Mo and occasionally contain as much as 100 mg Mo kg^{-1} (Thornton, 1977). There is little information on the distribution of Mo in soil profiles, but often Mo tends to be associated with OM (Adriano, 1986).

The Se in igneous rocks occurs mainly as metal selenides, which are usually associated with sulphides (Shamberger, 1983). Weathering results in the release of selenite, much of which is retained by hydrous Fe oxides and OM. Due to retention, sedimentary rocks sometimes have a higher concentration of Se than their precursor igneous rocks, and this is particularly so for shales (see Table 2.2; Gupta and Watkinson, 1985; Adriano, 1986). Concentrations of Se are generally higher in shales than in limestones and sandstones, and the Se is associated partly with ferrous sulphide and partly with OM (Fleming, 1988). The concentration of Se in soils is usually between 0.1 and 2 mg kg^{-1}, with an average of 0.2–0.3 mg kg^{-1} (Table 11.1; Fergusson, 1990; Fishbein, 1991). However, in some seleniferous soils, which are often derived from marine black shales, concentrations may be much higher, up to 30 mg kg^{-1} (Fleming, 1988; Mayland et al., 1989). High concentrations of Se may also occur in arid, alkaline soils.

Agricultural and Atmospheric Inputs of B, Mo and Se to Soils

Small amounts of each of these three elements may be added to soils as incidental constituents of fertilizers, but the amounts are usually negligible (Table 11.2). However, some phosphate rocks are enriched in Se, and the Se tends to persist in fertilizers derived from them (Adriano, 1986).

Concentrations of B, Mo and Se in sewage sludges are often rather higher than those in soils (as shown by comparing Table 2.8, p. 32, with Table 11.1). The concentration of B in sewage sludge is often enhanced by the widespread

Table 11.2. Reported concentrations of B, Mo and Se in fertilizers (mg kg⁻¹).

	B	Mo	Se	Reference
Calcium ammonium nitrate	9	56	–	Adriano, 1986
Superphosphate	132	335	–	
KCl	16	26	–	
Superphosphate	–	–	4.2–8.0	Gissel-Nielsen, 1971
Rock phosphate	–	–	0.03–25.0	
KCl	–	–	0.04–0.09	

use of B in detergents and bleaches (Mikkelsen and Camberato, 1995), while the high concentrations of Mo are due mainly to contamination from industry (Adriano, 1986). Unusually high concentrations of Mo (117–168 mg kg⁻¹) were reported in three Dutch sludges (Hemkes *et al.*, 1980). Concentrations of Se in sewage sludge are also sometimes higher than the range shown in Table 2.8, and may reach 15 mg kg⁻¹ as a result of contamination from industry (Fergusson, 1990; Haygarth, 1994). Where soils have been directly contaminated in the past by industrial and mine wastes, they sometimes contain relatively high concentrations of Mo (Mills and Davis, 1987).

Small amounts of B, Mo and Se are deposited from the atmosphere (Table 11.3), though there are few quantitative measurements. The presence of these elements in the atmosphere is due largely to the combustion of coal and oil, but additional small amounts of Se are transferred by the natural processes of volcanic activity and volatilization from soils and the oceans (Adriano, 1986). Coal often contains 1–20 mg Se kg⁻¹, and most of this enters the atmosphere during combustion (Mayland *et al.*, 1989). The importance of this source is confirmed by the observation that deposition of Se from the atmosphere is higher in industrial areas and is higher during the winter than during the summer (Haygarth, 1994). Globally, the wet deposition of Se is thought to amount, on average, to about 1.5 g ha⁻¹ year⁻¹ (Mayland, 1994).

Forms and Availability of B, Mo and Se in Soils

All three elements occur in both inorganic and organic forms in the soil, and all are subject to adsorption by the hydrous oxides of Fe and Al.

A large proportion of the B in some soils is present in the form of tourmaline, which is resistant to weathering. Some B also occurs in clay minerals, as a substituent for Al^{3+} and Si^{4+} (Mayland, 1994), and some is adsorbed by hydrous oxides of Al and Fe and by clay minerals, especially illites. Adsorption increases with increasing pH, particularly above pH 6.5 (Keren and Bingham, 1985; Harter, 1991). Some of the B in soils is associated with OM, partly by adsorption (Keren and Bingham, 1985) and partly through

Table 11.3. Amounts of B, Mo and Se deposited from the atmosphere (g ha^{-1} year^{-1}).

	B	Mo	Se	Reference
UK, seven sites	47–260	< 3–< 10	2.2–6.5	Cawse, 1980
UK, six sites	100–320	1–36	–	Wadsworth and Webber, 1980
Poland, field plots	71	–	–	Kabata-Pendais and Pendais, 1992
Czechoslovakia, two forests	1460; 4000	70; 100	–	Kabata-Pendais and Pendais, 1992

the reaction of boric acid with α-hydroxy aliphatic acids and with aromatic compounds containing o-dihydroxy groups (Stevenson, 1986). When a large proportion of the soil B is organically bound, the concentration of B tends to be highest in the surface horizons (Adriano, 1986). Although B is retained rather tightly by the OM, it is released slowly during decomposition, and OM acts as a major source of plant-available B in many soils. Small amounts of B, normally < 0.5% of the total, are present in the soil solution. In acid and neutral soils, the soluble B is present mainly as boric acid (H_3BO_3), but, at pH > 7, an increasing proportion is present as the borate anion (Moraghan and Mascagni, 1991). Boron is the only nutrient element that is present in the soil solution of most soils mainly as a non-ionized molecule. In humid regions, deficiencies of B are most likely to occur on light sandy soils, due to loss by leaching, but deficiencies may also occur on calcareous soils, due to poor availability. Availability is reduced when the soil pH is > 7, due to an increase in adsorption and a tendency for B to precipitate as the relatively insoluble Ca borate (McBride, 1994). The availability of B also tends to be low when conditions are dry, possibly due to microbial activity being curtailed and a slower release of B from plant residues (Moraghan and Mascagni, 1991). In contrast to the situation in humid regions, the soils of arid areas often contain high concentrations of borate, sometimes so high that they have toxic effects on plants (Keren and Bingham, 1985; McBride, 1994).

The Mo in soils is present partly in primary and secondary minerals, partly as molybdate (MoO_4^{2-}) adsorbed by hydrous Fe and Al oxides and clay minerals and partly as organically bound Mo (Harter, 1991). A small proportion occurs as molybdate in the soil solution. Molybdic acid, H_2MoO_4, is dissociated over the whole range of soil pH and, at pH > 4.3, most of the Mo in the soil solution is present as molybdate. In general, adsorption occurs most strongly under acid conditions (pH 3–5) and decreases with increasing pH (Harter, 1991; McBride, 1994). The amount of Mo in soils is usually low in sandy soils, particularly if they are strongly leached and/or calcareous. However, the availability of Mo is low, due to adsorption, in soils that are high in hydrous Fe oxides and which have a pH of < 6. Because the adsorption of Mo decreases with increasing pH, its availability, unlike that of other micronutrient elements (except Se), increases with an increase in soil pH (Moraghan and Mascagni, 1991). The concentration of molybdate in the soil solution often increases tenfold for each unit increase in soil pH (Tisdale *et al.*, 1985). The availability of Mo is also greater when drainage is poor (Berrow *et al.*, 1983), partly because chemically reducing conditions favour the release of Mo from minerals and partly because the complexes formed between molybdate and ferrous iron are more soluble than those formed with ferric iron (Allaway, 1986). Concentrations of Mo in herbage that are potentially toxic to grazing animals are, therefore, most likely to occur on soils that are high in Mo and also are poorly drained and have been limed.

Selenium occurs in soils in both inorganic and organic forms, and the total concentration is often greater in clay soils than in sandy soils. The inorganic Se may range in oxidation state from selenide (Se^{2-}) to selenate (SeO_4^{2-}), but, in most well-drained neutral and acid soils in humid regions, a large proportion occurs as selenite (SeO_3^{2-}), with a small proportion as selenate (Adriano, 1986). The selenite is often strongly adsorbed on to hydrous Fe oxides or clay minerals, or is precipitated as insoluble ferric selenite (Haygarth, 1994; McBride, 1994). As with Mo, adsorption decreases with increasing pH (Johnsson, 1991). In some soils, much of the Se present occurs in organic forms, partly as organic complexes and partly as Se-amino acids synthesized by microorganisms and plants (Gissel-Nielsen *et al.*, 1984; Mayland, 1994). The organic forms of Se are often dominant in the upper horizons of the soil, while Se associated with Fe oxides becomes increasingly important with depth (Fergusson, 1990). The concentration of Se in the soil solution is normally low ($< 1 \times 10^{-8}$ M), and depends on factors such as pH, redox conditions, the sorption capacity of the soil and microbial activity (Gissel-Nielsen *et al.*, 1984). It tends to be lowest when the soil pH is slightly acid to neutral (Gissel-Nielsen *et al.*, 1984). In soils with pH > 6.5, selenite is increasingly oxidized to selenate, which is only weakly adsorbed and is therefore more plant-available but also leached easily (Shamberger, 1983; Blaylock and James, 1994). The availability of Se is also increased by the presence of manganese oxides, which tend to oxidize selenite to selenate, and possibly by the secretion of exudates in the rhizosphere, which may promote this oxidation (Blaylock and James, 1994). On the other hand, the addition of OM was found to reduce the uptake of selenate by tall fescue, probably due to an enhanced reduction of selenate to selenite (Ajwa *et al.*, 1998). In arid regions, where leaching does not occur, selenate is often the dominant form in soils.

Losses of B, Mo and Se from Soils

Boron, Mo and Se can all be lost from soils by leaching, though all are subject to adsorption by soil constituents. The strength of adsorption varies with soil composition and pH, as discussed above, and these factors therefore influence the extent of leaching. With B, leaching occurs most readily under acid conditions, but, with Mo and Se, leaching increases with increasing pH (Adriano, 1986; Haygarth, 1994).

In addition to loss by leaching, Se can be lost from soils through the volatilization of compounds such as dimethylselenide, produced by the soil microorganisms. Many microorganisms, both bacteria and fungi, are able to methylate both selenite and selenate into the less toxic and volatile form, dimethylselenide. Small amounts of other volatile Se compounds, e.g. dimethyldiselenide and dimethylselenylsulphide, may also be produced (Frankenberger and Karlson, 1994). The quantitative importance of the volatilization of Se from soils is uncertain (Gissel-Nielsen et al., 1984; Lin et al., 1999), but it has been suggested that volatilization accounts for an appreciable proportion of the Se recycled in plant residues and animal excreta. Once in the atmosphere, dimethylselenide is probably subject to oxidation, with the products being sorbed on to aerosols, which may have an average residence time in the atmosphere of 7–9 days and are therefore liable to be transported over long distances (Frankenberger and Karlson, 1994).

Assessment of Plant-available B, Mo and Se in Soils

The most common procedure to extract the plant-available fraction of soil B is to boil with water or with dilute $CaCl_2$ (0.01–0.02 M): the latter has the advantage of providing a clear colourless extract (Sims and Johnson, 1991). The concentration of soluble B below which deficiency symptoms appear varies with plant species, and is lower for grasses than for clovers and lucerne (Keren and Bingham, 1985).

Although plant-available Mo is often assessed using acid ammonium oxalate solution, the concentrations of potentially soluble Mo in most soils are extremely low, and the effectiveness of the assessment is therefore limited by the sensitivity of the analysis (Gupta, 1993).

Extractants for plant-available Se include hot water, and solutions of ammonium bicarbonate-DTPA, calcium chloride, calcium nitrate and potassium sulphate (Adriano, 1986; Gupta, 1993). While these are useful in indicating soils where Se toxicity in animals may be a problem, they are less effective in identifying soils that are deficient in terms of the amounts of Se supplied via grassland herbage to grazing animals (Gupta, 1993).

Uptake of B, Mo and Se by Herbage Plants

The uptake of B by roots is thought to occur in the form of undissociated boric acid, by a process that is mainly passive and related to the uptake of water, though possibly with a small active component (Keren and Bingham, 1985; Kochian, 1991). Once absorbed into the root, some of the boric acid is bound by the cell walls, while some appears to diffuse into the symplasm and form esters with *cis*-diol groups. The formation of esters reduces the concentration of free boric acid, and allows more to diffuse into the symplasm (Kochian, 1991). Boron is thought to be transported to the leaves mainly as boric acid, but, once in the leaves, it becomes less mobile and shows relatively little movement in the phloem (Kochian, 1991).

Plants take up Mo as the molybdate anion from the soil solution, in which the concentration is typically in the range 2–8 µg Mo l^{-1}. There appears to be competition from sulphate at high concentrations (Mengel and Kirkby, 1987), but not from phosphate (Tisdale *et al.*, 1985). Molybdenum appears to be mobile in the plant, possibly moving as an organic complex with amino acids or sugars (Mengel and Kirkby, 1987; Romheld and Marschner, 1991).

The uptake of Se by plants occurs more readily in the form of selenate than as selenite (Mikkelsen *et al.*, 1989; Marschner, 1995), partly because selenite is strongly adsorbed by soil constituents. When soils are well supplied with Se, there is a close correlation between uptake and the water-soluble fraction, but the correlation is poor when soils are low in Se (Gissel-Nielsen *et al.*, 1984). The uptake of selenate into the root appears to require energy and to follow the same pathway as sulphate, which, at high concentrations, can reduce the uptake of selenate (Mikkelsen *et al.*, 1989). In addition to taking up Se through their roots, plants may also absorb gaseous forms, mainly dimethylselenide, from the atmosphere (Fishbein, 1991; Haygarth *et al.*, 1995).

Functions of B, Mo and Se in Herbage Plants

The functions of B in plants appear to be related to its tendency to form stable complexes with organic compounds that contain pairs of *cis* hydroxyl groups, compounds that include sugars, sugar alcohols and some *o*-diphenols. These reactions are probably involved in lignification and xylem development and in the transport of sugars (Marschner, 1995). Boron may also be essential for the development of pollen (Mengel and Kirkby, 1987; Mikkelsen and Camberato, 1995). However, its precise roles are not well understood, and it appears to be the only plant nutrient element that is not involved in the functioning of any enzyme system (Mikkelsen and Camberato, 1995). In general, dicots require about four times as much B as do monocots, and legumes appear to have an exceptionally large requirement (Adriano, 1986).

The functions of Mo in plants are related to electron transfer, and Mo is a constituent of several plant enzymes involved in oxidation and reduction. An important example is nitrate reductase. As a result of this role, less Mo is required by plants utilizing ammonium N than by plants utilizing nitrate N (Romheld and Marschner, 1991), and the symptoms of Mo deficiency resemble those of N deficiency (Mengel and Kirkby, 1987). Molybdenum is also a constituent of nitrogenase, and is therefore essential for N_2 fixation by the rhizobia in the root nodules of legumes (Mengel and Kirkby, 1987). When legumes are growing on soils deficient in Mo, the application of Mo often results in increased growth and a higher concentration of N in the herbage (Moraghan and Mascagni, 1991). Legumes that are not dependent on N_2 fixation still have a need for Mo as a constituent of nitrate reductase, but the requirement is less than for nitrogenase. In nitrogenase, which contains two Mo atoms per molecule, the Mo is involved in the transfer to N_2 of electrons supplied from Fe^{3+} via an Fe-S-protein (Romheld and Marschner, 1991).

Although Se is apparently not essential for plants, it is incorporated, at least partially, into organic compounds. When selenate is taken up, it is first reduced to selenite, which then forms various Se-amino acids, which are analogous to S-amino acids and are often incorporated into proteins (Mikkelsen et al., 1989). However, a small proportion of the Se absorbed by plants may be volatilized, mainly as dimethylselenide (Mayland et al., 1989), though the quantitative importance of this process is uncertain.

Distribution of B, Mo and Se within Herbage Plants

There is relatively little information on the distribution of B, Mo and Se in herbage plants, though some data on their distribution in various fractions of the shoot material of cocksfoot is shown in Table 3.3 (p. 53). In relation to the distribution between roots and shoots, limited evidence suggests that concentrations in roots and shoots are similar for B (Wilkinson and Gross, 1967; Banuelos et al., 1993), are higher in roots than shoots for Mo (John et al., 1972) and are higher in shoots than roots for Se (Wu and Huang, 1992). However, in tall fescue plants grown with a plentiful supply of Se, the concentrations in shoots and roots were similar, at about 0.95 mg kg^{-1} (Banuelos et al., 1993). In a later study, also with tall fescue, concentrations of Se were greater in the roots than in the shoots when the shoot concentrations were less than about 3 mg Se kg^{-1} (Ajwa et al., 1998). In ryegrass, there was little difference in B concentration between leaf, stem and inflorescence (Chiy and Phillips, 1999).

The concentration of B in white clover, grown in solution culture, was highest in the leaflets, while the roots, stolons and petioles were similar in concentration (Wilkinson and Gross, 1967), and, in field-grown white clover, the concentration of B was similar in leaf, stem and inflorescence (Chiy and Phillips, 1999). In lucerne also, the highest concentration of B was in the leaves, especially the oldest leaves, while the concentrations in stems and roots were similar to one another (Rominger et al., 1975). Boron is immobile within the plant, and there is no appreciable remobilization to young leaves (Mikkelsen and Camberato, 1995).

The concentration of Mo in three fractions of grasses decreased in the order: leaf > flowering head > stem (Fleming and Murphy, 1968). However, in red clover, the concentration of Mo was higher in the stem than in the leaf (Fleming, 1963; Gupta and Lipsett, 1981). Molybdenum is readily translocated within the plant (Romheld and Marschner, 1991).

Ranges of Concentration and Critical Concentrations of B, Mo and Se in Herbage Plants

Herbage concentrations of B, Mo and Se reported from various regions are shown in Table 11.4. For B, the usual range is about 4–10 mg kg^{-1}, though concentrations as high as 40 mg kg^{-1} have been reported. The usual range for Mo is about 0.1–4.0 mg kg^{-1}, and the concentration can vary widely in response to soil Mo content and pH, as well as plant species. Extremely high concentrations of more than 40 mg Mo kg^{-1} have been reported, especially in legumes, in some investigations (Mills and Davis, 1987). In a study of 12 nutrient elements in about 2000 samples of timothy in Finland, Mo showed a greater variation in concentration than did any of the other elements, from 0.02 to 16.6 mg kg^{-1} (Paasikallio, 1978).

  Concentrations of Se in herbage are usually between 0.05 and 1.0 mg kg^{-1} though, in areas associated with Se deficiency in animals, they may be as low as 0.02 mg Se kg^{-1} or even less (Davies and Watkinson, 1966; Levander, 1986; Fishbein, 1991). At the other extreme, in some areas of seleniferous soils, grasses may contain more than 50 mg Se kg^{-1}. In Ireland, ryegrass growing on a seleniferous soil was found to contain 11–21 mg kg^{-1}, with a steady increase in concentration as the grass matured over the period early

Table 11.4. Reported concentrations of B, Mo and Se in grassland herbage from different regions (mean and/or range values, mg kg^{-1}).

Region	Type of herbage	B	Mo	Se	Reference
USA	Grasses, normal range	5–15	0.1–2.0	0.01–0.5	Mayland and Wilkinson, 1996
Finland	Timothy	4.9 (1.4–34)	0.5 (0.02–16)	–	Paasikallio, 1978
Finland	Mixed grasses	–	0.14–0.80	0.007–0.028	Ettala and Kossila, 1979
Scotland	Grass, silage and hay	–	1.24	0.041	Scottish Agricultural Colleges, 1982
Europe	Grass	4–7	0.05–0.25	–	Kirchgessner *et al.*, 1960
Sweden	Red clover	14–40	–	–	Philipson, 1953
Manitoba	Grass/legume (1976)	–	0.4–20.0	–	Drysdale *et al.*, 1980
UK	Grass	–	–	0.013–0.230	Rickaby, 1981
New Zealand	Grass/clover	–	–	0.005–0.07	Grace *et al.*, 1995
Alberta	Grass/legume	–	–	0.002–2.0	Redshaw *et al.*, 1978
Denmark	Mixed pasture	–	–	0.005–0.280	Gissel-Nielsen, 1975

Table 11.5. Reported critical concentrations of B and Mo in the herbage of grasses and legumes (mg kg^{-1}).

	B	Mo	Reference
Grasses	< 3	< 0.1	Mayland and Wilkinson, 1996
Ryegrass	< 6	0.15	Marschner, 1995
Lucerne	10–30	0.5–0.9	Kelling and Matocha, 1990
White clover	c. 12	–	Sherrell, 1966
Subterranean clover	c. 10	–	Dear and Lipsett, 1987
Timothy	–	0.11	Gupta and Lipsett, 1981

June to late July (Lane and Fleming, 1967). And white clover grown on a seleniferous soil in pots contained > 100 mg Se kg^{-1} (Fleming, 1962).

The concept of critical concentration for plant growth applies only to B and Mo, and some values that have been proposed for these two elements are shown in Table 11.5. The critical concentration of B for grasses is low, probably < 3 mg kg^{-1} (Table 11.5; Sherrell, 1983). However, lucerne is particularly susceptible to B deficiency, and most assessments made for this species have been in the range from 10 to 30 mg B kg^{-1} (Table 11.5; Welch *et al.*, 1991).

For Mo, the critical concentration appears to be about 0.10–0.15 mg kg^{-1} for grasses and 0.5–0.9 mg kg^{-1} for legumes (Table 11.5).

Influence of Plant Species and Variety on Concentrations of B, Mo and Se in Herbage

Legumes generally have much higher concentrations of B than do grasses. Thus, in a comparison of herbage from West Virginia, USA, legumes contained, on average, 22 mg B kg^{-1} and grasses 7 mg kg^{-1} (Baker and Reid, 1977), and, in two comparisons involving grass–clover swards in New Zealand, average concentrations of B were 8 and 9 mg kg^{-1} in the grass and 33 and 36 mg kg^{-1} in the clover (McLaren *et al.*, 1990; Wheeler, 1998). Similarly, in a comparison involving several grass and legume species in the USA, concentrations of B were much higher in the legumes (Table 11.6).

Concentrations of Mo are sometimes higher in legumes than in grasses (Fleming, 1962, 1988; Jensen and Lesperance, 1971; Drysdale *et al.*, 1980; Kabata-Pendais and Pendais, 1992), and sometimes higher in grasses than in legumes (Beck, 1962; Fleming, 1963; Mills and Davis, 1987; Wheeler, 1998). The differences appear to be influenced by Mo supply. When there is a plentiful supply, clovers tend to have higher concentrations than grasses (Kabata-Pendais and Pendais, 1992).

With Se, the differences between grasses and legumes may also be influenced by supply. Several comparisons of grasses and clovers have shown no

Table 11.6. Concentrations of B in several grass and legume species (mg kg⁻¹), sampled at the flowering stage (from Smith *et al.*, 1974).

	B
Brome-grass	3
Timothy	3
Cocksfoot	5
Reed canarygrass	5
Red clover	29
Lucerne	26
Sainfoin	31

appreciable differences in Se concentration (Ehlig *et al.*, 1968; Gissel-Nielsen, 1975; Gupta and Winter, 1975; Kabata-Pendais and Pendais, 1992), especially with soils low in Se. However, when grown in culture solution supplemented with Se, the concentration in white clover was about three times that in tall fescue (Wu and Huang, 1992); and, when grown on a high-Se soil from Ireland, white and red clover contained three to five times as much Se as did cocksfoot and ryegrass (Fleming, 1965).

There are few data on differences in B concentration amongst individual species of the grasses and legumes. However, in a study in the USA, cocksfoot had a higher concentration than did timothy, brome-grass or reed canarygrass (Table 11.6). Although in this study there were only small differences between red clover, lucerne and sainfoin, another comparison of legume species showed lucerne and sainfoin to have higher concentrations of B than did white clover and red clover (Whitehead and Jones, 1969).

With Mo, there appear to be substantial species differences amongst both the grasses and the legumes. Amongst the grasses, the concentration of Mo was much higher in Yorkshire fog (0.46 mg kg⁻¹) and *Agrostis* (0.42 mg kg⁻¹) than in Italian ryegrass (0.17 mg kg⁻¹) (Fleming, 1973). In other comparisons, Mo concentration decreased in the order Yorkshire fog > meadow foxtail > cocksfoot (Kirchgessner *et al.*, 1960); and cocksfoot and timothy had higher concentrations of Mo than did perennial ryegrass and tall fescue (Patil and Jones, 1970). Amongst legume species, white clover had an average concentration of 0.64, red clover 0.44, lucerne 0.18 and sainfoin 0.18 mg kg⁻¹ (Whitehead and Jones, 1969). However, in another study, lucerne contained more Mo than did red clover (Gupta and Lipsett, 1981). Molybdenum has been reported to be particularly high in plants of the family *Compositae*, e.g. dandelion (Feely, 1990).

Comparisons of Se in grass species have shown *Agrostis tenuis* to contain more than cocksfoot and perennial ryegrass (Davies and Watkinson, 1966), and perennial ryegrass to contain rather more than Italian ryegrass and tall fescue (More and Coppenet, 1980). Amongst the legumes, lucerne often has a much higher concentration of Se than do other species (Ehlig *et al.*, 1968).

However, in one investigation, red clover contained more Se than lucerne (Heys and Hill, 1984). Differences amongst plant species in Se concentration are most apparent with a group of plants known as Se 'accumulators', which occur in seleniferous areas, especially in parts of the USA. These accumulator plants contain much higher concentrations of Se than do other plants growing in the vicinity (Anderson and Scarf, 1983; Welch et al., 1991), sometimes 100 times more than non-accumulator plants (Wu and Huang, 1992). Common 'accumulator' plants in the USA include species of Astragalus and Stanleya, and these can survive with concentrations of up to 20,000–30,000 mg Se kg^{-1}, whereas most non-accumulator plants suffer from toxicity at concentrations of more than about 50 mg kg^{-1} (Wu and Huang, 1992). More generally in seleniferous areas, the accumulator species may contain 50–500 mg Se kg^{-1}, while most grasses in the vicinity contain < 10 mg kg^{-1} (Welch et al., 1991). Tall fescue was relatively tolerant of high soil Se, and took up smaller amounts of Se than did four other grass species (Wu et al., 1988). Much of the Se in plants is incorporated into organic compounds, which are mainly derivatives of S-amino acids. In Se accumulator species, the compounds are mainly selenocystathione and Se-methylselenocysteine, while, in non-accumulator species, they are mainly selenomethionine and Se-methylselenomethionine (Adriano, 1986). All of these compounds, when consumed by grazing animals, are liable to cause Se toxicity (Levander, 1986). Although the non-accumulator species contain less Se than the accumulators, they vary in their uptake of Se and, when grown on high-Se soils, some may contain sufficient to cause toxic effects in animals (Hamilton and Beath, 1963).

Influence of Stage of Maturity on Concentrations of B, Mo and Se in Herbage

As B is not readily remobilized in plants, concentrations tend to be higher in old tissue than in young tissue, especially when the supply is limited. In white clover, for example, the concentration of B in the leaflets tended to increase with increasing maturity (Wilkinson and Gross, 1967). However, in four legume species grown with an adequate supply of B, the concentration in the whole shoot material showed little change with advancing maturity (Whitehead and Jones, 1969).

Molybdenum is generally a mobile element in plants (Mengel and Kirkby, 1987) and, although there was little change with maturity in a study with perennial ryegrass (Fleming and Murphy, 1968), a decline in Mo occurred in four legume species (Whitehead and Jones, 1969).

With Se, there was little change in concentration with increasing maturity in cocksfoot and timothy (Lessard et al., 1968), though an increase was reported in a study with ryegrass (Lane and Fleming, 1967). A factor that may possibly contribute to inconsistency in the change with maturity is volatilization. Both accumulator and non-accumulator plants may release Se in volatile form from intact plants, and also during the drying of the plant material. With lucerne, up to 3% of the total Se was lost during drying (Asher and Grundon, 1970).

Influence of Season of the Year and Weather Factors on Concentrations of B, Mo and Se in Herbage

Concentrations of B have been reported to be relatively low during the summer, both in ryegrass (Whitehead, 1966) and in lucerne (Stewart and Axley, 1956). The decline may be related to the soil moisture content, as the concentration of B in lucerne was found to decrease as soil moisture became limiting (Dible and Berger, 1952).

Seasonal effects on Mo concentration appear to be variable. There is sometimes a seasonal increase, with the concentration in autumn and early winter being higher than in spring or summer (Fleming, 1988), or higher in late summer than in early summer (Hemingway, 1962; Underwood and Suttle, 1999). However, there is sometimes little change in Mo concentration during the season (e.g. Fleming and Murphy, 1968). Some of the variation may be due to the effects of temperature and soil water status. Thus, increasing temperature has been reported to increase herbage Mo, with the concentration in subterranean clover being at least twice as high at 22°C as at 12°C (Reddy et al., 1981). Increasing soil water content also tends to increase the concentration of Mo in plants (Moraghan and Mascagni, 1991), though, in an investigation with subterranean clover, there was little effect (Reddy et al., 1981).

Herbage concentrations of Se tend to be lowest in summer. Thus, in Denmark, Se concentrations in the herbage of various grassland swards taken through the growing season were highest in March (0.20 mg kg^{-1}), declined to a minimum (about 0.02–0.03 mg kg^{-1}) in June, July and August, and increased somewhat in October–November (Gissel-Nielsen, 1975). In France, Se was also low in June, but it increased at the second cut in July and declined at the third cut in August (More and Coppenet, 1980). In Canada, there was little variation in the Se concentration of grasses and trefoil between June and September (Lessard et al., 1968).

Influence of Soil Type on Concentrations of B, Mo and Se in Herbage

The geographical distribution of B, Mo and Se in soils and plants has been studied most intensively in the USA. With B, there was no clear relationship between soil type and the occurrence of deficiency symptoms in plants (Welch et al., 1991). However, with Mo, the concentration in legumes tended to decrease from west to east, reflecting a decrease in the availability of soil Mo due to the change from calcareous to non-calcareous soils (Kubota, 1983b). Areas of Se deficiency and areas of Se toxicity both occur in the USA, with deficiency mainly in the north-east, north-west and south-east, and toxicity mainly on soils derived from shales in the midwest. Glaciation may have contributed to the incidence of deficiencies in the northern areas (Welch et al., 1991).

Geographical differences in the herbage concentrations of B, Mo and Se have also been studied in Finland. There, the concentration of B in timothy was related to the soluble B content of the soil and to soil pH: as pH increased, the herbage concentration of B decreased (Paasikallio, 1978). With both Mo and Se, differences in the concentrations in timothy between different soil types were generally small, but the lowest concentrations of both elements were associated with heavy clay soils and the highest concentrations with peat soils (Paasikallio, 1978).

Despite the small differences found in the Finnish study, some soil properties have large effects on the concentrations of Mo in herbage. Thus, concentrations of Mo tend to be increased by high soil pH, by impeded drainage and by the presence of large amounts of soil OM (Adriano, 1986). Liming often causes a large increase in the concentration of Mo in herbage, particularly when the soil Mo content is high and when the soil is poorly drained. The influence of soil pH on the concentration of Mo in grasses and clovers from a mixed sward is shown in Table 11.7. The effect of soil drainage was investigated in a study in which lucerne and birdsfoot trefoil were grown in large pots with a range of depths to the soil water-table: herbage concentrations of Mo increased as the depth to the soil water-table became shallower (Jensen and Lesperance, 1971), reflecting the increase in availability with anaerobic conditions. On a slightly acid soil high in Mo, variation in drainage conditions within a field resulted in a difference in the Mo concentration of mixed herbage, from 7 mg kg^{-1} on a freely drained area to 31 mg kg^{-1} on a poorly drained area (Reith, 1970). Excessively high concentrations of Mo in herbage, at least in the UK, occur mainly on soils derived from certain shales and slates – soils that contain more than 20 mg Mo kg^{-1} (Mitchell, 1963; Williams and Thornton, 1973; Leech and Thornton, 1987). When poorly drained, such soils may produce herbage with sufficient Mo to cause

Table 11.7. Influence of soil pH on the concentration of Mo (mg kg^{-1}) in grass and clover from a mixed sward (from Mills and Davis, 1987).

	pH					
	5.0	5.5	6.0	6.5	7.0	7.5
Grass	1.1	1.6	2.7	4.0	4.3	5.2
Clover	0.9	1.3	2.7	3.9	5.7	5.9

toxicity in cattle and sheep, toxicity that is due mainly to the Mo inhibiting the absorption of Cu (see p. 247).

With Se, areas of deficiency for animals are more widespread than areas of toxicity. Deficiencies, indicated by herbage containing < 0.05 mg Se kg^{-1}, are most likely to occur in cool humid regions (Gupta and Watkinson, 1985). Deficient areas are common in many parts of the USA, especially in the states adjoining the Great Lakes (Ammerman and Miller, 1975), and also in parts of Canada (Gupta and Watkinson, 1985). Deficiencies are also quite widespread in northern Europe, western and southern Australia, New Zealand and south-eastern China (Gissel-Nielsen *et al.*, 1984). Although soils deficient in Se are often sandy, the effects of soil texture on the supply of Se are ambiguous. Sandy soils generally have a low total concentration but a relatively high availability, whereas clay soils generally have a higher total content but low availability, due to strong adsorption (Gissel-Nielsen *et al.*, 1984; Gupta and Watkinson, 1985). Soil pH also has an appreciable effect, with availability being lowest in acid conditions. Toxicities of Se occur mainly in areas where the soils are derived from shales or from materials, such as loess, alluvium or glacial till, originating from shales. In contrast to Mo, the availability of Se is greater in well-drained than in poorly drained soils, and areas of toxicity are usually well drained and have a neutral to alkaline pH (Welch *et al.*, 1991).

Influence of Fertilizers and Lime on Concentrations of B, Mo and Se in Herbage

In general, the application of fertilizer N, P and K has little effect on herbage concentrations of B, Mo or Se, though a dilution effect is possible, particularly if the B, Mo or Se is in short supply. A dilution effect was probably responsible for reductions in herbage Mo following the application of ammonium nitrate (Hopkins *et al.*, 1994) and calcium ammonium nitrate (Reith *et al.*, 1984). Similarly, fertilizer N appeared to have a dilution effect in studies with Se (Gissel-Nielsen *et al.*, 1984). Examples of fertilizers having little effect include ammonium nitrate on B and Mo (Reid *et al.*, 1967b; Mudd, 1970b), superphosphate and KCl on Mo (Hemingway, 1962; Reid and Jung, 1965) and KCl on B (Smith, 1975).

In addition to a possible dilution effect, fertilizers may influence herbage concentrations of B, Mo or Se if they result in a change in soil pH or contain an ion that competes with the uptake of Mo or Se. In particular, fertilizers containing sulphate tend to reduce the herbage concentrations of both Mo and Se. For example, ammonium sulphate reduced both these elements in a pot experiment involving grass and clover (Williams and Thornton, 1972). A large reduction in herbage Mo also occurred following the application of ammonium sulphate at a rate of > 300 kg N ha^{-1} year^{-1} to long-term grassland on a peat soil (Feely, 1990). The influence of ammonium sulphate on herbage Mo was probably due partly to a pH effect and partly to the competitive effect of the sulphate ion (Gupta and Lipsett, 1981; Feely, 1990). Superphosphate was also found to reduce herbage Mo in permanent pasture (Reith *et al.*, 1984; Feely, 1990), apparently due to its sulphate content, since 'concentrated' superphosphate, not containing sulphate, increased the concentration of Mo in subterranean clover (Petrie and Jackson, 1982).

The effect of sulphate in reducing Se concentrations in herbage has been shown in field experiments in which ammonium sulphate was applied to grassland (Hemingway, 1962; More and Coppenet, 1980); and, in a pot experiment with grasses and clovers grown on a seleniferous soil, superphosphate reduced the herbage concentrations of Se to 20–30% of their original values (Fleming, 1962), an effect probably due to the sulphate component of the superphosphate (Fleming, 1965). Fertilization with sulphate, applied as gypsum, also significantly reduced Se in lucerne (Westermann and Robbins, 1974); and the addition of sodium sulphate reduced the concentration of Se in perennial ryegrass and subterranean clover (Pratley and McFarlane, 1974). Where the supply of Se is marginal, the application of sulphate, most commonly as superphosphate, is liable to induce Se deficiency in livestock. On the other hand, superphosphate had no effect on herbage Se in a study in France (Moré and Coppenet, 1980). Also phosphate added without sulphate increased Se in lucerne (Carter *et al.*, 1972). Sulphate may possibly have a competitive effect on the uptake of B, as the application of ammonium sulphate at 30 kg S ha^{-1} year^{-1} reduced the concentration of B in ryegrass from about 4.8 to 3.7 mg kg^{-1} (Chiy *et al.*, 1999), despite the tendency for B concentration to increase with increasing acidity.

Liming tends to reduce herbage concentrations of B and to increase concentrations of Mo and Se. With B, the effect is most apparent when the pH is raised above pH 6.5, as shown with tall fescue, in which there was little effect when the pH was raised from 4.3 to 6.3, but a substantial reduction in herbage B concentration when the pH was increased to 7.4 (Peterson and Newman, 1976). As noted above, soil pH has a major effect on herbage Mo, and a single application of lime to an acid soil may cause an increase to several times the original concentration (Reith and Mitchell, 1964; Reith, 1970; Burridge *et al.*, 1983). Although the availability of Se tends to increase with increasing pH, liming had no effect on the concentration of Se in timothy and red clover grown on a soil low in Se (Gupta and Winter, 1975).

Influence of the Addition of B, Mo and Se on Concentrations of these Elements in Herbage

Herbage concentrations of each of these three elements can be increased by the application of the element in an appropriate form. With B, the application of 0.5–2.5 kg B ha^{-1} as boric acid or borax ($Na_2B_4O_7.10H_2O$) is effective in greatly increasing the herbage concentration, especially in the first harvest after application (Dear and Lipsett, 1987; McLaren *et al.*, 1990). With Mo, effective sources include sodium molybdate and ammonium molybdate, and one or the other of these forms is sometimes incorporated into fertilizers, such as superphosphate (Adriano, 1986). The application of Mo as sodium molybdate at 0.4 kg ha^{-1} produced a large increase in herbage Mo, and the increase was greater in clover than in ryegrass – for example, from 1.9 to 6.9 mg kg^{-1} in ryegrass and from 1.4 to 11.9 mg kg^{-1} in clover (Reith *et al.*, 1984). In another study, sodium molybdate applied 1 year previously to a clover-based pasture was about half as effective as currently applied molybdate and after 2 years it was about 20% as effective (Barrow *et al.*, 1985).

Of the possible sources of Se, selenate is several times more effective than selenite in increasing herbage concentrations, at least in the first growth period after the application. In field experiments in the UK, involving the application of 70 and 140 g Se ha^{-1}, selenate produced much higher concentrations of Se in herbage than did selenite, but the effect of both forms on herbage Se declined rapidly, so that there was little effect by the end of the growing season (Archer, 1983). A similar decline was found in Ireland, with selenite applied at a lower rate of 25 g Se ha^{-1}, mixed with calcium ammonium nitrate (Culleton *et al.*, 1997). In the drier climate of Australia, sodium selenate, applied at a rate of 10 g Se ha^{-1} in either superphosphate or prills, is often effective in increasing herbage Se (Australian Agricultural Council, 1990). However, selenate is susceptible to leaching, a factor that explains the marked decline in its effect in some investigations, while selenite is subject to a progressive reduction in availability, due to adsorption. These disadvantages with both selenate and selenite illustrate the need for a slow-release form of Se and, in this context, selenate immobilized on pumice granules has been suggested as a possibility (Coutts and Atkinson, 1988). Another possible means of overcoming Se deficiency in grazing animals is through the direct application of selenite to herbage, thus avoiding immobilization in the soil, but this also gives variable results and the recovery of Se in the herbage is sometimes less than 10% (Haygarth, 1994).

Influence of Livestock Excreta on Concentrations of B, Mo and Se in Herbage

There appears to be no information of the influence of livestock excreta on herbage concentrations of B and Mo, but concentrations of Se have been reported to be up to 40% lower in urine patches than in the other parts of the sward, probably due mainly to a dilution effect (Joblin and Pritchard, 1983; Saunders, 1984). Excreted Se has a low availability for plant uptake (Maas, 1998), as that in urine is subject to adsorption in the soil, while that in faeces is mainly in insoluble forms, such as elemental Se and metal selenides (see p. 297). As a consequence, the supplementation of grazing animals with Se appears to have no effect on the concentration of Se in the subsequent growth of herbage (Maas, 1998).

Chemical Forms and Availability of B, Mo and Se in Herbage

The B in herbage plants is thought to occur almost entirely bound to *cis* hydroxyl groups of organic compounds, mainly components of the cell walls, but including soluble sugars (see p. 281). Its availability to animals appears not to have been studied.

The Mo in herbage occurs partly as water-soluble molybdates and partly as the insoluble oxide, MoO_3, and sulphide, MoS_2. The sulphide appears to be poorly absorbed by animals (Miller *et al.*, 1991).

The Se taken up by non-accumulator plants such as grasses and legumes is converted mainly into Se-amino acids, such as selenomethionine and selenocysteine, and these are incorporated, to a large extent, into proteins (Anderson and Scarf, 1983; Gissel-Nielsen *et al.*, 1984). The Se is readily available to animals (Mayland, 1994).

Occurrence and Metabolic Functions of B, Mo and Se in Ruminant Animals

Boron occurs in all animal tissues, usually at a concentration in the range between 0.05 and 0.6 mg kg^{-1} fresh weight, but several times greater in bones. Although B has not been shown to be essential for animals, it may possibly have a function in regulating the action of parahormone, and therefore in the metabolism of Ca, P and Mg (Nielsen, 1986). The concentration of B in cows' milk is usually within the range 0.3–1.0 mg B l^{-1} (see Table 4.4, p. 78; Nielsen, 1986; Anderson, 1992). It responds rapidly to changes in the dietary intake of B, but there is little variation due to breed or stage of lactation (Nielsen, 1986).

The Mo in animal tissues acts as a constituent of a number of enzymes involved in the metabolism of purines, pyrimidines, pteridines and aldehydes and in the oxidation of sulphite (Mills and Davis, 1987). During the biochemical reactions brought about by these enzymes, the Mo is involved in the transfer of electrons, and is subject to changes of oxidation state within the range Mo^{4+}–Mo^{6+}. The overall concentration in body tissue appears to show a wide variation within the range 0.03–4.0 mg kg^{-1} (see Table 4.1, p. 73; Mills and Davis, 1987), though it is often about 0.1–0.2 mg kg^{-1} (Pais and Benton Jones, 1997). Relatively high concentrations of Mo occur in the liver and kidneys (Miller *et al.*, 1991), though more than 50% of the Mo in the body may be present in the skeleton and there are substantial amounts in skin and wool (Mills and Davis, 1987). Concentrations of Mo in both blood

(Miller *et al.*, 1991) and milk (National Research Council, Subcommittee, 1989) reflect dietary intake. Cows' milk typically contains 0.02–0.05 mg Mo l^{-1} (see Table 4.4), though sometimes > 0.1 mg l^{-1} (National Research Council, Subcommittee, 1989). Almost all the Mo in cows' milk is associated with the enzyme xanthine oxidase (Mills and Davis, 1987).

When the concentration of Mo in grazed herbage is high, the Mo may have a toxic effect on ruminant animals by interfering with their metabolism of Cu (see p. 247). The disorder of cattle known as 'teart' is an example of this toxic effect, and occurs when herbage contains between about 20 and 100 mg Mo kg^{-1}. Such concentrations are restricted to areas where there is a combination of high soil Mo and a soil pH of > 7. The soils are often developed from black shales or granite alluviums and are poorly drained.

Selenium occurs in the animal body at an average concentration of 0.2–0.5 mg kg^{-1} live weight (see Table 4.1). It is widely distributed in the body, but concentrations are highest in the liver and kidneys (Ullrey, 1987; Australian Agricultural Council, 1990). Some Se is incorporated into proteins as Se-amino acids, such as selenocystine and selenomethionine (Levander, 1986). One Se-containing protein, the enzyme glutathione peroxidase, has a specific function as part of an enzyme system that chemically reduces H_2O_2 and other peroxides, and thus protects cells from damage by oxidation (Miller *et al.*, 1991; Maas, 1998). Glutathione peroxidase contains four Se atoms per molecule, with selenocysteine as an essential component (Miller *et al.*, 1991). Selenium is also a constituent of deiodinase enzymes and is therefore required for the metabolism of the thyroid hormones: a deficiency of Se accentuates the effects of I deficiency (Lee *et al.*, 1999). The concentration of Se in the blood is related to dietary intake over a wide range (Miller *et al.*, 1991), and concentrations in the plasma of > 0.08–0.12 mg l^{-1} indicate that the supply is adequate (Mayland, 1994). The measurement of glutathione peroxidase in the blood can also be used to assess Se status (Sinclair *et al.*, 1993). The concentration of Se in milk also varies with Se intake (Levander, 1986), though, with diets that are high in Se, the increase in milk concentration is proportionately less. In a study of milk from eight different areas of New Zealand, the mean concentration from the individual areas ranged from 2.9 to 9.7 µg Se l^{-1} (Millar *et al.*, 1973), but much higher concentrations (160–1270 µg ml^{-1}) have been reported from seleniferous areas in the USA (Rosenfeld and Beath, 1964). Normally, with concentrations in milk of up to about 10 µg Se l^{-1}, 60–75% of the Se occurs in the protein fraction (Miller, 1979).

Absorption of Dietary B, Mo and Se by Ruminant Animals

As B has not been shown to be essential for animals, its absorption from the diet appears not to have been studied.

Most dietary Mo is readily absorbed, though the extent of absorption and retention are influenced by the amount of sulphate and, to a lesser extent, by other constituents in the diet (Mills and Davis, 1987; Miller et al., 1991).

Selenium is absorbed mainly from the duodenum (National Research Council, Subcommittee, 1989; Miller et al., 1991), with the Se-amino acids being absorbed more readily than selenite (Miller et al., 1991; Lee et al., 1999). The extent of absorption from grassland herbage is thought to be about 40–50% (Underwood and Suttle, 1999), though it is influenced by the microbial conversion of soluble forms of Se to insoluble inorganic forms in the rumen (Australian Agricultural Council, 1990). Relatively large amounts of Se are secreted into the small intestine, and most of this is reabsorbed subsequently and transported in the blood, mainly in association with protein. A high concentration of sulphate in the diet interferes with the absorption of selenite and may therefore increase the risk of Se deficiency (Goodrich and Garrett, 1986). On the other hand, ingested soil can supply Se to grazing animals (Healy et al., 1970), and soil ingestion was found to increase the concentrations of Se in the plasma and liver of lambs (Grace et al., 1996a).

Nutritional Requirements for Mo and Se in Ruminant Animals

The requirement for Mo is sufficiently low for it to be met by almost all diets. On the basis of the few data available, the minimum dietary concentration is probably between 0.025 and 0.35 mg kg^{-1} (Miller et al., 1991).

The concentration of Se in the diet that is necessary to prevent deficiency in ruminant animals is probably no more than 0.1 mg kg^{-1}, though the value is not well defined. During a temporary shortage, there may be some mobilization of Se from the liver (Lee et al., 1999). In Australia and New Zealand, the incidence of Se-responsive conditions has suggested that a dietary concentration of 0.03 mg kg^{-1} is adequate for sheep and a concentration of 0.05 mg kg^{-1} for cattle (Australian Agricultural Council, 1990). In the UK, the dietary requirement for Se has been estimated at 0.05 mg Se kg^{-1} (Agricultural Research Council, 1980). The actual value depends on the chemical form of Se, the previous Se status of the animal and the amounts of other constituents of the diet that may influence the need for Se. Some constituents, particularly vitamin E and, to a lesser extent, S-amino acids, have a sparing effect on Se, while others, particularly polyunsaturated fatty acids, exacerbate Se deficiency (National Research Council, Subcommittee, 1989; Miller et al., 1991). The requirement for Se is also related to the physiological state of the animal. For example, in the USA, it is recommended that 0.01 mg kg^{-1} may be adequate for growing and finishing steers and heifers, while the diets of breeding bulls and pregnant and lactating cows should contain 0.05–0.10 mg kg^{-1} (National Research Council, 1984). Although animals require at least 0.03–0.1 mg Se kg^{-1} in their diets to prevent deficiency, they are liable to suffer from Se toxicity when the dietary concentration exceeds about 2 mg kg^{-1} (National Research Council, Subcommittee, 1989; Miller *et al.*, 1991).

Effects of Deficiencies and Toxicities of B, Mo and Se on Ruminant Animals

There appear to be no instances of either deficiency or toxicity of B in ruminant animals consuming normal diets.

Deficiencies of Mo are also unknown in animals on herbage diets (Miller *et al.*, 1991), though Mo toxicity is prevalent in some areas. The toxicity is due primarily to an induced deficiency of Cu (see p. 247), and results in diarrhoea, anorexia, lack of pigmentation in hair and wool and, eventually, premature death (Miller *et al.*, 1991). Other symptoms include stunted growth and deformed bones. The maximum concentration of Mo in the diet that can be tolerated by cattle without toxic effects is usually about 5–10 mg kg^{-1}, but a higher concentration may be acceptable if the concentrations of Cu and sulphate are also high (Miller *et al.*, 1991). In the UK, Mo toxicity is associated with the so-called 'teart' pastures, which typically have herbage concentrations of Mo in the range 20–100 mg kg^{-1} DM. In areas where the herbage concentration of Mo is potentially toxic to livestock, the risk can be reduced by the application of sulphate (Gupta and Lipsett, 1981).

With Se, areas of both deficiency and toxicity are widespread. Areas of deficiency include parts of the USA, Canada, New Zealand and northern Europe (Welch *et al.*, 1991). A UK survey of Se concentration in the blood of sheep during the 1970s indicated that a deficient or marginal supply of Se was rather widespread (Anderson *et al.*, 1979), but this situation has now been counteracted by the inclusion of Se in many dietary supplements. A major sign of Se deficiency, especially in young lambs and calves, is a myopathy, sometimes known as white muscle disease or nutritional muscular dystrophy, in which the animals collapse with a rapid heartbeat and respiration rate. The symptoms are typically associated with an excessive peroxidation of lipids, resulting in chalky white striations and degeneration and necrosis of the muscles (Miller *et al.*, 1991). When the heart muscle is seriously affected, the effects may be fatal. Deaths are most likely in young lambs up to the age of weaning and in young calves up to 3–4 months of age (Australian Agricultural Council, 1990), but symptoms may also occur in cattle up to 2 years of age. However, not all animals that have a low Se status develop myopathy of the muscles, and it is possible that symptoms develop only when vitamin E is also deficient and/or when there are other predisposing factors, such as unaccustomed exercise, inclement weather or an increased intake of polyunsaturated fatty acids (Price, 1989; MacPherson, 1994). In addition to causing myopathy, a deficiency of Se may reduce immunocompetence and cause unthriftiness in sheep and cattle. Such Se-deficiency unthriftiness is widespread in New Zealand and is characterized by a reduction in growth rate and sometimes by diarrhoea. It may reflect a reduction in voluntary feed intake and is most prevalent during the autumn and winter months (Minson, 1990; Maas, 1998). When lambs affected by unthriftiness are treated with Se, they may produce up to 30% more wool than those that are untreated (Shamberger, 1984). When Se is deficient, the supply to grazing animals can be augmented by treatment of the sward with selenate or selenite, or, for individual animals, by Se given orally (e.g. as an intraruminal Se bolus) or by annual injections of barium selenate (Culleton *et al.*, 1997; Grace *et al.*, 1997).

As with deficiency, Se toxicity occurs in limited areas in many countries. In the USA, these are mainly areas of well-drained and alkaline soils, derived from shales, which occur in parts of the western plains and Rocky Mountain region (Welch *et al.*, 1991). Selenium toxicity can be classified into three types: acute toxicity, chronic blind staggers and chronic alkali disease. Acute Se toxicity results from the consumption of large amounts of accumulator plants or from the accidental provision of excess supplementary Se. It is characterized by abnormal movement and posture, difficulty in breathing and rapid death. Chronic blind staggers results from the more prolonged consumption of smaller amounts of Se accumulator plants, and the symptoms include blindness and slight ataxia (Miller *et al.*, 1991). Chronic alkali disease results from the long-term consumption of diets containing 5–40 mg Se kg^{-1}, and produces symptoms such as loss of appetite, ill thrift, poor coat condition and reduced wool growth, lameness and infertility (Australian Agricultural Council, 1990).

Excretion of B, Mo and Se by Ruminant Animals

Boron is excreted partly in the faeces and partly in the urine, the proportion in the urine increasing with increasing dietary intake (Nielsen, 1986). The faeces of dairy cattle grazing a grass–clover sward contained an average of 15 mg B kg^{-1} DM (Kirchmann and Witter, 1992).

Molybdenum, like B, is excreted partly in the faeces and partly in the urine, but the proportion in the urine depends largely on the S content of the diet. On low-S diets, most of the excreted Mo is in the faeces, whereas, on high-S diets, an increasing proportion of the excreted Mo is in the urine (Barrow, 1987). Cattle slurry has been reported to have an average concentration of < 1 mg Mo kg^{-1} on a DM basis (Verloo and Willaert, 1990).

Selenium is excreted partly in the faeces and partly in the urine, but also partly in expired air, with the proportions being influenced by the Se intake and the nature of the diet. In ruminant animals, most Se is excreted in the faeces, which often contain more than twice as much as the urine (Shamberger, 1983; Langlands *et al.*, 1986; Levander, 1986; Mayland, 1994). However, excretion in the urine increases with increasing intake (Langlands *et al.*, 1986). The excretion of Se in expired air is important only when the intake of Se is excessive (Barrow, 1987) and, in this situation, dimethyl selenide is the main product (Mayland, 1994). Most of the Se in the faeces consists of unabsorbed Se from the diet, and results partly from the conversion, by microorganisms in the rumen, of organic forms of Se into insoluble and unavailable forms. These include elemental Se and metal selenides (Mayland, 1994; Maas, 1998). The faeces also contain some endogenous Se from saliva and bile. Much of the Se excreted in the urine appears to be in the form of the trimethyl selenium ion, $(CH_3)_3Se^+$ (Miller, 1979; Mayland, 1994). Cattle slurry was reported to have an average concentration of 0.2 mg Se kg^{-1} DM (Verloo and Willaert, 1990).

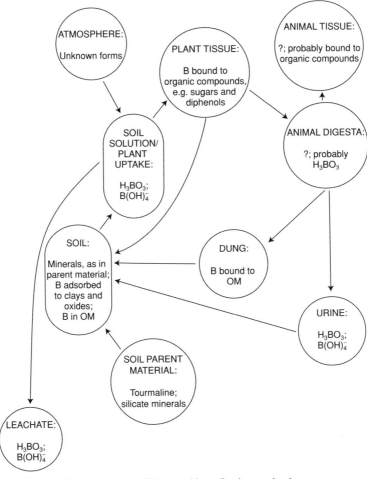

Fig. 11.1. Outline of the main forms of B involved in cycling in grassland.

Transformations of B, Mo and Se in Grassland Systems

The three elements differ in the type of compound in which they are present in soil parent materials, but, in soils, each occurs partly as an anion adsorbed to Fe and Al oxides and partly associated with OM, as well as partly as residual minerals (Figs 11.1, 11.2 and 11.3). In plant tissues, each is bound, at least partly, by organic constituents, B by carbohydrates and phenolic compounds, Mo and Se by proteins. As B is not an essential element or a particularly toxic element, little is known of its transformations within the animal. Both Mo and Se are associated with proteins, with Se forming analogues of the S-amino acids.

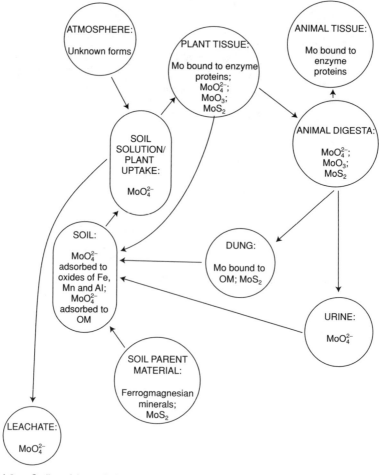

Fig. 11.2.   Outline of the main forms of Mo involved in cycling in grassland.

Quantitative Balances of B, Mo and Se in Grassland Systems

The estimated balances (Table 11.8) indicate that, in some situations, atmospheric inputs make a substantial contribution to the amounts of B, Mo and Se in circulation. The atmospheric input is particularly important for B, and it appears likely that most of the input accumulates in the soil.

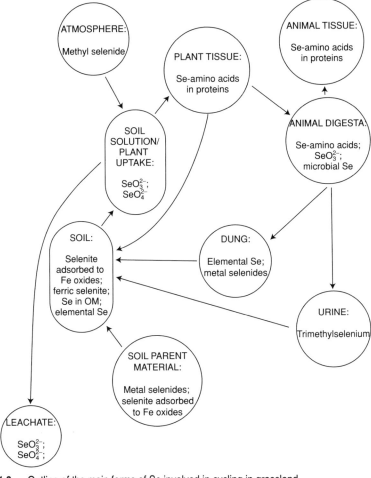

Fig. 11.3. Outline of the main forms of Se involved in cycling in grassland.

Table 11.8. Estimated balances of B, Mo and Se (g ha⁻¹ year⁻¹) typical of two systems of grassland management, intensively managed grass and extensively managed grass–clover, both grazed by cattle (except during the winter) in UK conditions.

	Intensive grass (dairy cows)			Extensive grass–clover (beef cattle)		
	B	Mo	Se	B	Mo	Se
Inputs						
Fertilizer	6	7	0.3	0	0	0
Atmosphere	150	2	3	150	2	3
Aspects of recycling						
Uptake into herbage	150	40	10	9	0.3	0.15
Consumption of herbage by animals	120	32	8	4.5	0.15	0.08
Dead stubble + herbage to soil	60	16	4	12	0.4	0.2
Dead roots to soil	60	20	3	12	0.5	0.15
Excreta to soil of grazed area	96	26	6.5	4.4	0.05	0.04
Outputs						
Milk/live-weight gain	7	0.7	0.2	0.06	0.1	0.04
Leaching/runoff	60	1	1	70	1	1
Loss through excreta off sward	17	5	1.2	0	0	0
Gain/loss to soil	82	2	0.6	80	1	2

Glossary

Apoenzyme: the portion of an enzyme (which contains a non-protein component) that is a protein.

Arbuscules: specialized fungal structures formed within the cortical cells of roots by many mycorrhizal fungi.

Cation exchange capacity: the potential of a soil (or other solid material) to adsorb cations, often expressed in milligram equivalents per 100 g of soil.

Chelate: a chemical complex in which two or more atoms of an organic compound (acting as a ligand) bind to a single metal ion, thus forming a ring structure.

Coenzyme: a molecule that acts with an enzyme in catalysing a biochemical reaction, usually by accepting or donating a particular chemical group.

Complex: a compound in which a metallic atom or ion is bound by the donation of a lone pair of electrons.

Concentrate feed: a feed that is high in energy and/or protein and low in fibre.

Critical concentration: the concentration of a nutrient element in a plant tissue that occurs when the supply of the nutrient is just adequate, or slightly less than adequate, for maximum growth.

Dry deposition: the transfer of material from the atmosphere to land (or sea) surfaces by the processes of gravitational settling, turbulent transport, molecular diffusion and impaction.

Endogenous: from within the body.

Erythrocyte: red blood cell.

Forage (species): plants of which the foliage is used for feeding to ruminant animals.

Goitre: enlargement of the thyroid gland (in the neck), often caused by a deficiency of I and/or the presence of goitrogens in the diet.

Homeostasis (in animals): (i) the state of equilibrium in the chemical composition and functions of the body and body fluids; and (ii) the processes through which such equilibria are maintained.

301

Hypomagnesaemia: a disorder in animals due to a low concentration of Mg in the blood plasma; it occurs particularly in recently calved cows, and less frequently in ewes.

Inositol: a cyclic sugar alcohol, $C_6H_6(OH)_6$.

Iodophor: a compound comprising iodine linked to a surface-active compound, which has both antimicrobial and detergent properties.

Isomorphous replacement: replacement of one atom by another of similar size in the lattice of a mineral, especially a clay mineral, without disrupting the lattice structure.

Kaolinite: a clay mineral with a 1 : 1 crystal lattice, i.e. single Si tetrahedral sheets alternating with single Al octahedral sheets.

Leaching: the removal of soluble constituents (usually from soil), by water moving downwards.

Ley: a sown grass or grass–clover sward intended to remain *in situ* for between 1 and 5 years, often alternating with a similar period under arable cultivation.

Ligand: a molecule, atom or ion that acts as the electron donor in one or more coordination bonds. If the ligand is organic and a heterocyclic ring is formed with a metal ion, the product is a chelate.

Macropore (in soil): a pore or crack of sufficiently large diameter for water to flow readily, often formed by alternate wetting and drying of the soil or through the decay of roots or the action of soil fauna.

Mass flow: the movement of nutrient ions in the flow of water in soils, often caused by the plant uptake of water.

Mineralization (in soils): the processes by which the organic forms of elements in plant, animal and microbial residues and in humified organic matter are converted into inorganic or 'mineral' forms.

Molar: the concentration resulting from 1 mole (see below) dissolved in 1 l of solution.

Mole: the amount of a substance that has a weight in grams numerically equal to the molecular weight of the substance.

Mycorrhiza: an association between fungal hyphae and plant roots, with the hyphae extending into the root cortex and also into the soil for a distance of several centimetres or more.

Occlusion: the adsorption, usually of a soluble ion, in a non-exchangeable form, e.g. in the interior of a mineral particle.

Oxidation: in a chemical reaction, the loss of electrons, often by transfer to an oxidizing agent, which is thus reduced.

Pasture: grassland grazed, on a long-term basis, by ruminant animals.

Prosthetic group (of enzyme): a non-amino acid group (e.g. haem) linked to the protein component of the enzyme.

Redox condition: the balance between chemical oxidation and reduction in a given situation.

Reduction: in a chemical reaction, the addition of electrons, often by the addition of hydrogen; reduction is the opposite of oxidation.

Residence time: the mean period for which an element or compound remains in a specified form.

Rhizosphere: the zone of the soil immediately surrounding a plant root, where there is usually a large microbiological population and where soil properties are modified by the presence of the root.

Ruminant (animals): herbivorous mammals, such as cattle and sheep, that have complex stomachs containing microorganisms that are able to break down the cellulose and hemicellulose in plant cell walls.

Shale: a sedimentary rock formed largely from compacted clay deposits, with variable amounts of organic material.

Siderophore (or phytosiderophore): a naturally occurring compound that will chelate with trivalent or divalent metal ions, especially Fe^{3+}, thus rendering the ion(s) more readily available for uptake.

Slurry: a mixture of dung and urine, with a variable amount of water, from housed livestock.

Smectite: a group of clay minerals with a 2 : 1 lattice structure (i.e. two Si tetrahedral sheets on each side of a single Al octahedral sheet) and with isomorphous replacement, producing a negative charge. Montmorillonite is an example.

Stocking rate (or density): the number of grazing animals per unit area of land.

Stolon: a stem growing horizontally at or just below the soil surface, capable of rooting and sending up new shoots at the nodes.

Symplasm (in plants): the continuum of cytoplasm that results from the cytoplasm in different cells being connected by plasmodesmata (see Fig. 3.2, p. 47).

Wet deposition: the transfer to the land (or sea) surface of gaseous and particulate matter from the atmosphere by rain or snow.

References

Aarts, H.F.M., Biewinga, E.E. and Van Keulen, H. (1992) Dairy farming systems based on efficient nutrient management. *Netherlands Journal of Agricultural Science* 40, 285–299.

Adams, C.A. and Sheard, R.W. (1966) Alterations in the nitrogen metabolism of *Medicago sativa* and *Dactylis glomerata* as influenced by potassium and sulfur nutrition. *Canadian Journal of Plant Science* 46, 671–680.

Adams, R.S. (1975) Variability in mineral and trace element content of dairy cattle feeds. *Journal of Dairy Science* 58, 1538–1548.

Adams, S.N. (1973) The response of pastures in Northern Ireland to N, P and K fertilizers and to animal slurries II. Effects on mineral composition. *Journal of Agricultural Science, Cambridge* 81, 419–428.

Adams, S.N. (1984) Some effects of lime, nitrogen, and soluble and insoluble phosphate on the yield and mineral composition of established grassland. *Journal of Agricultural Science, Cambridge* 102, 219–226.

Adams, S.N. and Honeysett, J.L. (1964) Some effects of soil waterlogging on the cobalt and copper status of pasture plants grown in pots. *Australian Journal of Agricultural Research* 15, 357–367.

Adriano, D.C. (1986) *Trace Elements in the Terrestrial Environment.* Springer-Verlag, New York.

Agricultural and Food Research Council (1991) Technical Committee on Responses to Nutrients, Report No. 6. A re-appraisal of the calcium and phosphorus requirements of sheep and cattle. *Nutrition Abstracts and Reviews* 61B, 573–612.

Agricultural and Food Research Council (1992) Technical Committee on Responses to Nutrients, Report No. 9. Nutritive requirements of ruminant animals: protein. *Nutrition Abstracts and Reviews* 62B, 787–835.

Agricultural Research Council (1980) *The Nutrient Requirements of Ruminant Livestock.* Commonwealth Agricultural Bureaux, Farnham Royal, 351 pp.

Aguilar, R. and Heil, R.D. (1988) Soil organic carbon, nitrogen and phosphorus quantities in northern Great Plains rangeland. *Soil Science Society of America Journal* 52, 1076–1081.

Ajwa, H.A., Banuelos, G.S. and Mayland, H.F. (1998) Selenium uptake by plants amended with inorganic and organic materials. *Journal of Environmental Quality* 27, 1218–1227.

Alderman, G. and Jones, D.I.H. (1967) The iodine content of pastures. *Journal of the Science of Food and Agriculture* 18, 197–199.

Alderman, G. and Stranks, M.H. (1967) The iodine content of bulk herd milk in summer in relation to estimated dietary iodine intake of cows. *Journal of the Science of Food and Agriculture* 18, 151–153.

Allaway, W.H. (1986) Soil–plant–animal and human interrelationships in trace element nutrition. In: Mertz, W. (ed.) *Trace Elements in Human and Animal Nutrition*, 5th edn, Vol. 2. Academic Press, Orlando, pp. 465–488.

Amiro, B.D. and Johnston, F.L. (1989) Volatilization of iodine from vegetation. *Atmospheric Environment* 23, 533–538.

Ammerman, C.B. and Goodrich, R.D. (1983) Advances in mineral nutrition in ruminants. *Journal of Animal Science* 57 (Suppl. 2), 519–533.

Ammerman, C.B. and Miller, S.M. (1975) Selenium in ruminant nutrition: a review. *Journal of Dairy Science* 58, 1561–1577.

Anderson, J.W. and Scarf, A.R. (1983) Selenium and plant metabolism. In: Robb, D.A. and Pierpoint, W.S. (eds) *Metals and Micronutrients: Uptake and Utilization by Plants*. Academic Press, London, pp. 241–275.

Anderson, P.H., Berrett, S. and Patterson, D.S.P. (1979) The biological selenium status of livestock in Britain as indicated by sheep erythrocyte glutathione peroxidase activity. *Veterinary Record* 104, 235–238.

Anderson, R.R. (1992) Comparison of trace elements in milk of four species. *Journal of Dairy Science* 75, 3050–3055.

Andrew, C.S. (1960) The effect of phosphorus, potassium and calcium on the growth, chemical composition and symptoms of deficiency of white clover in a sub-tropical environment. *Australian Journal of Agricultural Research* 11, 149–161.

Andrew, C.S. (1977) The effect of sulphur on the growth, sulphur and nitrogen concentrations, and critical sulphur concentrations of some tropical and temperate pasture legumes. *Australian Journal of Agricultural Research* 28, 807–820.

Andrew, C.S. and Hegarty, M.P. (1969) Comparative responses to manganese excess of eight tropical and four temperate pasture legume species. *Australian Journal of Agricultural Research* 20, 687–696.

Andrew, C.S. and Robins, M.F. (1969) The effect of phosphorus on the growth and chemical composition of some tropical pasture legumes. 1. Growth and critical percentages of phosphorus. *Australian Journal of Agricultural Research* 20, 665–674.

Andrews, E.D. (1966) Cobalt concentrations in some New Zealand fodder plants grown in Co-sufficient and Co-deficient soils. *New Zealand Journal of Agricultural Research* 9, 829–838.

Ap Griffith, G. and Walters, R.J.K. (1966) The sodium and potassium content of some grass genera, species and varieties. *Journal of Agricultural Science, Cambridge* 67, 81–89.

Ap Griffith, G., Jones, D.I.H. and Walters, R.J.K. (1965) Specific and varietal differences in sodium and potassium in grasses. *Journal of the Science of Food and Agriculture* 16, 94–98.

Archer, F.C. (1983) The uptake of applied selenium by grassland herbage. *Journal of the Science of Food and Agriculture* 34, 49.

Archer, F.C. and Hodgson, I.H. (1987) Total and extractable trace element contents of soils in England and Wales. *Journal of Soil Science* 38, 421–431.

Archer, J. (1988) *Crop Nutrition and Fertilizer Use*, 2nd edn. Farming Press, Ipswich.

Arnold, P.W. and Close, B.M. (1961) Release of non-exchangeable potassium from some British soils cropped in the glasshouse. *Journal of Agricultural Science, Cambridge* 57, 295–303.

Asher, C.J. (1991) Beneficial elements, functional nutrients and possible new essential elements. In: Mortvedt, J.J. (ed.) *Micronutrients in Agriculture*. Soil Science Society of America, Madison, pp. 703–723.

Asher, C.J. and Grundon, N.J. (1970) Volatile losses of mineral constituents from forage plants. In: *Proceedings of the 11th International Grassland Congress, Surfer's Paradise, Queensland*. University of Queensland Press, St Lucia, pp. 329–332.

Australian Agricultural Council, Ruminants Subcommittee (1990) *Feeding Standards for Australian Livestock: Ruminants*. CSIRO, Melbourne, 266 pp.

Bailey, J.S. and Laidlaw, A.S. (1999) The interactive effects of phosphorus, potassium, lime and molybdenum on the growth and morphology of white clover (*T. repens* L.) at establishment. *Grass and Forage Science* 54, 69–76.

Bailey, R.W. (1964) Pasture quality and ruminant nutrition. 1. *New Zealand Journal of Agricultural Research* 7, 496–507.

Baker, B.S. and Reid, R.L. (1977) *Mineral Concentration of Forage Species Grown in Central West Virginia on Various Soil Series*. Bulletin 657, West Virginia University Agricultural and Forestry Experiment Station, Morgantown, West Virginia.

Baker, R.D., Doyle, C.J. and Lidgate, H. (1991) Grass production. In: Thomas, C., Reeve, A. and Fisher, G.E.J. (eds) *Milk from Grass*, 2nd edn. ICI, Billingham, pp. 1–26.

Balasko, J.A. (1977) Effects of N, P, and K fertilization on yield and quality of tall fescue forage in winter. *Agronomy Journal* 69, 425–428.

Banuelos, G.S., Cardon, G.E., Phene, C.J., Wu, L., Akohone, S. and Zambrzuski, S. (1993) Soil boron and selenium removal by three plant species. *Plant and Soil* 148, 253–263.

Banwart, W.L. and Pierre, W.H. (1975) Cation–anion balance of field-grown crops. II. Effect of P and K fertilization and soil pH. *Agronomy Journal* 67, 20–25.

Barak, P. and Helmke, P.A. (1993) The chemistry of zinc. In: Robson, A.D. (ed.) *Zinc in Soils and Plants*. Kluwer Academic Publishers, Dordrecht, pp. 1–13.

Barber, S.A. (1980) Soil–plant interactions in the phosphorus nutrition of plants. In: Khasawneh, F.E., Sample, E.C. and Kamprath, E.J. (eds) *The Role of Phosphorus in Agriculture*. American Society of Agronomy, Madison, pp. 591–615.

Barber, S.A. (1984) *Soil Nutrient Bioavailability*. Wiley-Interscience, New York, 398 pp.

Barraclough, D. (1995) 15N isotope dilution techniques to study soil nitrogen transformations and plant uptake. *Fertilizer Research* 42, 185–192.

Barraclough, P.B. and Leigh, R.A. (1993) Grass yield in relation to potassium supply and the concentration of cations in tissue water. *Journal of Agricultural Science, Cambridge* 121, 157–168.

Barrow, N.J. (1975) Chemical form of inorganic phosphate in sheep faeces. *Australian Journal of Soil Research* 13, 63–67.

Barrow, N.J. (1987) Return of nutrients by animals. In: Snaydon, R.W. (ed.) *Managed Grasslands: Analytical Studies*. Elsevier, Amsterdam, pp. 181–186.

Barrow, N.J. and Lambourne, L.J. (1962) Partition of excreted nitrogen, sulphur and phosphorus between the faeces and urine of sheep being fed pasture. *Australian Journal of Agricultural Research* 13, 461–471.

Barrow, N.J., Leahy, P.J., Southey, I.N. and Purser, D.B. (1985) Initial and residual effectiveness of molybdate fertilizer in two areas of southwestern Australia. *Australian Journal of Agricultural Research* 36, 579–587.

Batey, T. (1982) Nitrogen cycling in upland pastures of the UK. *Philosophical Transactions of the Royal Society, London*, B 296, 551–556.

Beard, W.E. and Guenzi, W.D. (1983) Volatile sulphur compounds from a redox-controlled cattle-manure slurry. *Journal of Environmental Qualtiy* 12, 113–116.

Beaton, J.D. and Soper, R.J. (1986) Plant responses to sulfur in western Canada. In: Tabatabai, M.A. (ed.) *Sulfur in Agriculture*. American Society of Agronomy, Madison, pp. 375–403.

Beauchemin, S., Simard, R.R. and Cluis, D. (1996) Phosphorus sorption–desorption kinetics of soil under contrasting land use. *Journal of Environmental Quality* 25, 1317–1325.

Beck, A.B. (1962) The levels of copper, molybdenum and inorganic sulphate in some Western Australian pastures. *Australian Journal of Experimental Agriculture and Animal Husbandry* 2, 40–45.

Begg, J.E. and Freney, J.R. (1960) *Chemical Composition of Some Grazed Native Pasture Species in the New England Region of New South Wales*. CSIRO Division of Plant Industry Report no. 18. Canberra.

Belesky, D.P. and Jung, G.A. (1982) Seasonal variation of water-soluble and total zinc in cool-season grasses. *Agronomy Journal* 74, 1009–1012.

Benacchio, S.S., Baumgardner, M.F. and Mott, G.O. (1970) Residual effect of grass-pasture feeding systems on the fertility of the soil under a pasture sward. *Proceedings of the Soil Science Society of America* 34, 621–624.

Bentley, O.G. and Phillips, P.H. (1951) The effect of low manganese rations upon dairy cattle. *Journal of Dairy Science* 34, 396–403.

Berrow, M.L. and Burridge, J.C. (1980) Trace element levels in soils: effects of sewage sludge. In: *Inorganic Pollution and Agriculture*. MAFF Reference Book 326, HMSO, London, pp. 159–183.

Berrow, M.L. and Mitchell, R.L. (1980) Location of trace elements in soil profiles: total and extractable contents of individual horizons. *Transactions of the Royal Society of Edinburgh: Earth Sciences* 71, 103–121.

Berrow, M.L. and Webber, J. (1972) Trace elements in sewage sludges. *Journal of the Science of Food and Agriculture* 23, 93–100.

Berrow, M.L., Burridge, J.C. and Reith, J.W.S. (1983) Soil drainage conditions and related trace element contents. *Journal of the Science of Food and Agriculture* 34, 53–54.

Betteridge, K., Andrews, W.G.K. and Sedcole, J.R. (1986) Intake and excretion of nitrogen, potassium and phosphorus by grazing steers. *Journal of Agricultural Science, Cambridge* 106, 393–404.

Black, A.L. and Wight, J.R. (1979) Range fertilization: nitrogen and phosphorus uptake and recovery over time. *Journal of Range Management* 32, 349–353.

Blaser, R.E. and Kimbrough, E.L. (1968) Potassium nutrition of forage crops with perennials. In: Kilmer, V.J., Younts, S.E. and Brady, N.C. (eds) *The Role of Potassium in Agriculture*. American Society of Agronomy, Madison, pp. 423–445.

Blaxter, K. (1980) Soils, plant and animals. In: *Annual Report, 1979–80*. Macaulay Institute for Soil Research, Aberdeen, UK, pp. 138–157.

Blaxter, K.L., Wainman, F.W., Dewey, P.J.S., Davidson, J., Denerley, H. and Gunn, J.B. (1971) The effects of nitrogenous fertilizer on the nutritive value of artificially dried grass. *Journal of Agricultural Science, Cambridge* 76, 307–319.

Blaylock, M.J. and James, B.R. (1994) Redox transformations and plant uptake of selenium resulting from root-soil interactions. *Plant and Soil* 158, 1–12.

Blevins, D.G. (1994) Uptake, translocation and function of essential mineral elements in crop plants. In: Boote, K.J. *et al.* (eds) *Physiology and Determination of Crop Yield*. American Society of Agronomy, Madison, pp. 259–275.

Bohn, H.L., McNeal, B.L. and O'Connor, G.A. (1979) *Soil Chemistry*. Wiley, New York, 329 pp.

Boila, R.J., Derlin, T.J., Drysdale, R.A. and Lillie, L.E. (1984) Geographical variation in the copper and molybdenum contents of forages grown in northwestern Manitoba. *Canadian Journal of Animal Science* 64, 899–918.

Bolan, N.S., Syers, J.K., Tillman, R.W. and Scotter, D.R. (1988) Effect of liming and phosphate additions on sulphate leaching in soils. *Journal of Soil Science* 39, 493–504.

Bolton, J., Nowakowski, T.Z. and Lazarus, W. (1976) Sulphur–nitrogen interaction effects on the yield and composition of the protein-N, non-protein-N and soluble carbohydrates in perennial ryegrass. *Journal of the Science of Food and Agriculture* 27, 553–560.

Boswell, C.C. and Gregg, P.E.H. (1998) Sulfur fertilizers for grazed pasture systems. In: Maynard, D.G. (ed.) *Sulfur in the Environment*. Marcel Dekker, New York, pp. 95–134.

Botkin, D.B., Jordan, P.A., Dominski, A.S., Lowendorf, H.S. and Hutchinson, G.E. (1973) Sodium dynamics in a northern ecosystem. *Proceedings of the National Academy of Science, USA* 70, 2745–2748.

Bowen, H.D. (1981) Alleviating mechanical impedance. In: Arkin, G.F. and Taylor, H.M. (eds) *Modifying the Root Environment to Reduce Crop Stress*. American Society of Agricultural Engineers, St Joseph, pp. 21–57.

Bradford, G.R., Arkley, R.J., Pratt, P.F. and Blair, F.L. (1967) Total content of nine mineral elements in fifty selected benchmark soil profiles of California. *Hilgardia* 38, 541–556.

Brady, N.C. and Weil, R.R. (1999) *The Nature and Properties of Soils*, 12th edn. Prentice Hall, Upper Saddle River, New Jersey, 881 pp.

Braithwaite, G.D. (1976) Calcium and phosphorus metabolism in ruminants with special reference to parturient paresis. *Journal of Dairy Research* 43, 501–520.

Brauer, F.P. and Ballou, N.E. (1975) Isotope ratios of iodine and other radionuclides as nuclear power pollution inidicators. In: *Isotope Ratios as Pollutant Source and Behaviour Indicators*. International Atomic Energy Agency, Vienna, pp. 215–229.

Breenbroek, B., Poppii, K.J. and Wossink, G.A.A. (1996) Environmental farm accounting: the case of the Dutch nutrients accounting system. *Agricultural Systems* 51, 29–40.

Bremner, I. and Knight, A.H. (1970) The complexes of zinc, copper and manganese present in ryegrass. *British Journal of Nutrition* 24, 279–289.

Bremner, I. and Mills, C.F. (1981) Absorption, transport and tissue storage of essential trace elements. *Philosophical Transactions of the Royal Society* B294, 75–89.

Brennan, R.F., Armour, J.D. and Reuter, D.J. (1993) Diagnosis of zinc deficiency. In: Robson, A.D. (ed.) *Zinc in Soils and Plants*. Kluwer Academic Publishers, Dordrecht, pp. 167–181.

Bristow, A.W. and Garwood, E.A. (1984) Deposition of sulphur from the atmosphere and the sulphur balance in four soils under grass. *Journal of Agricultural Science* 103, 463–468.

Brockman, J.S., Shaw, P.G. and Wolton, K.M. (1970) The effect of phosphate and potash fertilizers on cut and grazed grassland. *Journal of Agricultural Science, Cambridge* 74, 397–407.

Bromfield, S.M. and Jones, O.L. (1970) The effect of sheep on the recycling of phosphorus in hayed-off pastures. *Australian Journal of Agricultural Research* 21, 699–711.

Brooks, P.C., Powlson, D.S. and Jenkinson, D.S. (1984) Phosphorus in the soil microbial biomass. *Soil Biology and Biochemistry* 16, 169–175.

Brown, J.R. and Sleper, D.A. (1980) Mineral concentration in two tall fescue genotypes grown under variable soil nutrient levels. *Agronomy Journal* 72, 742–745.

Brown, K.A. (1982) Sulphur in the environment: a review. *Environmental Pollution series B* 3, 47–80.

Brown, L., Jewkes, E.C. and Scholefield, D. (1999) The effect of sulphur application on efficiency of nitrogen use in grassland: some preliminary results. In: Corrall, A.J. (ed.) *Accounting for Nutrients*. Occasional Symposium no. 33, British Grassland Society, Reading, pp. 63–67.

Brown, P.H., Calmak, I. and Zhang, Q. (1993) Form and function of zinc in plants. In: Robson, A.D. (ed.) *Zinc in Soils and Plants*. Kluwer Academic Publishers, Dordrecht, pp. 93–106.

Burford, J.R. and Bremner, J.M. (1972) Is phosphate reduced to phosphine in waterlogged soils? *Soil Biology and Biochemistry* 4, 489–495.

Burkhalter, D.L., Nethery, M.W., Miller, W.J., Whitlock, R.H. and Allen, J.C. (1979) Effects of low chloride intake on performance, clinical characteristics, and chloride, sodium, potassium and nitrogen metabolism in dairy calves. *Journal of Dairy Science* 62, 1895–1901.

Burmester, C.H., Adams, F. and Haaland, R.L. (1981) Effects of nitrogen and sulphur fertilizers on sulphur content of tall fescue and *Phalaris*. *Agronomy Journal* 73, 614–618.

Burns, R.H., Johnston, A., Hamilton, J.W., McColloch, R.J., Duncan, W.E. and Fisk, H.G. (1964) Minerals in domestic wools. *Journal of Animal Science* 23, 5–11.

Burridge, J.C. (1970) Vegetational factors affecting the trace element content of plants. In: Mills, C.F. (ed.) *Trace Element Metabolism in Animals*. E. & S. Livingstone, Edinburgh, pp. 412–415.

Burridge, J.C., Reith, J.W.S. and Berrow, M.L. (1983) Soil factors and treatments affecting trace elements in crops and herbage. In: Suttle, N.F., Gunn, R.G., Allen, W.M., Linklater, K.A. and Wiener, G. (eds) *Trace Elements in Animal Production and Veterinary Practice*. British Society of Animal Production Occasional Publication no. 7, British Society of Animal Production, Edinburgh, pp. 77–85.

Burton, G.W., Jackson, J.E. and Knox, F.E. (1959) The influence of light reduction upon the production, persistence and chemical composition of Coastal Bermudagrass, *Cynodon dactylon*. *Agronomy Journal* 51, 537–542.

Bussink, D.W. (1994) Relationships between ammonia volatilization and nitrogen fertilizer application rate, intake and excretion of herbage nitrogen by cattle on grazed swards. *Fertilizer Research* 38, 111–121.

Butler, G.W. and Jones, D.I.H. (1973) Mineral biochemistry of herbage. In: Butler, G.W. and Bailey, R.W. (eds) *Chemistry and Biochemistry of Herbage*. Academic Press, London, pp. 127–162.

Buxton, D.R., Mertens, D.R. and Fisher, D.S. (1996) Forage quality and ruminant utilization. In: Moser, L.E., Buxton, D.R. and Casler, M.D. (eds) *Cool-Season Forage Grasses*. American Society of Agronomy, Madison, pp. 229–266.

Campbell, G.W., Atkins, D.H.F., Bower, J.S., Irwin, J.G., Simpson, D. and Williams, M.L. (1990) The spatial distribution of the deposition of sulphur and nitrogen in the UK. *Journal of the Science of Food and Agriculture* 53, 427–428.

Campos, M.L.A.M., Nightingale, P.D. and Jickells, T.D. (1996) A comparison of methyl iodide emissions from seawater and wet depositional fluxes of iodine over the southern North Sea. *Tellus* 48B, 106–114.

Caradus, J.R. (1992) Heritability of, and relationships between phosphorus and nitrogen concentration in shoot, stolon and root of white clover (*Trifolium repens* L.). *Plant and Soil* 146, 209–217.

Carran, R.A. (1988) Influence of soil N transformations on cation uptake by urine-affected pastures. *New Zealand Journal of Agricultural Research* 31, 65–70.

Carter, D.L., Robbins, C.W. and Brown, M.J. (1972) Effect of phosphorus fertilization on the selenium concentration in alfalfa (*Medicago sativa*). *Soil Science Society of America Proceedings* 36, 624–628.

Casler, M.D., Collins, M. and Reich, J.M. (1987) Location, year, maturity and alfalfa competition effects on mineral element concentrations in smooth bromegrass. *Agronomy Journal* 79, 774–778.

Cawse, P.A. (1980) Deposition of trace elements from the atmosphere in the UK. In: *Inorganic Pollution and Agriculture*, MAFF Reference Book 326, HMSO, London, pp. 22–46.

Cawse, P.A. (1987) Trace and major elements in the atmosphere at rural locations in Great Britain, 1972–81. In: Coughtrey, P.J., Martin, M.H. and Unsworth, M.H. (eds) *Pollutant Transfer and Fate in Ecosystems*. Blackwell Scientific Publications, Oxford, pp. 89–122.

Ceccotti, S.P. (1996) Plant nutrient sulphur – a review of nutrient balance, environmental impact and fertilizers. *Fertilizer Research* 43, 117–125.

Challa, J., Braithwaite, G.D. and Dhanoa, M.S. (1989) Phosphorus homeostasis in growing calves. *Journal of Agricultural Science, Cambridge* 112, 217–226.

Chamberlain, A.T. and Wilkinson, J.M. (1996) *Feeding the Dairy Cow*. Chalcombe Publications, Lincoln.

Chambers, B.J. and Garwood, T.W.D. (1998) Lime loss rates from arable and grassland soils. *Journal of Agricultural Science, Cambridge* 131, 455–464.

Chaney, R.L. (1993) Zinc phytotoxicity. In: Robson, A.D. (ed.) *Zinc in Soils and Plants*. Kluwer Academic Publishers, Dordrecht, pp. 135–150.

Chen, Y. and Barak, P. (1982) Iron nutrition of plants in calcareous soils. *Advances in Agronomy* 35, 217–240.

Cherney, J.H., Cherney, D.J.R. and Braulsema, T.W. (1998) Potassium management. In: Cherney, J.H. and Cherney, D.J.R. (eds) *Grass for Dairy Cattle*. CAB International, Wallingford, pp. 137–160.

Chesworth, W. (1991) Geochemistry of micronutrients. In: Mortvedt, J.J. (ed.) *Micronutrients in Agriculture*, 2nd edn. Soil Science Society of America, Madison, pp. 593–662.

Chiy, P.C., Avezinius, J.A. and Phillips, C.J.C. (1999) Sodium fertilizer application to pasture. 9. The effects of combined or separate applications of sodium and sulphur fertilizers on herbage composition and dairy cow production. *Grass and Forage Science* 54, 312–321.

Chiy, P.C. and Phillips, C.J.C. (1993) Sodium fertilizer application to pasture. 4. Effects on mineral uptake and the sodium and potassium status of steers. *Grass and Forage Science* 48, 260–270.

Chiy, P.C. and Phillips, C.J.C. (1995a) Sodium in forage crops. In: Phillips, C.J.C. and Chiy, P.C. (eds) *Sodium in Agriculture*. Chalcombe Publications, Canterbury, pp. 43–69.

Chiy, P.C. and Phillips, C.J.C. (1995b) Sodium in ruminant nutrition, production, reproduction and health. In: Phillips, C.J.C. and Chiy, P.C. (eds) *Sodium in Agriculture*. Chalcombe Publications, Canterbury, pp. 105–143.

Chiy, P.C. and Phillips, C.J.C. (1996a) Effects of sodium fertilizer on the chemical composition of grass and clover leaves, stems and inflorescences. *Journal of the Science of Food and Agriculture* 72, 501–510.

Chiy, P.C. and Phillips, C.J.C. (1996b) Sodium nutrition of dairy cows. In: Phillips, C.J.C. (ed.) *Progress in Dairy Science*. CAB International, Wallingford, pp. 29–44.

Chiy, P.C. and Phillips, C.J.C. (1997) Effects of sodium fertilizer on the chemical composition of perennial ryegrass and white clover leaves of different physiological ages. *Journal of the Science of Food and Agriculture* 73, 337–348.

Chiy, P.C. and Phillips, C.J.C. (1998) Sodium fertilizer application to pasture. 6. Effects of combined applications with sulphur on herbage production and chemical composition in the season of application. *Grass and Forage Science* 53, 1–10.

Chiy, P.C. and Phillips, C.J.C. (1999) Effects of sodium fertilizer on the distribution of trace elements, toxic metals and water-soluble carbohydrates in grass and clover fractions. *Journal of the Science of Food and Agriculture* 79, 2017–2024.

Chiy, P.C., Phillips, C.J.C. and Ajele, C.L. (1994) Sodium fertilizer application to pasture. 5. Effects on herbage digestibility and mineral availability in sheep. *Grass and Forage Science* 49, 25–33.

Chowdhury, A.K., McLaren, R.G. and Swift, R.S. (1997) Effects of phosphate and lime applications on pasture zinc status. *New Zealand Journal of Agricultural Research* 40, 417–424.

Christie, P. (1987) Some long-term effects of slurry on grassland. *Journal of Agricultural Science, Cambridge* 108, 529–541.

Christie, P. (1990) Accumulation of potentaially toxic metals in grassland from long-term slurry application. In: Merckx, R., Vereeken, H. and Vlassak, K. (eds) *Fertilization and the Environment*. Leuven University Press, Leuven, pp. 124–130.

Clark, F.E. (1977) Internal cycling of [15]nitrogen in shortgrass prairie. *Ecology* 58, 1322–1333.

Clark, F.E. and Woodmansee, R.G. (1992) Nutrient cycling. In: Coupland, R.G. (ed.) *Natural Grasslands: Introduction and Western Hemisphere*, Elsevier, Amsterdam, pp. 137–146.

Clarkson, D.T. (1985) Factors affecting mineral acquisition by plants. *Annual Review of Plant Physiology* 36, 77–115.

Clarkson, D.T. and Hanson, J.B. (1980) The mineral nutrition of higher plants. *Annual Review of Plant Physiology* 31, 239–298.

Clement, C.R. and Hopper, M.J. (1968) The supply of potassium to high yielding cut grass. *NAAS Quarterly Review* 79, 101–109.

Clement, C.R., Jones, L.H.P. and Hopper, M.J. (1972) The leaching of some elements from herbage plants by simulated rain. *Journal of Applied Ecology* 9, 249–260.

Clough, T.J., Sherlock, R.R., Cameron, K.C. and Ledgard, S.F. (1996) Fate of urine nitrogen on mineral and peat soils in New Zealand. *Plant and Soil* 178, 141–152.

Collins, M. (1983) Changes in composition of alfalfa, red clover and birdsfoot trefoil during autumn. *Agronomy Journal* 75, 287–291.

Collins, M. (1985) Wetting and maturity effects on mineral concentrations in legume hay. *Agronomy Journal* 77, 779–782.

Conroy, E. (1961) Effects of heavy applications of nitrogen on the composition of herbage. *Irish Journal of Agricultural Research* 1, 67–71.

Coppenet, M. (1964) [Changes in the mineral composition of pasture grasses under a grazing regime.] *Comptes Rendus Séances Académie d'Agriculture (France)* 50, 330–342.

Coppock, C.E. (1986) Mineral utilization by the lactating cow – chlorine. *Journal of Dairy Science* 69, 595–603.

Cornforth, I.S. (1984) Mineral nutrients in pasture species. *Proceedings of the New Zealand Society of Animal Production* 44, 135–137.

Coutts, G. and Atkinson, D. (1988) Selenium supply to herbage. In: *Annual Report, 1987*. Macaulay Land Use Research Institute, Aberdeen, pp. 63–64.

Cowling, D.W. and Jones, L.H.P. (1978) Sulphur and amino acid fractions in perennial ryegrass as influenced by soil and atmospheric supplies of sulphur. In: Brogan, J.C. (ed.) *Sulphur in Forages*. An Foras Taluntais, Dublin, pp. 15–31.

Cowling, D.W. and Lockyer, D.R. (1967) A comparison of the reaction of different grass species to fertilizer nitrogen and to growth in association with white clover. 2. Yield of nitrogen. *Journal of the British Grassland Society* 22, 53–61.

Cowling, D.W. and Lockyer, D.R. (1978) The effect of SO_2 on *Lolium perenne* L. grown at different levels of sulphur and nitrogen nutrition. *Journal of Experimental Botany* 29, 257–265.

Crush, J.R. (1983) Variation in the magnesium concentration of ryegrass plants. *New Zealand Journal of Agricultural Research* 26, 337–340.

Crush, J.R. and Caradus, J.R. (1995) Cyanogenesis potential and iodine concentration in white clover (*Trifolium repens* L.) cultivars. *New Zealand Journal of Agricultural Research* 38, 309–316.

Crush, J.R., Evans, J.P.M. and Cosgrove, G.P. (1989) Chemical composition of ryegrass and prairie grass pastures. *New Zealand Journal of Agricultural Research* 32, 461–468.

Culleton, N. and Fleming, G.A. (1983) Mineral composition of ryegrass cultivars. *Irish Journal of Agricultural Research* 22, 21–29.

Culleton, N., Parle, P.J., Rogers, P.A.M., Murphy, W.E. and Murphy, J. (1997) Selenium supplementation for dairy cows. *Irish Journal of Agricultural and Food Research* 36, 23–29.

Cunningham, R.K. (1963) Cation–anion relationships in crop nutrition. In: *Annual Report, 1962*. Rothamsted Experimental Station, Harpenden, Hertfordshire, p. 65.

Cunningham, R.K. (1964) Cation–anion relationships in crop nutrition. 1. Factors affecting cations in Italian ryegrass. *Journal of Agricultural Science, Cambridge* 63, 97–101.

Currier, C.G., Haaland, R.L., Hoveland, C.S., Elkins, C.B. and Odom, J.W. (1983) Breeding for higher magnesium content in orchardgrass (*D. glomerata* L.). In: *Proceedings of the 14th International Grassland Congress, Lexington*, Westview Press, Boulder, Colorado, 127–129.

Cushnahan, A., Gordon, F.J., Bailey, J.S. and Mayne, C.S. (1996) A note on the effect of sodium fertilization of pasture on the performance of lactating dairy cows. *Irish Journal of Agricultural and Food Research* 35, 43–47.

Dahlman, R.C. and Kucera, C.L. (1965) Root productivity and turnover in native prairie. *Ecology* 46, 84–89.

Dam Kofoed, A. (1980) Copper and its utilisation in Danish agriculture. *Fertilizer Research* 1, 63–71.

Dampney, P.M.R. (1992) The effect of timing and rate of potash application on the yield and herbage composition of grass grown for silage. *Grass and Forage Science* 47, 280–289.

Dampney, P.M.R. and Unwin, R.J. (1993) More efficient and effective use of NPK in today's conditions. In: Hopkins, A. and Younie, D. (eds) *Forward with Grass into Europe*. British Grassland Society Occasional Symposium 27, British Grassland Society, Reading, 62–72.

Davey, B.G. and Mitchell, R.L. (1968) The distribution of trace elements in cocksfoot at flowering. *Journal of the Science of Food and Agriculture* 19, 425–431.

Davies, B.E. (1994) Soil chemistry and bioavailability with special reference to trace elements. In: Farago, M.E. (ed.) *Plants and the Chemical Elements*. VCH, Weinheim, pp. 1–30.

Davies, E.B. and Watkinson, J.H. (1966) Uptake of native and applied selenium by pasture species. 1. *New Zealand Journal of Agricultural Research* 9, 317–327.

Davies, R.O., Jones, D.I.H. and Milton, W.E.J. (1959) Factors affecting the composition and nutritive value of herbage from fescue and *Molinia* areas. *Journal of Agricultural Science, Cambridge* 53, 268–285.

Davies, W.E., Ap Griffith, G. and Ellington, A. (1966) The assessment of herbage legume varieties II. The *in vitro* digestibility, water soluble carbohydrate and mineral content of primary growth of clover and lucerne. *Journal of Agricultural Science, Cambridge* 66, 351–357.

Davies, W.E., Thomas, T.A. and Young, N.R. (1968) The assessment of herbage legume varieties III. Annual variation in chemical composition of eight varieties. *Journal of Agricultural Science, Cambridge* 71, 233–241.

Davis, G.K. and Mertz, W. (1987) Copper. In: Mertz, W. (ed.) *Trace Elements in Human and Animal Nutrition*, 5th edn, Vol. 1. Academic Press, San Diego, pp. 301–364.

Davis, M.R. (1991) The comparative phosphorus requirements of some temperate perennial legumes. *Plant and Soil* 133, 17–30.

Davis, R.D. and Beckett, P.H.T. (1978) Upper critical levels of toxic elements in plants. 2. Critical levels of copper in young barley, wheat, rape, lettuce and ryegrass, and of nickel and zinc in young barley and ryegrass. *New Phytologist* 80, 23–32.

Davis, R.D., Carlton-Smith, C.H., Stark, J.H. and Campbell, J.A. (1988) Distribution of metals in grassland following surface applications of sewage sludge. *Environmental Pollution* 49, 99–115.

Day, T.A. and Detling, J.K. (1990) Grassland patch dynamics and herbivore grazing preference following urine deposition. *Ecology* 71, 180–183.

Dear, B.S. and Lipsett, J. (1987) The effect of boron supply on the growth and seed production of subterranean clover (*Trifolium subterraneum* L.). *Australian Journal of Agricultural Research* 38, 537–546.

Deenen, P.J.A.G. and Middelkoop, N. (1992) Effects of dung and urine in nitrogen uptake and yield of perennial ryegrass. *Netherlands Journal of Agricultural Research* 40, 469–482.

Deighton, N. and Goodman, B.A. (1995) The speciation of metals in biological systems. In: Ure, A.M. and Davidson, C.M. (eds) *Chemical Speciation in the Environment*. Blackie Academic and Professional, London, pp. 307–334.

Deinum, B. (1966) Influence of some climatological factors on the chemical composition and feeding value of herbage. In: *Proceedings of the 10th International Grassland Congress, Helsinki*, Valtioneuvoston Kirjapaino, Helsinki, 415–418.

Deinum, B. (1985) Root mass of grass swards in different grazing systems. *Netherlands Journal of Agricultural Science* 33, 377–384.

Deinum, B. and Sibma, L. (1980) Nitrate content of herbage in relation to nitrogen fertilization and management. In: Prins, W.H. and Arnold, G.H. (eds) *The Role of Nitrogen in Intensive Grassland Production*. Pudoc, Wageningen, pp. 95–102.

Department of the Environment (UK) (1997) *The United Kingdom National Air Quality Strategy* The Stationery Office Ltd, London, pp. 189.

De Wit, C.T., Dijkshoorn, W. and Noggle, J.C. (1963) Ionic balance and growth of plants. *Verslagen van Landbouwkundige Onderzoekingen* 69(15), 69.

Diaz-Fierros, F., Villar, M.C., Gil, F., Leirós, M.C., Carballas, M., Carballas, T. and Cabaneiro, A. (1987) Laboratory study of the availability of nutrients in physical fractions of cattle slurry. *Journal of Agricultural Science, Cambridge* 108, 353–359.

Dible, W.T. and Berger, K.C. (1952) Boron content of alfalfa as influenced by boron supply. *Proceedings of the Soil Science Society of America* 16, 60–62.

Dickinson, C.H. and Craig, G. (1990) Effects of water on the decomposition and release of nutrients from cow pats. *New Phytologist*, 115, 139–147.

Dickinson, N.M. (1983) Decomposition of grass litter in a successional grassland. *Pedobiologia* 25, 117–126.

Dickinson, N.M. (1984) Seasonal dynamics and compartmentation of nutrients in a grassland meadow in lowland England. *Journal of Applied Ecology* 21, 695–701.

Dickson, I.A. and Macpherson, A. (1976) The effects on ewes and lambs of grazing pasture containing differing levels of nitrate-nitrogen. *Journal of the British Grassland Society* 31, 129–134.

Dijkshoorn, W. (1958) Nitrogen, chlorine and potassium in perennial ryegrass and their relation to the mineral balance. *Netherlands Journal of Agricultural Science* 6, 131–138.

Dijkshoorn, W., Lampe, J.E.M. and Van Burg, P.F.J. (1960) A method of diagnosing the sulphur nutrition status of herbage. *Plant and Soil* 13, 227–241.

Dinkelaker, B., Romheld, V. and Marschner, H. (1989) Citric acid excretion and precipitation of calcium citrate in the rhizosphere of white lupin (*Lupinus albus* L.). *Plant, Cell and Environment* 12, 285–292.

Drysdale, R.A., Devlin, T.J., Lillie, L.E., Fletcher, W.K. and Clark, K.W. (1980) Nutrient concentrations in grass and legume forages of northwestern Manitoba. *Canadian Journal of Animal Science* 60, 991–1002.

Dua, K. and Care, A.D. (1995) Impaired absorption of magnesium in the aetiology of grass tetany. *British Veterinary Journal* 151, 413–426.

Dunlop, J. and Hart, A.L. (1987) Mineral nutrition. In: Baker, M.J. and Williams, W.M. (eds) *White Clover*. CAB International, Wallingford, pp. 153–183.

Durand, M. and Komisarczuk, S. (1988) Influence of major minerals on rumen microbiota. *Journal of Nutrition* 118, 249–260.

During, C. and Weeda, W.C. (1973) Some effects of cattle dung on soil properties, pasture production and nutrient uptake I. Dung as a source of phosphorus. *New Zealand Journal of Agricultural Research* 16, 423–430.

Early, M.S.B., Cameron, K.C. and Fraser, P.M. (1998) The fate of potassium, calcium and magnesium in simulated urine patches on irrigated dairy pasture soil. *New Zealand Journal of Agricultural Research* 41, 117–124.

Eason, W.R. and Newman, E.I. (1990) Rapid cycling of nitrogen and phosphorus from dying roots of *Lolium perenne*. *Oecologia* 82, 432–436.

Easton, H.S., Mackay, A.D. and Lee, J. (1997) Genetic variation for macro- and micro-nutrient concentrations in perennial ryegrass (*Lolium perenne* L.). *Australian Journal of Agricultural Research* 48, 657–666.

Edmeades, D.C., Smart, C.E. and Wheeler, D.M. (1983) Effects of lime on the chemical composition of ryegrass and white clover grown on a yellow-brown loam. *New Zealand Journal of Agricultural Research* 26, 473–481.

Eghball, B., Binford, G.D. and Baltensperger, D.D. (1996) Phosphorus movement and adsorption in a soil receiving long-term manure and fertilizer application. *Journal of Environmental Quality* 25, 1339–1343.

Ehlig, C.F., Allaway, W.H., Cary, E.E. and Kubota, J. (1968) Differences among plant species in selenium accumulation from soils low in available selenium. *Agronomy Journal* 60, 43–47.

Elkins, C.B. and Hoveland, C.S. (1977) Soil oxygen and temperature effects on tetany potential of three annual forage species. *Agronomy Journal* 69, 626–628.

Emanuele, S.M., Staples, C.R. and Wilcox, C.J. (1991) Extent and site of mineral release from six forage species incubated in mobile Dacron bags. *Journal of Animal Science* 69, 801–810.

Ettala, E. and Kossila, V. (1979) Mineral content in heavily nitrogen fertilized grass and its silage. *Annales Agriculturae Fenniae* 18, 252–262.

Evangelou, V.P., Wang, J. and Phillips, R.E. (1994) New developments and perspectives on soil potassium quantity/intensity relationships. *Advances in Agronomy* 52, 173–227.

Evans, C. (1985) The effect of applying cobalt sulphate to soil on the cobalt content of herbage. *Soil Use and Management* 1, 50–53.

Evans, D.R., Thomas, T.A., Williams, T.A. and Ellis Davies, W. (1986) Effect of fertilizers on the yield and chemical composition of pure sown white clover and on soil nutrient status. *Grass and Forage Science* 41, 295–302.

Evans, P.S. (1973) The effect of repeated defoliation to three different levels on root growth of five pasture species. *New Zealand Journal of Agricultural Research* 16, 31–34.

Evans, P.S. (1977) Comparative root morphology of some pasture grasses and clovers. *New Zealand Journal of Agricultural Research* 20, 331–335.

Evans, P.S. (1978) Plant root distribution and water use patterns of some pasture and crop species. *New Zealand Journal of Agricultural Research* 21, 261–265.

Fairbourn, M.L. and Batchelder, A.R. (1980) Factors influencing magnesium in High Plains forage. *Journal of Range Management* 33, 435–438.

Fangmeier, A., Hadwinger-Fangmeier, A., Van der Eerden, L. and Jager, H-J. (1994) Effects of atmospheric ammonia on vegetation – a review. *Environmental Pollution* 86, 43–82.

Farwell, S.O., Sherrard, A.E., Pack, M.E. and Adams, D.F. (1979) Sulfur compounds volatilized from soils at different moisture contents. *Soil Biology and Biochemistry* 11, 411–415.

Feely, L. (1990) Agronomic effectiveness of nitrogen, sulphur and phosphorus for reducing molybdenum uptake by herbage grown on peatland. *Irish Journal of Agricultural Research* 29, 129–139.

Fergusson, J.E. (1990) *The Heavy Elements: Chemistry, Environmental Impact and Health Effects.* Pergamon Press, Oxford, 614 pp.

Fincher, G.T., Menson, W.G. and Burton, G.W. (1981) Effects of cattle feces rapidly buried by dung beetles on yield and quality of Coastal Bermudagrass. *Agronomy Journal* 73, 775–779.

Fishbein, L. (1991) Selenium. In: Merian, E. (ed.) *Metals and their Compounds in the Environment.* VCH Publishers, New York, pp. 1153–1190.

Fisher, E., Thornton, B., Hudson, G. and Edwards, A.C. (1998) The variability in total and extractable soil phosphorus under a grazed pasture. *Plant and Soil* 203, 249–255.

Fixen, P.E. (1993) Crop responses to chloride. *Advances in Agronomy* 50, 107–150.

Fixen, P.E. and Grove, J.H. (1990) Testing soils for phosphorus. In: Westermann, R.L. (ed.) *Soil Testing and Plant Analysis*, 3rd edn. Soil Society of America, Madison, pp. 141–180.

Fleming, G.A. (1962) Selenium in Irish soils and plants. *Soil Science* 94, 28–35.

Fleming, G.A. (1963) Distribution of major and trace elements in some common pasture species. *Journal of the Science of Food and Agriculture* 14, 203–208.

Fleming, G.A. (1965) Trace elements in plants with particular reference to pasture species. *Outlook on Agriculture* 4, 270–285.

Fleming, G.A. (1973) Mineral composition of herbage. In: Butler, G.W. and Bailey, R.W. (eds) *Chemistry and Biochemistry of Herbage.* Academic Press, London, pp. 529–566.

Fleming, G.A. (1980) Essential micronutrients II. Iodine and selenium. In: Davies, B.E. (ed.) *Applied Soil Trace Elements.* John Wiley and Sons, Chichester, pp. 199–234.

Fleming, G.A. (1988) Aspects of agricultural and environmental trace element research in the Republic of Ireland. In: Thornton, I. (ed.) *Geochemistry and Health.* Science Reviews, Northwood, pp. 123–137.

Fleming, G.A. and Coulter, B.S. (1963) Mineral elements in pasture plants. In: *Proceedings, 1st Regional Conferencce of the International Potash Institute, Wexford, 1963*, International Potash Institute, Berne, pp, 63–70.

Fleming, G.A. and Murphy, W.E. (1968) The uptake of some major and trace elements by grass as affected by season and stage of maturity. *Grass and Forage Science* 23, 174–185.

Floate, M.J.S. (1970) Decomposition of organic materials from hill soils and pastures. II. *Soil Biology and Biochemistry* 2, 173–185.

Follett, R.F. and Wilkinson, S.R. (1995) Nutrient management of forages. In: Barnes, R.F., Miller, D.A. and Nelson, C.J. (eds) *Forages*, 5th edn, Vol. II. Iowa State University Press, Ames, pp. 55–82.

Follett, R.F., Power, J.F., Grunes, D.L. and Klein, C.A. (1977) Effect of N, K and P fertilization, N source and clipping on potential tetany hazard of bromegrass. *Plant and Soil* 48, 485–508.

Fontenot, J.P. and Church, D.C. (1979) The macro (major) minerals. In: Church, D.C. (ed.) *Digestive Physiology and Nutrition of Ruminants*, Vol. 2 – *Nutrition*, 2nd edn. O & B Books, Corvallis, Oregon, pp. 56–99.

Fontenot, J.P., Allen, V.G., Bunce, G.E. and Goff, J.P. (1989) Factors influencing magnesium absorption and metabolism in ruminants. *Journal of Animal Science* 67, 3445–3455.

Forbes, J.C. and Gelman, A.L. (1981) Copper and other minerals in herbage species and varieties on copper-deficient soils. *Grass and Forage Science* 36, 25–30.

Forbes, P.J., Black, K.E. and Hooker, J.E. (1997) Temperature-induced alteration to root longevity in *Lolium perenne. Plant and Soil* 190, 87–90.

Fowler, D. (1986) Transfer to terrestrial surfaces. *Philosophical Transactions of the Royal Society, London*, B 305, 281–297.

Frame, J., Fisher, G.E.J. and Tiley, G.E.D. (1994) Wildflowers in grassland systems. In: Haggar, R.J. and Peel, S. (eds) *Grassland Management and Nature Conservation*. British Grassland Society Occasional Symposium no. 28, British Grassland Society, Reading, 104–114.

Frankenberger, W.T. and Karlson, U. (1994) Microbial volatilization of selenium from soils and sediments. In: Frankenburg, W.T. and Benson, S. (eds) *Selenium in the Environment*, Marcel Dekker, New York, pp. 369–387.

Fransen, B., Blijjenberg, J. and de Kroon, H. (1999) Root morphological and physiological plasticity of perennial grass species and the exploitation of spatial and temporal heterogenous nutrient patches. *Plant and Soil* 211, 179–189.

Freney, J.R. (1986) Forms and reactions of organic sulfur compounds in soils. In: Tabatabai, M.A. (ed.) *Sulfur in Agriculture*. American Society of Agronomy, Madison, pp. 207–232.

Froslie, A. and Norheim, G. (1983) Copper, molybdenum, zinc and sulphur in Norwegian forages and their possible role in chronic copper poisoning in sheep. *Acta Agriculturae Scandinavica* 33, 97–104.

Frossard, E., Brossard, M., Hedley, M.J. and Metherell, A. (1995) Reactions controlling the cycling of P in soils. In: Tiessen, H. (ed.) *Phosphorus in the Global Environment*. John Wiley and Sons, Chichester, pp. 107–138.

Frossard, E., Condron, L.M., Oberson, A., Sinaj, S. and Fardeau, J.C. (2000) Processes governing phosphorus availability in temperate soils. *Journal of Environmental Quality* 29, 15–23.

Fuge, R. (1988) Sources of halogens in the environment, influences on human and animal health. *Environmental Geochemistry and Health* 10, 51–61.

Fuge, R. (1996) Geochemistry of iodine in relation to iodine deficiency diseases. In: Appleton, J.D., Fuge, R. and McCall, G.J.H. (eds) *Environmental Geochemistry and Health*. Geological Society, London, pp. 201–211.

Fuge, R. and Johnson, C.C. (1986) The geochemistry of iodine – a review. *Environmental Geochemistry and Health* 8, 31–54.

Fulkerson, R.S., Mowat, D.N., Tossell, W.E. and Winch, J.E. (1967) Yield of dry matter, *in vitro* digestible dry matter and crude protein of forages. *Canadian Journal of Plant Science* 47, 683–690.

Furr, A.K., Lawrence, A.W., Tong, S.S.C., Grandolfo, M.C., Hofstader, R.A., Bache, C.A., Gutenmann, W.H. and Lisk, D.J. (1976) Multielement and chlorinated hydrocarbon analysis of municipal sewage sludges of American cities. *Environmental Science and Technology* 10, 683–687.

Gardner, E.H., Jackson, T.L., Webster, G.R. and Turley, R.H. (1960) Some effects of fertilization on the yield, botanical and chemical composition of irrigated grass and grass/clover pasture swards. *Canadian Journal of Plant Science* 40, 546–562.

Garsed, S.G. and Read, D.J. (1977) The uptake and metabolism of $^{35}SO_2$ in plants of differing sensitivity to sulphur dioxide. *Environmental Pollution* 13, 177–186.

Garwood, E.A. (1967a) Studies on the roots of grasses. In: *Annual Report, 1966*, Grassland Research Institute, Hurley, pp. 72–79.

Garwood, E.A. (1967b) Some effects of soil water conditions and soil temperature on the roots of grasses. 1. The effect of irrigation on the weight of root material under various swards. *Journal of the British Grassland Society* 22, 176–181.

Gastal, F. and Nelson, C.J. (1994) Nitrogen use within the growing leaf blade of tall fescue. *Plant Physiology* 105, 191–197.

Gerritse, R.G. and Vriesema, R. (1984) Phosphate distribution in animal waste slurries. *Journal of Agricultural Science, Cambridge* 102, 159–161.

Ghani, A., McLaren, R.G. and Swift, R.S. (1990) Seasonal fluctuations of sulphate and soil microbial biomass-S in the surface of a Wakanui soil. *New Zealand Journal of Agricultural Research* 33, 467–472.

Ghani, A., McLaren, R.G. and Swift, R.S. (1992) Sulphur mineralisation and transformations in soils as influenced by additions of carbon, nitrogen and sulphur. *Soil Biology and Biochemistry* 24, 331–341.

Gilbert, M.A. and Robson, A.D. (1984a) Sulphur nutrition of temperate pasture species I. Effects of N-supply on the external and internal sulphur requirement of subterranean clover and ryegrass. *Australian Journal of Agricultural Research* 35, 379–388.

Gilbert, M.A. and Robson, A.D. (1984b) Sulfur nutrition of temperate pasture species II. A comparison of subterranean clover cultivars, medics and grasses. *Australian Journal of Agricultural Research* 35, 389–398.

Gilkes, R.J. and McKenzie, R.M. (1988) Geochemistry of manganese in soil. In: Graham, R.D., Hannan, R.J. and Uren, N.C. (eds) *Manganese in Soils and Plants*. Kluwer Academic Publishers, Dordrecht, pp. 23–35.

Gill, K., Jarvis, S.C. and Hatch, D.J. (1995) Mineralization of nitrogen in long-term pasture soils: effects of management. *Plant and Soil* 172, 153–162.

Gillingham, A.G. (1987) Phosphorus cycling in managed grasslands. In: Snaydon, R.W. (ed.) *Managed Grasslands: Analytical Studies*. Elsevier, Amsterdam, pp. 173–180.

Gillingham, A.G., Syers, J.K. and Gregg, P.E.H. (1980) Phosphorus uptake and return in grazed steep hill pastures. II. Above-ground components of the phosphorus cycle. *New Zealand Journal of Agricultural Research* 23, 323–330.

Gillingham, A.G., Sheath, G.W. and Sutton, M.M. (1987) Soil contamination and sample washing effects on major nutrient and soluble sugar concentrations of pasture. *New Zealand Journal of Agricultural Research* 30, 281–284.

Giordano, P.M. and Mortvedt, J.J. (1980) Zinc uptake and accumulation by agricultural crops. In: Nriagu, J.O. (ed.) *Zinc in the Environment, Part 11: Health Effects.* Wiley, New York, pp. 401–413.

Gissel-Nielsen, G. (1971) Selenium content of some fertilizers and their influence on uptake of selenium in plants. *Journal of Agricultural and Food Chemistry* 19, 564–566.

Gissel-Nielsen, G. (1975) Selenium concentration in Danish forage crops. *Acta Agriculturae Scandinavica* 25, 216–220.

Gissel-Nielsen, G., Gupta, U.C., Lamand, M. and Westermarck, T. (1984) Selenium in soils and plants and its importance in livestock and human nutrition. *Advances in Agronomy* 37, 397–460.

Gladstones, J.S. (1962) The mineral composition of lupins. 2. A comparison of the copper, manganese, molybdenum and cobalt contents of lupins and other species at one site. *Australian Journal of Experimental Agriculture and Animal Husbandry* 2, 213–220.

Gladstones, J.S. and Loneragan, J.F. (1967) Mineral elements in temperate crop and pasture plants. 1. Zinc. *Australian Journal of Agricultural Research* 18, 427–446.

Gladstones, J.S., Loneragan, J.F. and Simmonds, W.J. (1975) Mineral elements in temperate crop and pasture plants. 3. Copper. *Australian Journal of Agricultural Research* 26, 113–126.

Goff, J.P. and Horst, R.L. (1997) Effects of the addition of potassium or sodium, but not calcium, to prepartum rations on milk fever in dairy cows. *Journal of Dairy Science* 80, 176–186.

Goh, K.M. and Gregg, P.E.H. (1980) Re-utilization by perennial ryegrass (*Lolium perenne* L.) of labelled fertilizer sulphur incorporated in field-grown tops and roots of pasture plants added to soils. *Fertilizer Research* 1, 73–85.

Goh, K.M. and Kee, K.K. (1978) Effects of nitrogen and sulphur fertilization on the digestibility and chemical composition of perennial ryegrass (*Lolium perenne*, L). *Plant and Soil* 50, 161–177.

Goh, K.M. and Nguyen, M.L. (1997) Estimating net annual soil sulfur mineralisation in New Zealand grazed pastures using mass balance methods. *Australian Journal of Agricultural Research* 48, 477–484.

Goh, K.M. and Williams, P.H. (1999) Comparative nutrient budgets of temperate grazed pastoral systems. In: Smaling, E.M.A., Oenema, O. and Fresco, L.O. (eds) *Nutrient Disequilibria in Agroecosystems: Concepts and Case Studies.* CAB International, Wallingford, pp. 75–97.

Goh, K.M., Haynes, R.J. and Kee, K.K. (1979) Ionic balance and composition of perennial ryegrass (*Lolium perenne* L.) as influenced by nitrogen and sulphur fertilization. *New Zealand Journal of Agricultural Research* 22, 319–327.

Goodrich, R.D. and Garrett, J.E. (1986) Sulfur in livestock nutrition. In: Tabatabai, M.A. (ed.) *Sulfur in Agriculture.* American Society of Agronomy, Madison, pp. 617–633.

Gooneratne, S.R., Buckley, W.T. and Christensen, D.A. (1989) Review of copper deficiency and metabolism in ruminants. *Canadian Journal of Animal Science* 69, 819–845.

Gostik, K.G. (1981) Accumulation of organic nitrogen in soils after application to grassland: some studies in England and Wales. In: Brogan, J.C. (ed.) *Nitrogen Losses and Surface Run-off from Landspreading of Manures*. Martinus Nijhoff, The Hague, pp. 369–372.

Goswami, A.K. and Willcox, J.S. (1969) Effect of applying increasing levels of nitrogen to ryegrass. *Journal of the Science of Food and Agriculture* 20, 592–599.

Goulding, K.W.T. (1990) Nitrogen deposition to land from the atmosphere. *Soil Use and Management* 6, 61–62.

Goulding, K.W.T. and Loveland, P.J. (1986) The classification and mapping of potassium reserves in soils of England and Wales. *Journal of Soil Science* 37, 555–565.

Goulding, K.W.T., Poulton, P.R., Thomas, V.H. and Williams, R.J.B. (1986) Atmospheric deposition at Rothamsted, Saxmundham and Woburn Experimental Stations, England, 1969–1984. *Water, Air and Soil Pollution* 29, 27–49.

Goulding, K.W.T., Bailey, N.J. and Bradbury, N.J. (1998) A modelling study of nitrogen deposited to arable land from the atmosphere and its contribution to nitrate leaching. *Soil Use and Management* 14, 70–77.

Grace, N.D. (1983a) Amounts and distribution of mineral elements associated with fleece-free body weight gains in the grazing sheep. *New Zealand Journal of Agricultural Research* 26, 59–70.

Grace, N.D. (1983b) Animal requirements for sulphur. In: Gregg, P.E.H. and Syers, J.K. (eds) *Sulphur in New Zealand Agriculture*. Proceedings of Technical Workshop, Massey University, Palmerston North, pp. 85–95.

Grace, N.D. (1984) The determination of mineral requirements of sheep and cattle. *Proceedings of the New Zealand Society of Animal Production* 44, 139–141.

Grace, N.D. and Clark, R.G. (1991) Trace element requirements, diagnosis and prevention of deficiencies in sheep and cattle. In: Tsuda, T., Sasaki, Y. and Kawashima, R. (eds) *Physiological Aspects of Digestion and Metabolism in Ruminants*. Academic Press, San Diego, pp. 321–346.

Grace, N.D. and Healy, W.B. (1974) Effect of ingestion of soil on faecal loss and retention of Mg, Ca, P, K and Na in sheep fed two levels of dried grass. *New Zealand Journal of Agricultural Research* 17, 73–78.

Grace, N.D. and Lee, J. (1992) Influence of high zinc intakes, season and staple site on the elemental composition of wool and fleece quality in grazing sheep. *New Zealand Journal of Agricultural Research* 35, 367–377.

Grace, N.D., Davies, E. and Monro, J. (1977) Association of Mg, Ca, P and K with various fractions in the diet, digesta and faeces of sheep fed fresh pasture. *New Zealand Journal of Agricultural Research* 20, 441–448.

Grace, N.D., Mills, A.R. and Death, A.F. (1995) The relationship between daily Se intakes and blood Se concentrations in pregnant dairy cows. *Proceedings of the New Zealand Society of Animal Production* 55, 174–175.

Grace, N.D., Rounce, J.R. and Lee, J. (1996a) Effect of soil ingestion on the storage of Se, vitamin B_{12}, Cu, Cd, Fe, Mn and Zn in the liver of sheep fed lucerne pellets. *New Zealand Journal of Agricultural Research* 39, 325–331.

Grace, N.D., Waghorn, G.C., Mills, A.R. and Death, A.F. (1996b) Effect of iodine supplementation on milk iodine concentrations and productivity of dairy cows. *Proceedings of New Zealand Society of Animal Production* 56, 143–145.

Grace, N.D., Lee, J., Mills, R.A. and Death, A.F. (1997) Influence of Se status on milk concentrations in dairy cows. *New Zealand Journal of Agricultural Research* 40, 75–78.

Grady, T.M., Heisler, C.R. and Blincoe, C. (1978) Chemical properties of manganese in lucerne. *Journal of the Science of Food and Agriculture* 29, 207–212.

Graven, E.H., Attoe, O.J. and Smith, D. (1965) Effect of liming and flooding on manganese toxicity in alfalfa. *Soil Science Society of America Proceedings* 29, 702–706.

Greene, L.W., Pinchak, W.E. and Heitschmidt, R.K. (1987) Seasonal dynamics of minerals in forages at the Texas Experimental Ranch. *Journal of Range Management* 40, 502–506.

Gregg, P.E.H., Goh, K.M. and Brash, D.W. (1977) Isotope studies of the uptake of sulphur by pasture plants. II. Uptake from various soil depths at several field sites. *New Zealand Journal of Agricultural Research* 20, 229–233.

Gregory, P.J. (1988) Growth and functioning of plant roots. In: Wild, A. (ed.) *Russell's Soil Conditions and Plant Growth*. Longman, Harlow, pp. 113–167.

Gressel, N. and McColl, J.G. (1997) Phosphorus mineralization and organic matter decomposition. A critical review. In: Cadisch, G. and Giller, K.E. (eds) *Driven by Nature: Plant Litter Quality and Decomposition*. CAB International, Wallingford, pp. 297–309.

Grewal, H.S. and Williams, R. (1999) Alfalfa genotypes differ in their ability to tolerate zinc deficiency. *Plant and Soil* 214, 39–48.

Griffith, W.K., Teel, M.R. and Parker, H.E. (1964) Influence of nitrogen and potassium on the yield and chemical composition of orchardgrass. *Agronomy Journal* 56, 473–475.

Grime, J.P. and Hutchinson, T.C. (1967) The incidence of lime-chlorosis in the natural vegetation of England. *Journal of Ecology* 55, 557–566.

Gross, C.F. and Jung, G.A. (1978) Magnesium, Ca and K concentration in temperate-origin forage species as affected by temperature and Mg fertilization. *Agronomy Journal* 70, 397–403.

Gross, C.F. and Jung, G.A. (1981) Season, temperature, soil pH and Mg fertilizer effects on herbage Ca and P levels and ratios of grasses and legumes. *Agronomy Journal* 73, 629–634.

Grunes, D.L. (1983) Uptake of magnesium by different plant species. In: Fontenot, J.P., Bunce, G.E., Webb, K.E. and Allen, V.G. (eds) *Role of Magnesium in Animal Nutrition*. Virginia State University, Blacksburg, pp. 23–38.

Grunes, D.L. and Welch, R.M. (1989) Plant contents of magnesium, calcium and potassium in relation to animal nutrition. *Journal of Animal Science* 67, 3485–3494.

Grunes, D.L., Stout, P.R. and Brownell, J.R. (1970) Grass tetany of ruminants. *Advances in Agronomy* 22, 331–374.

Gueguen, L., Lamand, M. and Mischy, F. (1989) Mineral requirements. In: Jarrige, R. (ed.) *Ruminant Nutrition*. John Libbey Eurotext, London/Paris, pp. 49–59.

Gupta, U.C. (1979) Boron nutrition of crops. *Advances in Agronomy* 31, 273–307.

Gupta, U.C. (1993) Boron, molybdenum and selenium. In: Carter, M.R. (ed.) *Soil Sampling and Methods of Analysis*. Lewis Publishers, Boca Raton, Florida, pp. 91–99.

Gupta, U.C. and Lipsett, J. (1981) Molybdenum in soils, plants and animals. *Advances in Agronomy* 34, 73–115.

Gupta, U.C. and Watkinson, J.H. (1985) Agricultural significance of selenium. *Outlook on Agriculture* 14, 183–189.

Gupta, U.C. and Winter, K.A. (1975) Selenium content of soils and crops and the effects of lime and sulphur on plant selenium. *Canadian Journal of Soil Science* 55, 161–166.

Haas, H.J. (1958) Effects of fertilizers, age of sward and decomposition on weight of grass roots and of grass and alfalfa on soil nitrogen and carbon. *Agronomy Journal* 50, 5–9.

Hagstrom, G.R. (1986) Fertilizer sources of sulfur and their use. In: Tabatabai, M.A. (ed.) *Sulfur in Agriculture*. American Society of Agronomy, Madison, pp. 567–581.

Hambidge, K.M., Casey, C.E. and Krebs, N.F. (1986) Zinc. In: Mertz, W. (ed.) *Trace Elements in Human and Animal Nutrition*, 5th edn, Vol. 2. Academic Press, Orlando, pp. 1–137.

Hamilton, J.W. and Beath, O.A. (1963) Uptake of available selenium by certain range plants. *Journal of Range Management* 16, 261–265.

Handreck, K.A. and Riceman, D.S. (1969) Cobalt distribution in several pasture species grown in culture solution. *Australian Journal of Agricultural Research* 20, 213–216.

Harrison, A.F. (1978) Phosphorus cycles of forest and upland grassland ecosystems and some effects of land management practices. In: Porter, R. and Fitzsimmons, D.W. (eds) *Phosphorus in the Environment: its Chemistry and Biochemistry*. Elsevier, Amsterdam, pp. 175–199.

Hart, A.L. (1989) Nodule phosphorus and nodule activity in white clover. *New Zealand Journal of Agricultural Research* 32, 145–149.

Harter, R.D. (1991) Micronutrient adsorption-desorption reactions in soils. In: Mortvedt, J.J. (ed.) *Micronutrients in Agriculture*. Soil Science Society of America, Madison, pp. 59–87.

Harter, R.D. and Naidu, R. (1995) Role of metal–organic complexation in metal sorption by soils. *Advances in Agronomy* 55, 219–263.

Hartmans, J. (1974) Factors affecting the herbage iodine content. *Netherlands Journal of Agricultural Science* 22, 195–206.

Hassink, J. (1995) Effect of the non-fertilizer N supply of grassland soils on the response of herbage to N fertilization under mowing conditions. *Plant and Soil* 175, 159–166.

Hassink, J. and Neeteson, J.J. (1991) Effect of grassland management of the amounts of soil organic N and C. *Netherlands Journal of Agricultural Science* 39, 225–236.

Havre, G.N. and Dishington, I.E. (1962) The mineral composition of pasture as influenced by various types of heavy nitrogen dressings. *Acta Agriculturae Scandinavica* 12, 298–308.

Haydock, K.P. and Norris, D.O. (1966) Opposed curves for nitrogen per cent on dry weight by *Rhizobium*-dependent and nitrate-dependent legumes. *Australian Journal of Science* 29, 426–427.

Haygarth, P.M. (1994) Global importance and global cycling of selenium. In: Frankenberger, W.T. and Benson, S. (eds) *Selenium in the Environment*. Marcel Dekker, New York, pp. 1–27.

Haygarth, P.M., Harrison, A.F. and Jones, K.C. (1995) Plant selenium from soil and the atmosphere. *Journal of Environmental Quality* 24, 768–771.

Haygarth, P.M., Chapman, P.J., Jarvis, S.C. and Smith, R.V. (1998) Phosphorus budgets for two contrasting farming systems in the UK. *Soil Use and Management* 14, 160–167.

Haynes, R.J. (1980) Ion exchange properties of roots and interactions within the root apoplasm: their role in ion accumulation by plants. *Botanical Review* 46, 75–99.

Haynes, R.J. (1986) Uptake and assimilation of mineral nitrogen by plants. In: Haynes, R.J. (ed.) *Mineral Nitrogen and the Plant–Soil System*. Academic Press, London, pp. 303–378.

Haynes, R.J. and Swift, R.S. (1984) Amounts and forms of micronutrient cations in a group of loessial grassland soils of New Zealand. *Geoderma* 33, 53–62.

Haynes, R.J. and Williams, P.H. (1992a) Long-term effect of superphosphate on accumulation of soil phosphorus and exchangeable cations on a grazed, irrigated pasture site. *Plant and Soil* 142, 123–133.

Haynes, R.J. and Williams, P.H. (1992b) Accumulation of soil organic matter and the forms, mineralization potential and plant-availability of accumulated organic sulphur: effects of pasture improvement and intensive cultivation. *Soil Biology and Biochemistry* 24, 209–217.

Haynes, R.J. and Williams, P.H. (1993) Nutrient cycling and soil fertility in the grazed pasture ecosystem. *Advances in Agronomy* 49, 119–199.

Healy, W.B. (1973) Nutritional aspects of soil ingestion by grazing animals. In: Butler, G.W. and Bailey, R.W. (eds) *Chemistry and Biochemistry of Herbage*. Academic Press, London, pp. 567–588.

Healy, W.B., McCabe, W.J. and Wilson, G.F. (1970) Ingested soil as a source of microelements for grazing animals. *New Zealand Journal of Agricultural Research* 13, 503–512.

Healy, W.B., Crouchley, G., Gillett, R.L., Rankin, P.C. and Watts, H.M. (1972) Ingested soil and iodine deficiency in lambs. *New Zealand Journal of Agricultural Research* 15, 778–782.

Heddle, R.G. (1967) Long-term effects of fertilizers on herbage production. I. Yields and botanical composition. *Journal of Agricultural Science, Cambridge* 69, 425–431.

Heddle, R.G. and Crooks, P. (1967) Long-term effects of fertilizers on herbage production. II. Chemical composition. *Journal of Agricultural Science, Cambridge* 69, 433–441.

Helmke, P.A. (1999) The chemical composition of soils. In: Sumner, M.E. (ed.) *Handbook of Soil Science*. CRC Press, Boca Raton, Florida, pp. B-3–B-24.

Hemingway, R.G. (1961) Magnesium, potassium, sodium and calcium contents of herbage as influenced by fertilizer treatments over a three year period. *Journal of the British Grassland Society* 16, 106–116.

Hemingway, R.G. (1962) Copper, molybdenum, manganese and iron contents of herbage as influenced by fertilizer treatments over a three year period. *Journal of the British Grassland Society* 17, 182–187.

Hemingway, R.G. (1963) Soil and herbage potassium levels in relation to yield. *Journal of the Science of Food and Agriculture* 14, 188–195.

Hemingway, R.G., MacPherson, A., Duthie, A.K. and Brown, N.A. (1968) The mineral composition of hay and silage grown in Scotland in relation to the ARC standards for mineral requirements of dairy cattle. *Journal of Agricultural Science, Cambridge* 71, 53–59.

Hemken, R.W. (1979) Factors that influence the iodine content of milk and meat: a review. *Journal of Animal Science* 48, 981–985.

Hemkes, O.J., Kemp, A. and van Broekhoven, L.W. (1980) Accumulation of heavy metals in the soil due to annual dressings with sewage sludge. *Netherlands Journal of Agricultural Science* 28, 228–237.

Heng, L.K., White, R.E., Bolan, N.S. and Scotter, D.R. (1991) Leaching losses of major nutrients from a mole-drained soil under pasture. *New Zealand Journal of Agricultural Research* 34, 325–334.

Henzell, E.F. and Ross, P.J. (1973) The nitrogen cycle of pasture ecosystems. In: Butler, G.W. and Bailey, R.W. (eds) *Chemistry and Biochemistry of Herbage*, Vol. 2. Academic Press, London, pp. 227–246.

Herriott, J.B.D. and Wells, D.A. (1963) The grazing animal and sward productivity. *Journal of Agricultural Science, Cambridge* 61, 89–99.

Hetrick, B.A.D., Kitt, D.G. and Wilson, G.T. (1988) Mycorrhizal dependence and growth habit of warm-season and cool-season tallgrass prairie plants. *Canadian Journal of Botany* 66, 1376–1380.

Hetzel, B.S. and Maberly, G.F. (1986) Iodine. In: Mertz, W. (ed.) *Trace Elements in Human and Animal Nutrition*, 5th edn, Vol. 2. Academic Press, Orlando, pp. 139–208.

Heys, V. and Hill, R. (1984) The selenium concentration of cereal grain and conserved and fresh herbage from farms in England and Wales. *Journal of Agricultural Science, Cambridge* 102, 367–369.

Hidiroglou, M. (1979) Trace element deficiencies and fertility in ruminants: a review. *Journal of Dairy Science* 62, 1195–1206.

Hinedi, Z.R., Chang, A.C. and Lee, R.W.K. (1989) Characterization of phosphorus in sludge extracts using P-31 nuclear magnetic resonance spectroscopy. *Journal of Environmental Quality* 18, 323–329.

Hinsinger, P. (1998) How do plants acquire mineral nutrients? Chemical processes involved in the rhizosphere. *Advances in Agronomy* 64, 225–265.

Hodgson, J. and Spedding, C.R.W. (1966) The health and performance of the grazing animal in relation to fertilizer nitrogen usage. 1. Calves. *Journal of Agricultural Science, Cambridge* 67, 155–167.

Hodgson, J.F., Leach, R.M. and Allaway, W.H. (1962) Micro-nutrients in soils and plants in relation to animal nutrition. *Journal of Agricultural and Food Chemistry* 10, 171–174.

Hogg, D.E. (1981) A lysimeter study of mineral nutrient losses from urine and dung applications on pasture. *New Zealand Journal of Experimental Agriculture* 9, 39–46.

Hogg, D.E. and Karlovsky, J. (1968) The relative effectiveness of various magnesium fertilizers on a magnesium-deficient pasture. *New Zealand Journal of Agricultural Research* 11, 171–183.

Hojjati, S.M., Templeton, W.C. and Taylor, T.H. (1977) Changes in chemical composition of Kentucky bluegrass and tall fescue herbages following N fertilization. *Agronomy Journal* 69, 264–268.

Holecheck, J.L., Galyean, M.L., Wallace, J.D. and Wofford, H. (1985) Evaluation of faecal indices for predicting phosphorus status of cattle. *Grass and Forage Science* 40, 489–492.

Holford, I.C.R. and Crocker, G.J. (1994) Long-term effects of lime on pasture yields and responses to phosphate fertilizers on eight acidic soils. *Australian Journal of Agricultural Research* 45, 1051–1062.

Holter, P. (1979) Effects of dung beetles (*Aphodius* spp.) and earthworms on the disappearance of cattle dung. *Oikos* 32, 393–402.

Hooda, P.S., Moynagh, M., Svoboda, I.F., Edwards, A.C., Anderson, H.A. and Sym, G. (1999) Phosphorus loss in drainflow from intensively managed grassland soils. *Journal of Environmental Quality* 28, 1235–1242.

Hopkins, A., Adamson, A.H. and Bowling, P.J. (1994) Response of permanent and reseeded grassland to fertilizer nitrogen. 2. Effects on concentrations of Ca, Mg, K, Na, S, P, Mn, Zn, Cu, Co and Mo in herbage at a range of sites. *Grass and Forage Science* 49, 9–20.

Hopper, M.J. and Clement, C.R. (1967) The supply of potassium to grassland: an integration of field, pot and laboratory investigations. In: *Transactions of Meeting of Commissions II and IV of the International Society of Soil Science, Aberdeen, 1966*, International Society of Soil Science, Aberdeen, 237–246.

Hoque, S., Heath, S.B. and Killham, K. (1987) Evaluation of methods to assess adequacy of potential soil S supply to crops. *Plant and Soil* 101, 3–8.

Horn, F.P., Reid, R.L. and Jung, G.A. (1974) Iodine nutrition and thyroid function of ewes and lambs on orchardgrass under different levels of nitrogen and microelement fertilization. *Journal of Animal Science* 38, 968–974.

Horst, R.L. (1986) Regulation of calcium and phosphorus homeostasis in the dairy cow. *Journal of Dairy Science* 69, 604–616.

Houba, V.J.G., Novozamsky, I. and van der Lee, J.J. (1986) Inorganic chemical analysis of plant tissue: possibilities and limitations. *Netherlands Journal of Agricultural Science* 34, 449–456.

Houba, V.J.G., Novozamsky, I. and van der Lee, J.J. (1997) Quality aspects in laboratories for soil and plant analysis. In: Hood, T.M. and Benton Jones, J. (eds) *Soil and Plant Analysis in Sustainable Agriculture and Environment*. Marcel Dekker, New York, pp. 31–52.

Hu, S.T., Hannaway, D.B. and Youngberg, H.W. (1992) *Forage Resources of China*. Pudoc, Wageningen, 327 pp.

Hué, N.V. (1995) Sewage sludge. In: Rechcigl, J.E. (ed.) *Soil Amendments and Environmental Quality*. Lewis Publishers, Boca Raton, pp. 199–247.

Hume, I.D. and Bird, P.R. (1970) Synthesis of microbial protein in the rumen IV. The influence of the level and form of dietary sulphur. *Australian Journal of Agricultural Research* 21, 315–322.

Hunt, I.V., Alexander, R.H. and Rutherford, A.A. (1964) The effect of various manuring practices on the magnesium status of spring herbage. *Journal of the British Grassland Society* 19, 224–230.

Hunt, I.V., Frame, J. and Harkess, R.D. (1976) Removal of mineral nutrients by red clover varieties. *Journal of the British Grassland Society* 31, 171–179.

Hunter, B.A., Johnson, M.S. and Thompson, D.J. (1987) Ecotoxicology of copper and cadmium in a contaminated grassland ecosystem. I. Soil and vegetation contamination. *Journal of Applied Ecology* 24, 573–586.

Hurley, L.S. and Keen, C.L. (1987) Manganese. In: Mertz, W. (ed.) *Trace Elements in Human and Animal Nutrition*, 5th edn, Vol. 1. Academic Press, San Diego, pp. 185–223.

Hurley, W.L. and Doane, R.M. (1989) Recent developments in the roles of vitamins and minerals in reproduction. *Journal of Dairy Science* 72, 784–804.

Hylton, L.O., Ulrich, A., Cornelius, D.R. and Okhi, K. (1965) Phosphorus nutrition of Italian ryegrass relative to growth, moisture content and mineral constituents. *Agronomy Journal* 57, 505–508.

Hylton, L.O., Ulrich, A. and Cornelius, D.R. (1967) Potassium and sodium inter-relations in growth and mineral content of Italian ryegrass. *Agronomy Journal* 59, 311–314.

Jackson, M.L. (1964) Chemical composition of soil. In: Bear, F.E. (ed.) *Chemistry of the Soil*, 2nd edn. Reinhold, New York, pp. 71–141.

Jacoby, B. (1995) Nutrient uptake by plants. In: Pessarakli, M. (ed.) *Handbook of Plant and Crop Physiology*. Marcel Dekker, New York, pp. 1–22.

James, D.W., Dhumal, S.S., Rumbaugh, M.D. and Tindall, T.A. (1995) Inheritance of potassium–sodium nutritional traits in alfalfa. *Agronomy Journal* 87, 681–686.

Janzen, H.H. and Ellert, B.H. (1998) Sulfur dynamics in cultivated, temperate agroecosystems. In: Maynard, D.G. (ed.) *Sulfur in the Environment*. Marcel Dekker, New York, pp. 11–43.

Jarrige, R. (1989) *Ruminant Nutrition*. John Libbey Eurotext, London/Paris, 389 pp.

Jarvis, S.C. (1997) Emission processes and their interactions in grassland soils. In: Jarvis, S.C. and Pain, B.F. (eds) *Gaseous Nitrogen Emissions from Grassland Soils*. CAB International, Wallingford, pp. 1–17.

Jarvis, S.C. (1998) Nitrogen management and sustainability. In: Cherney, J.H. and Cherney, D.J.R. (eds) *Grass for Dairy Cattle*. CAB International, Wallingford, pp. 161–192.

Jarvis, S.C. and Austin, A.R. (1983) Soil and plant factors limiting the availability of copper to a beef suckler herd. *Journal of Agricultural Science* 101, 39–46.

Jarvis, S.C. and Hatch, D.J. (1985) Rates of hydrogen ion efflux by nodulated legumes grown in flowing solution culture with continuous pH monitoring and adjustment. *Annals of Botany* 55, 41–51.

Jarvis, S.C. and Whitehead, D.C. (1981) The influence of some soil and plant factors on the concentration of copper in perennial ryegrass. *Plant and Soil* 60, 275–286.

Jarvis, S.C., Hatch, D.J. and Lockyer, D.R. (1989) Ammonia fluxes from grazed grassland: annual losses from cattle production systems and their relation to nitrogen inputs. *Journal of Agricultural Science, Cambridge* 113, 99–108.

Jarvis, S.C., Scholefield, D. and Pain, B. (1995) Nitrogen cycling in grazing systems. In: Bacon, P.E. (ed.) *Nitrogen Fertilization in the Environment*. Marcel Dekker, New York, pp. 381–419.

Jarvis, S.C., Stockdale, A.E., Shepherd, M.A. and Powlson, D.S. (1996a) Nitrogen mineralization in temperate agricultural soils: processes and measurement. *Advances in Agronomy* 57, 187–235.

Jarvis, S.C., Wilkins, R.J. and Pain, B.F. (1996b) Opportunites for reducing the environmental impact of dairy farming managements: a systems approach. *Grass and Forage Science* 51, 21–31.

Jenkinson, D.S. (1969) Radiocarbon dating of soil organic matter. In: *Annual Report, 1968*, Part I. Rothamsted Experimental Station, Harpenden, Hertfordshire, p. 73.

Jenkinson, D.S. (1981) The fate of plant and animal residues in soil. In: Greenland, D.J. and Hayes, M.H.B. (eds) *The Chemistry of Soil Processes*. John Wiley, Chichester, pp. 505–561.

Jensen, E.H. and Lesperance, A.L. (1971) Molybdenum accumulation by forage plants. *Agronomy Journal* 63, 201–204.

Joblin, K.N. (1981) Effect of urine on the elemental composition of spring regrowth herbage in a ryegrass pasture. *New Zealand Journal of Agricultural Research* 24, 293–298.

Joblin, K.N. and Keogh, R.G. (1979) The element composition of herbage at urine patch sites in a ryegrass pasture. *Journal of Agricultural Science, Cambridge* 92, 571–574.

Joblin, K.N. and Pritchard, M.W. (1983) Urinary effect on variations in the selenium and sulphur contents of ryegrass from pasture. *Plant and Soil* 70, 69–76.

John, M.K., Eaton, G.W., Case, V.W. and Chuah, H.H. (1972) Liming of alfalfa (*Medicago sativa* L.) II. Effect on mineral composition. *Plant and Soil* 37, 363–374.

Johnsson, L. (1991) Selenium uptake by plants as a function of soil type, organic matter content and pH. *Plant and Soil* 133, 57–64.

Jones, D.J.C. (1960) The effect of sulphate of ammonia applications on the sulphur content of various grass and clover mixtures. *Journal of Agricultural Science, Cambridge* 54, 188–194.

Jones, D.I.H. and Thomas, T.A. (1987) Minerals in pastures and supplements. In: Snaydon, R.W. (ed.) *Ecosystems of the World: 17B Managed Grasslands.* Elsevier, Amsterdam, pp. 145–153.

Jones, D.L. (1998) Organic acids in the rhizosphere – a critical review. *Plant and Soil* 205, 25–44.

Jones, D.L. and Darrah, P.R. (1994) Role of root derived organic acids in the mobilization of nutrients from the rhizosphere. *Plant and Soil* 166, 247–257.

Jones, D.L., Darrah, P.R. and Kochian, L.V. (1996) Critical evaluation of organic acid mediated iron dissolution in the rhizosphere and its potential role in root iron uptake. *Plant and Soil* 180, 57–66.

Jones, E. (1963) Studies on the magnesium content of mixed herbage and some individual grass and clover species. *Journal of the British Grassland Society* 18, 131–138.

Jones, E. and Sparrow, P.E. (1977) Magnesium uptake by herbage. *Journal of the Science of Food and Agriculture* 28, 905–910.

Jones, J.B., Jr. (1991) Plant tissue analysis in micronutrients. In: Mortvedt, J.J. (ed.) *Micronutrients in Agriculture*, 2nd edn. Soil Science Society of America, Madison, pp. 477–521.

Jones, M.B. (1986) Sulfur availability indexes. In: Tabatabai, M.A. (ed.) *Sulfur in Agriculture*. American Society of Agronomy, Madison, pp. 549–566.

Jones, M.B. and Woodmansee, R.G. (1979) Biogeochemical cycling in annual grassland systems. *Botanical Research* 45, 111–144.

Jones, M.B., Ruckman, J.E., Williams, W.A. and Koenigs, R.L. (1980) Sulfur diagnostic criteria as affected by age and defoliation of subclover. *Agronomy Journal* 72, 1043–1046.

Jones, M.B., Rendig, V.V., Torell, D.T. and Inouye, T.S. (1982) Forage quality for sheep and chemical composition associated with sulfur fertilization on a sulfur deficient site. *Agronomy Journal* 74, 775–780.

Jones, O.L. and Bromfield, S.M. (1969) Phosphorus changes during the leaching and decomposition of hayed-off pasture plants. *Australian Journal of Agricultural Research* 20, 653–663.

Jungk, A. and Claassen, N. (1997) Ion diffusion in the soil–root system. *Advances in Agronomy* 61, 53–110.

Kabata-Pendais, A. and Adriano, D.C. (1995) Trace metals. In: Rechcigl, J.E. (ed.) *Soil Amendments and Environmental Quality*. Lewis Publishers, Boca Raton, pp. 139–167.

Kabata-Pendais, A. and Pendais, H. (1992) *Trace Elements in Soils and Plants*, 2nd edn. CRC Press, Boca Raton, Florida, 365 pp.

Kähäri, J. and Nissinen, H. (1978) The mineral element contents of timothy (*Phleum pratense*) in Finland. 1. The elements calcium, magnesium, phosphorus, potassium, chromium, cobalt, iron, copper, manganese, sodium and zinc. *Acta Agriculturae Scandinavica* Suppl. 20, 26–39.

Kamprath, E.J. and Jones, U.S. (1986) Plant responses to sulphur in the southwestern United States. In: Tabatabai, M.A. (ed.) *Sulfur in Agriculture*. American Society of Agronomy, Madison, pp. 323–343.

Kappel, L.C., Morgan, E.B., Kilgore, L., Ingraham, R.H. and Babcock, D.K. (1985) Seasonal changes of mineral content of southern forages. *Journal of Dairy Science* 68, 1822–1827.

Kelling, K.A. and Matocha, J.E. (1990) Plant analysis as an aid in fertilizing forage crops. In: Westerman, R.L. (ed.) *Soil Testing and Plant Analysis*, 3rd edn. Soil Science Society of America, Madison, pp. 603–643.

Kemp, A. (1960) Hypomagnesaemia in milking cows: the response of serum magnesium to alterations in herbage composition resulting from potash and nitrogen fertilizer applications on pasture. *Netherlands Journal of Agricultural Science* 8, 281–304.

Kemp, A. (1983) The effect of fertilizer treatment of grassland on the biological availability of magnesium to ruminants. In: Fontenot, J.P., Bunce, G.E., Webb, K.E. and Allen, V.G. (eds) *Role of Magnesium in Animal Nutrition*. Virginia Polytechnic Institute, Blacksburg, pp. 143–157.

Kemp, A. and Geurink, J.H. (1978) Grassland farming and minerals in cattle. *Netherlands Journal of Agricultural Science* 26, 161–169.

Kennedy, A.P. and Till, A.R. (1981) The distribution in soil and plant of ^{35}S from sheep excreta. *Australian Journal of Agricultural Research* 32, 339–351.

Keren, R. and Bingham, F.T. (1985) Boron in water, soils and plants. *Advances in Soil Science* 1, 229–276.

Kershaw, E.S. and Banton, C.L. (1965) The mineral content of S22 ryegrass on calcareous loam soil in response to fertilizer treatments. *Journal of the Science of Food and Agriculture* 16, 698–701.

Kershaw, K.A. (1963) Pattern in vegetation and its causality. *Ecology* 44, 377–385.

Keywood, M.D., Chivas, A.R., Fifield, L.K., Cresswell, R.G. and Ayers, G.P. (1997) The accession of chloride to the western half of the Australian continent. *Australian Journal of Soil Research* 35, 1177–1189.

Kidambi, S.P., Matches, A.G. and Bolger, T.P. (1990) Mineral concentrations in alfalfa and sainfoin as influenced by soil moisture level. *Agronomy Journal* 82, 229–236.

Kiekens, L. (1995) Zinc. In: Alloway, B.J. (ed.) *Heavy Metals in Soils*, 2nd edn. Blackie Academic and Professional, London, pp. 284–305.

Killham, K. (1985) Vesicular-arbuscular mycorrhizal mediation of trace and minor element uptake in perennial grasses: relation to livestock herbage. In: Fitter, A.H.

(ed.) *Ecological Interaction in Soil*. Blackwell Scientific Publications, Oxford, pp. 225–232.

Kilmer, V.J. (1979) Minerals and agriculture. In: Trudinger, P.A. amd Swaine, D.J. (eds) *Biogeochemical Cycling of Mineral-forming Elements*. Elsevier, Amsterdam, pp. 515–558.

Kilmer, V.J., Bennett, O.L., Stahly, V.F. and Timmons, D.R. (1960) Yield and mineral composition of eight forage species grown at four levels of soil moisture. *Agronomy Journal* 52, 282–285.

Kincaid, R. (1988) Macro elements for ruminants. In: Church, D.C. (ed.) *The Ruminant Animal: Digestive Physiology and Nutrition*. Prentice Hall, Englewood Cliffs, New Jersey, pp. 326–341.

Kirchgessner, M., Merz, G. and Oelschlager, W. (1960) Der Einfluss der Vegetationsstadiums auf den Mengen- und Spurenelementgehalt drier Grasarten. *Archiv für Tierernährung* 10, 414–427.

Kirchmann, H. and Witter, E. (1992) Composition of fresh, aerobic and anaerobic farm animal dungs. *Bioresource Technology* 40, 137–142.

Kirkman, J.H., Basker, A., Surapaneni, A. and MacGregor, A.N. (1994) Potassium in the soils of New Zealand – a review. *New Zealand Journal of Agricultural Research* 37, 207–227.

Knauer, N. (1966) The use of plant analysis for determining the P and K needs of grasslands. In: *Proceedings of the 10th International Grassland Congress, Helsinki, 1966*, Valtioneuvoston Kirjapaino, Helsinki, 171–174.

Koch, J.T., Rachar, D.B. and Kay, B.D. (1989) Microbial participation in iodide removal from solution by organic soils. *Canadian Journal of Soil Science* 69, 127–135.

Kochian, L.V. (1991) Mechanisms of micronutrient uptake and translocation in plants. In: Mortvedt, J.J. (ed.) *Micronutrients in Agriculture*, 2nd edn. Soil Science Society of America, Madison, pp. 229–296.

Kochian, L.V. (1993) Zinc absorption from hydroponic solutions by plant roots. In: Robson, A.D. (ed.) *Zinc in Soils and Plants*. Kluwer Academic Publishers, Dordrecht, pp. 45–57.

Kochian, L.V. and Lucas, W.J. (1988) Potassium transport in roots. *Advances in Botany* 15, 93–178.

Kowalenko, C.G. (1993) Extraction of available sulfur. In: Carter, M.R. (ed.) *Soil Sampling and Methods of Analysis*. Lewis Publishers, Boca Raton, pp. 65–74.

Krauskopf, K.B. and Bird, D.K. (1995) *Introduction to Geochemistry*, 3rd edn. McGraw-Hill, New York, 647 pp.

Kreil, W., Wacker, G., Kaltofen, H. and Hey, E. (1966) Heavy nitrogen fertilizing to pasture. In: *Proceedings of 9th International Grassland Congress, São Paulo, 1965*, Agricultural Secretariat, São Paulo, 1093–1098.

Kubota, J. (1964) Cobalt content of New England soils in relation to cobalt levels in forage for ruminant animals. *Soil Science Society of America Proceedings* 28, 246–251.

Kubota, J. (1983a) Copper status of United States soils and forage plants. *Agronomy Journal* 75, 913–918.

Kubota, J. (1983b) Soils and plants and the geochemical environment. In: Thornton, I. (ed.) *Applied Environmental Geochemistry*. Academic Press, London, pp. 103–122.

Kubota, J., Lemon, E.R. and Allaway, W.H. (1963) The effect of soil moisture content upon the uptake of molybdenum, copper and cobalt by alsike clover. *Proceedings of the Soil Science Society of America* 27, 679–683.

Kubota, J., Oberley, G.H. and Naphan, E.A. (1980) Magnesium in grasses at three selected regions in the United States and its relation to grass tetany. *Agronomy Journal* 72, 907–914.

Kubota, J., Welch, R.M. and Van Campen, D. (1987) Soil-related nutritional problem areas for grazing animals. *Advances in Soil Science* 6, 189–215.

Kucey, R.M.N., Janzen, H.H. and Leggett, M.E. (1989) Microbially mediated increases in plant-available phosphorus. *Advances in Agronomy* 42, 199–228.

Låg, J. (1987) Soil properties of special interest in connection with health problems. *Experientia* 43, 63–67.

Laidlaw, A.S., Christie, P. and Lee, H.W. (1996) Effect of white clover cultivar on apparent transfer of nitrogen from clover to grass and estimation of relative turnover rates of nitrogen in roots. *Plant and Soil* 179, 243–253.

Lambert, J. and Toussaint, B. (1978) [An investigation of the factors influencing the phosphorus content of herbage.] *Phosphore et Agriculture* 73, 1–12.

Lambert, T.L. and Blincoe, C. (1971) High concentrations of cobalt in wheat grasses. *Journal of the Science of Food and Agriculture* 22, 8–9.

Lamond, R.E., Whitney, D.A. and Marsh, B.H. (1995) Sulfur fertilization of smooth bromegrass in Kansas. *Agronomy Journal* 87, 13–16.

Lane, J.C. and Fleming, G.A. (1967) The effect of stage of maturity and season on the uptake of selenium, molybdenum and copper by perennial ryegrass. In: *Transactions International Society of Soil Science, Commissions II and IV, Aberdeen, 1966*, International Society of Soil Science, Aberdeen, 289–297.

Langlands, J.P. (1987) Assessing the nutrient status of herbivores. In: Hacker, J.B. and Ternouth, J.H. (eds) *The Nutrition of Herbivores*. Academic Press, Sydney, pp. 363–390.

Langlands, J.P. and Sutherland, H.A.M. (1973) Sulphur as a nutrient for Merino sheep. I. *British Journal of Nutrition* 30, 529–535.

Langlands, J.P., Bowles, J.E., Donald, G.E. and Smith, A.J. (1982) The nutrition of ruminants grazing native and improved pasture V. Effects of stocking rate and soil ingestion on the copper and molybdenum status of grazing sheep. *Australian Journal of Agricultural Research* 30, 565–575.

Langlands, J.P., Bowles, J.E., Donald, G.E. and Smith, A.J. (1986) Selenium excretion in sheep. *Australian Journal of Agricultural Research* 37, 201–209.

Lantinga, E.A., Keuning, J.A., Groenwold, J. and Deenen, P.A.G. (1987) Distribution of excreted nitrogen by grazing cattle and its effects on sward quality, herbage production and utilization. In: Van der Meer, H.G., Unwin, R.J., van Dijk, T.A. and Ennik, G.C. (eds) *Animal Manure on Grassland and Fodder Crops: Fertilizer or Waste*. Martinus Nijhoff, Dordrecht, pp. 103–117.

Lanyon, L.E. and Smith, P.W. (1985) Potassium nutrition of alfalfa and other forage legumes. In: Munson, R.D. (ed.) *Potassium in Agriculture*. American Society of Agronomy, Madison, pp. 861–893.

Laredo, M.A. and Minson, D.J. (1975) The voluntary intake and digestibility by sheep of leaf and stem fractions of *Lolium perenne*. *Journal of the British Grassland Society* 30, 73–77.

Larvor, P. (1983) The pools of cellular nutrients: minerals. In: Riis, P.M. (ed.) *Dynamic Biochemistry of Animal Production*. Elsevier, Amsterdam, pp. 281–317.

Laughlin, W.M., Martin, P.F. and Smith, G.R. (1973) Potassium rate and source influences on yield and composition of bromegrass forage. *Agronomy Journal* 65, 85–87.

Laurie, S.H. and Manthey, J.A. (1994) The chemistry and role of metal ion chelation in plant uptake processes. In: Manthey, J.A., Crowley, D.E. and Luster, D.G. (eds) *Biochemistry of Metal Micronutrients in the Rhizosphere.* Lewis Publishers, Boca Raton, pp. 165–182.

Ledgard, S.F. and Giller, K.E. (1995) Atmospheric N_2 fixation as an alternative N source. In: Bacon, P.E. (ed.) *Nitrogen Fertilizer in the Environment.* Marcel Dekker, New York, pp. 443–486.

Ledgard, S.F. and Saunders, W.M.H. (1982) Effects of nitrogen fertilizer and urine on pasture performance and the influence of soil phosphorus and potassium status. *New Zealand Journal of Agricultural Research* 25, 541–547.

Ledgard, S.F. and Upsdell, M.P. (1991) Sulphur inputs from rainfall throughout New Zealand. *New Zealand Journal of Agricultural Research*, 34, 105–111.

Ledgard, S.F., Steele, K.W. and Saunders, W.M.H. (1982) Effects of cow urine and its major constituents on pasture properties. *New Zealand Journal of Agricultural Research* 25, 61–68.

Ledgard, S.F., Sprosen, M.S. and Steele, K.W. (1996) Nitrogen fixation by nine white clover cultivars in grazed pasture as affected by nitrogen fertilization. *Plant and Soil* 178, 193–203.

Lee, J., Masters, D.G., White, C.L., Grace, N.D. and Judson, G.J. (1999) Current issues in trace element nutrition of grazing livestock in Australia and New Zealand. *Australian Journal of Agricultural Research* 50, 1341–1364.

Leech, A.F. and Thornton, I. (1987) Trace elements in soils and pasture herbage on farms with bovine hypocupraemia. *Journal of Agricultural Science, Cambridge* 108, 591–597.

Lemaire, G. and Chapman, D. (1996) Tissue flows in grazed plant communities. In: Hodgson, J. and Illius, A.W. (eds) *The Ecology and Management of Grazing Systems.* CAB International, Wallingford, pp. 3–36.

Lessard, J.R., Hidiroglou, M., Carson, R.B. and Dermine, P. (1968) Intra-seasonal variations in the selenium content of various forage crops at Kapuskasing, Ontario. *Canadian Journal of Plant Science* 48, 581–585.

Levander, O.A. (1986) Selenium. In: Mertz, W. (ed.) *Trace Elements in Human and Animal Nutrition*, 5th edn, Vol. 2. Academic Press, Orlando, pp. 209–279.

Levi-Minzi, R., Riffaldi, R. and Saviozzi, A. (1986) Organic matter and nutrients in fresh and mature farmyard manure. *Agricultural Wastes* 16, 225–236.

Lewis, D.C. (1992) Effect of plant age on the critical inorganic and total phosphorus concentrations in selected tissues of subterranean clover. *Australian Journal of Agricultural Research* 43, 215–223.

Li, M., Osaki, M., Rao, I.M. and Tadano, T. (1997) Secretion of phytase from the roots of several plant species under phosphorus deficient conditions. *Plant and Soil* 195, 161–169.

Li, X.-L., Marschner, H. and George, E. (1991) Acquisition of phosphorus and copper by VA-mycorrhizal hyphae and root-to-shoot transport in white clover. *Plant and Soil* 136, 49–57.

Lightner, J.W., Rhykerd, C.L., Van Scoyoc, G.E., Mengel, D.B., Noller, C.H. and Blair, B.O. (1981) Effect of N-P-K fertilization on mineral composition of *Poa pratensis* L. grown on a shallow muck soil. In: *Proceedings of the 14th*

International Grassland Congress, Lexington, Westview Press, Boulder, Colorado, pp. 294–296.

Lightner, J.W., Rhykerd, C.L., Mengel, D.B., Van Scoyoe, G.E., Hood, E.L. and Noller, C.H. (1983) Influence of NPK fertilization on calcium and magnesium in *Poa pratensis* L. with reference to dietary requirements of grazing cattle. *Communications in Soil Science and Plant Analysis* 14, 121–130.

Lin, Z.-Q., Hansen, D., Zayed, A. and Terry, N. (1999) Biological selenium volatilization: method of measurement under field conditions. *Journal of Environmental Quality* 28, 309–315.

Lindsay, W.L. (1991) Inorganic equilibria affecting micronutrients in soils. In: Mortvedt, J.J. (ed.) *Micronutrients in Agriculture.* Soil Science Society of America, Madison, pp. 89–112.

Lipton, D.S., Blanchar, R.W. and Blevins, D.G. (1987) Citrate, malate and succinate concentrations in exudates from P-sufficient and P-stressed *Medicago sativa* L. seedlings. *Plant Physiology* 85, 315–317.

Little, D.A. (1981) Utilization of minerals. In: Hacker, J.B. (ed.) *Nutritional Limits to Animal Production from Pastures.* Commonwealth Agricultural Bureaux, Farnham Royal, pp. 259–283.

Lloyd, A. (1994) Effectiveness of cattle slurry as a sulphur source for grass cut for silage. *Grass and Forage Science* 49, 203–208.

Lockyer, D.R. (1985) The effect of sulphur dioxide on the growth of four grasses. *Journal of Experimental Botany* 36, 1851–1859.

Loneragan, J.F. (1988) Distribution and movement of manganese in plants. In: Graham, R.D., Hannan, R.J. and Uren, N.C. (eds) *Manganese in Soils and Plants.* Kluwer Academic Publishers, Dordrecht, pp. 113–124.

Loneragan, J.F. and Webb, M.J. (1993) Interactions between zinc and other nutrients affecting the growth of plants. In: Robson, A.D. (ed.) *Zinc in Soils and Plants.* Kluwer Academic Publishers, Dordrecht, pp. 119–134.

Longnecker, N.E. and Robson, A.D. (1993) Distribution and transport of zinc in plants. In: Robson, A.D. (ed.) *Zinc in Soils and Plants.* Kluwer Academic Publishers, Dordrecht, pp. 79–91.

Loper, G.M. and Smith, D. (1961) *Changes in Micronutrient Composition of the Herbage of Alfalfa, Medium Red Clover, Ladino Clover and Bromegrass with Advance in Maturity.* Research Report no. 8, Agricultural Experiment Station, University of Wisconsin, Madison, Wisconsin, USA.

Lotero, J., Woodhouse, W.W. and Petersen, R.G. (1966) Local effect on fertility of urine voided by grazing cattle. *Agronomy Journal* 58, 262–265.

Lowrance, R.R., Leonard, R.A. and Asmussen, L.E. (1985) Nutrient budgets for agricultural watersheds in the southeastern coastal plain. *Ecology* 66, 287–296.

Lunt, O.R., Branson, R.L. and Clark, S.B. (1966) Response of five grass species to phosphorus on six soils. In: *Proceedings of the 9th International Grassland Congress, São Paulo, 1965,* Agricultural Secretariat, São Paulo, pp. 1697–1701.

Lyttleton, J.W. (1973) Proteins and nucleic acids. In: Butler, G.W. and Bailey, R.W. (eds) *Chemistry and Biochemistry of Herbage,* Vol. 1. Academic Press, London, pp. 63–103.

Maas, J. (1998) Selenium metabolism in grazing ruminants: deficiency, supplementation and environmental implications. In: Frankenberger, W.T. and Engberg, R.A. (eds) *Environmental Chemistry of Selenium.* Marcel Dekker, New York, pp. 113–128.

McAdam, P.A. and O'Dell, G.D. (1982) Mineral profile of blood plasma of lactating dairy cows. *Journal of Dairy Science* 65, 1219–1226.

McBride, M.B. (1989) Reactions controlling heavy metal solubility in soils. *Advances in Soil Science* 10, 1–56.

McBride, M.B. (1994) *Environmental Chemistry of Soils*. Oxford University Press, New York, 406 pp.

McConaghy, S., McAllister, J.S.V., Todd, J.R., Rankin, J.E.F. and Kerr, J. (1963) The effects of magnesium compounds and of fertilizers on the mineral composition of herbage and on the incidence of hypomagnesaemia in dairy cows. *Journal of Agricultural Science, Cambridge* 60, 313–328.

McCoy, M.A., Smyth, J.A., Ellis, W.A., Arthur, J.R. and Kennedy, D.G. (1997) Experimental reproduction of iodine deficiency in cattle. *Veterinary Record* 141, 544–547.

MacDiarmid, B.N. and Watkin, B.R. (1971) The cattle dung patch. I. Effect of dung patches on yield and botanical composition of surrounding and underlying pasture. *Journal of the British Grassland Society* 26, 239–245.

MacDiarmid, B.N. and Watkin, B.R. (1972) The cattle dung patch. 3. Distribution and rate of decay of dung patches and their influence on grazing behaviour. *Journal of the British Grassland Society* 27, 48–54.

McDonald, P., Edwards, R.A., Greenhalgh, J.F.D. and Morgan, C.A. (1995) *Animal Nutrition*, 5th edn. Longman, Harlow, 607 pp.

McDowell, L.R. (1992) *Minerals in Animal and Human Nutrition*. Academic Press, San Diego, 524 pp.

McFarlane, J.D. (1989) The effect of copper supply on vegetative and seed yield of pasture legumes and the field calibration of a tissue test for detecting copper deficiency. 1. Subterranean clover. *Australian Journal of Agricultural Research* 40, 817–832.

McGrath, D. and Fleming, G.A. (1988) Iodine levels in Irish soils and grasses. *Irish Journal of Agricultural Research* 27, 75–81.

McGrath, D., Poole, D.B.R., Fleming, G.A. and Sinnott, J. (1982) Soil ingestion by grazing sheep. *Irish Journal of Agricultural Research* 21, 135–145.

McGrath, S.P. and Goulding, K.W.T. (1990) Deposition of sulphur and nitrogen from the atmosphere. *Journal of the Science of Food and Agriculture* 53, 426–427.

McGrath, S.P. and Jarvis, S.C. (1993) Recent considerations of grassland 'soil quality' in temperate regions. In: Baker, M.J. (ed.) *Grasslands for Our World*. SIR Publishing, Wellington, New Zealand, pp. 512–521.

McGrath, S.P. and Loveland, P.J. (1992) *The Soil Geochemical Atlas of England and Wales*. Blackie Academic and Professional, London, 101 pp.

McIntosh, S., Crooks, P. and Simpson, K. (1973a) Sources of magnesium for grassland. *Journal of Agricultural Science, Cambridge* 81, 507–511.

McIntosh, S., Crooks, P. and Simpson, K. (1973b) The effects of applied N, K and Mg on the distribution of magnesium in the plant. *Plant and Soil* 39, 389–397.

Mackay, A.D., Saggar, S., Trolove, S.N. and Lambert, M.G. (1995) Use of an unsorted pasture sample in herbage testing for sulphur, phosphorus and nitrogen. *New Zealand Journal of Agricultural Research* 38, 483–493.

Macklon, A. (1988) Trace element uptake by roots. In: *Annual Report, 1987*, Macaulay Land Use Research Institute, Aberdeen, pp. 73–74.

Macklon, A.E.S., Mackie-Dawson, L.A., Sim, A., Shand, C.A. and Lilly, A. (1994) Soil P resources, plant growth and rooting characteristics in nutrient-poor upland grasslands. *Plant and Soil* 163, 257–266.

McLaren, R.G. (1975) Marginal S supplies for grassland herbage in south-east Scotland. *Journal of Agricultural Science, Cambridge* 85, 571–573.

McLaren, R.G. (1976) Effect of fertilizers on the S content of herbage. *Grass and Forage Science* 31, 99–104.

McLaren, R.G., Lawson, D.M., Swift, R.S. and Purves, D. (1985) The effects of cobalt additions on soil and herbage cobalt concentrations in some southeast Scotland pastures. *Journal of Agricultural Science, Cambridge* 105, 347–363.

McLaren, R.G., McLenaghan, R.D. and Swift, R.S. (1990) Boron applications to pastures: effect on herbage and soil boron concentrations. *New Zealand Journal of Agricultural Research* 33, 277–284.

McLaren, R.G., McLenaghan, R.D. and Swift, R.S. (1991) Zinc applications to pastures: effect on herbage and soil zinc concentrations. *New Zealand Journal of Agricultural Research* 34, 113–118.

MacLeod, L.B. (1965) Effect of nitrogen and potassium on the yield and chemical composition of alfalfa, bromegrass, orchardgrass and timothy grown as pure species. *Agronomy Journal* 57, 261–266.

McManus, W.R., Robinson, V.N.E. and Grout, L.L. (1977) The physical distribution of mineral material on forage plant cell walls. *Australian Journal of Agricultural Research* 28, 651–662.

McNaught, K.J. (1958) Potassium deficiency in pastures. 1. Potassium content of legumes and grasses. *New Zealand Journal of Agricultural Research* 1, 148–181.

McNaught, K.J. (1959) Effect of potassium fertilizer on sodium, magnesium and calcium in plant tissues. *New Zealand Journal of Agriculture* 99, 442.

McNaught, K.J. (1970) Diagnosis of mineral deficiencies in grass-legume pastures by plant analysis. In: *Proceedings, 11th International Grassland Congress, Surfers' Paradise, Queensland*, University of Queensland Press, St. Lucia, 333–338.

McNaught, K.J. and Christoffels, P.J.E. (1961) Effect of sulphur deficiency on sulphur and nitrogen levels in pastures and lucerne. *New Zealand Journal of Agricultural Research* 4, 177–196.

McNaught, K.J. and Karlovsky, J. (1964) Sodium chloride responses in pasture on a potassium-deficient soil. *New Zealand Journal of Agricultural Research* 7, 386–404.

McNaught, K.J., Dorofaeff, F.D., Ludecke, T.E. and Cottier, K. (1973) Effect of potassium fertilizer, soil magnesium status and soil type on uptake of magnesium by pasture plants from magnesium fertilizers. *New Zealand Journal of Experimental Agriculture* 1, 329–347.

MacNicol, R.D. and Beckett, P.H.T. (1985) Critical tissue concentrations of potentially toxic elements. *Plant and Soil* 85, 107–129.

MacPherson, A. (1989) Trace element deficiency in ruminants. *Outlook on Agriculture* 18, 124–132.

MacPherson, A. (1994) Selenium, vitamin E and biological oxidation. In: Garnsworthy, P.C. and Cole, D.J.A. (eds) *Recent Advances in Animal Nutrition, 1994.* Nottingham University Press, Nottingham, pp. 3–30.

McSweeney, C.S., Cross, R.B., Wholohan, B.T. and Murphy, M.R. (1988) Diagnosis of sodium status in small ruminants. *Australian Journal of Agricultural Research* 39, 935–942.

Mäkälä, A. (1967) On the water-solubility of plant minerals. *Journal of the Scientific Agricultural Society of Finland* 39, 166–181.

Mansell, G.P., Syers, J.K. and Gregg, P.E.H. (1981) Plant availability of phosphorus in dead herbage ingested by surface-casting earthworms. *Soil Biology and Biochemistry* 13, 163–167.

Manthey, J.A., Luster, D.G. and Crowley, D.E. (1994) Biochemistry of metal micronutrients in the rhizosphere: an introduction. In: Manthey, J.A., Crowley, D.E. and Luster, D.G. (eds) *Biochemistry of Metal Micronutrients in the Rhizosphere*. Lewis Publishers, Boca Raton, pp. 1–13.

Markus, D.K. and Battle, W.R. (1965) Soil and plant responses to long-term fertilization of alfalfa (*Medicago sativa* L.). *Agronomy Journal* 57, 613–616.

Marriott, C. (1988) The transfer of fixed N from clover to grass. In: *Annual Report, 1987*. Macaulay Land Use Research Institute, Aberdeen, p. 81.

Marschner, H. (1988) Mechanisms of manganese acquisition by roots from soils. In: Graham, R.D., Hannan, R.J. and Uren, N.C. (eds) *Manganese in Soils and Plants*. Kluwer Academic Publishers, Dordrecht, pp. 191–204.

Marschner, H. (1993) Zinc uptake from soils. In: Robson, A.D. (ed.) *Zinc in Soils and Plants*. Kluwer Academic Publishers, Dordrecht, pp. 59–77.

Marschner, H. (1995) *Mineral Nutrition of Higher Plants*, 2nd edn. Academic Press, London, 889 pp.

Marschner, H. and Dell, B. (1994) Nutrient uptake in mycorrhizal symbiosis. *Plant and Soil* 159, 89–102.

Marsh, K.B., Tillman, R.W. and Syers, J.K. (1992) Effect of changes in pH on sulphate and phosphate retention in soils. *New Zealand Journal of Agricultural Research* 35, 93–100.

Marsh, R. and Campling, R.C. (1970) Fouling of pastures by dung. *Herbage Abstracts* 40, 123–130.

Martens, H., Kubel, O.W. and Gabel, G. (1987) Effects of low sodium intake on magnesium metabolism of sheep. *Journal of Agricultural Science, Cambridge* 108, 237–243.

Mason, B. and Moore, C.E. (1982) *Principles of Geochemistry*, 4th edn. Wiley, New York, 344 pp.

Mason, R.W. (1976) Milk iodine content as an estimate of the dietary iodine status of sheep. *British Veterinary Journal* 132, 374–379.

Mason, V.C., Kessank, P., Ononiwu, J.C. and Narang, M.P. (1981) Factors influencing faecal nitrogen excretion in sheep. 2. *Zeitschrift für Tierphysiologie, Tierernährung und Futtermittelkunde* 45, 174–184.

Mathews, B.W., Sollenberger, L.E., Nair, V.D. and Staples, C.R. (1994) Impact of grazing management on soil nitrogen, phosphorus, potassium and sulphur distribution. *Journal of Environmental Quality* 23, 1006–1013.

Mathews, B.W., Tritschler, J.P. and Miyasaka, S.C. (1998) Phosphorus management and sustainability. In: Cherney, J.H. and Cherney, D.J.R. (eds) *Grass for Dairy Cattle*. CAB International, Wallingford, pp. 193–222.

Mayland, H.F. (1994) Selenium in plant and animal nutrition. In: Frankenberger, W.T. and Benson, S. (eds) *Selenium in the Environment*. Marcel Dekker, New York, pp. 29–45.

Mayland, H.F. and Wilkinson, S.R. (1989) Soil factors affecting magnesium availability in plant–animal systems: a review. *Journal of Animal Science* 67, 3437–3444.

Mayland, H.F. and Wilkinson, S.R. (1996) Mineral nutrition. In: Mosier, L.E., Buxton, D.R. and Casler, M.D. (eds) *Cool Season Forage Grasses*. American Society of Agronomy, Madison, pp. 165–191.

Mayland, H.F., Florence, A.R., Rosenau, R.C., Lazar, V.A. and Turner, H.A. (1975) Soil ingestion by catttle on semi-arid range as reflected by titanium analysis of feces. *Journal of Range Management* 28, 448–452.

Mayland, H.F., James, F., Panter, K.E. and Sonderegger, J.L. (1989) Selenium in seleniferous environments. In: Jacobs, L.W. (ed.) *Selenium in Agriculture and the Environment*. American Society of Agronomy, Madison, pp. 15–50.

Mays, D.A., Wilkinson, S.R. and Cole, C.V. (1980) Phosphorus nutrition of forages. In: Khasawneh, F.E., Sample, E.C. and Kamprath, E.J. (eds) *The Role of Phosphorus in Agriculture*. American Society of Agronomy, Madison, pp. 805–846.

Mazzolini, A.P., Jeffery, J.J., Uren, N.C. and Legge, G.J.F. (1983) Distribution of copper and other elements in ryegrass roots determined with a scanning proton microprobe. *Australian Journal of Plant Physiology* 10, 153–166.

Mehra, A. and Farago, M.E. (1994) Metal ions and plant nutrition. In: Farago, M.E. (ed.) *Plants and the Chemical Elements*. VCH, Weinheim, pp. 31–66.

Melsted, S.W., Motto, H.L. and Peck, T.R. (1969) Critical plant nutrient composition values useful in interpreting plant analysis data. *Agronomy Journal* 61, 17–20.

Mengel, K. and Kirkby, E.A. (1987) *Principles of Plant Nutrition*, 4th edn. International Potash Institute, Worblaufen-Bern, 687 pp.

Mengel, K. and Steffens, D. (1985) Potassium uptake of ryegrass (*Lolium perenne*) and red clover (*Trifolium pratense*) as related to root parameters. *Biology and Fertility of Soils* 1, 53–58.

Metson, A.J. (1974) Magnesium in New Zealand soils I. *New Zealand Journal of Experimental Agriculture* 2, 277–319.

Metson, A.J. (1979) Sulphur in New Zealand soils. 1. *New Zealand Journal of Agricultural Research* 22, 95–114.

Metson, A.J. and Saunders, W.M.H. (1978a) Seasonal variations in chemical composition of pasture I. Calcium, magnesium, potassium, sodium, and phosphorus. *New Zealand Journal of Agricultural Research* 21, 341–353.

Metson, A.J. and Saunders, W.M.H. (1978b) Seasonal variations in chemical composition of pasture II. Nitrogen, sulphur, and soluble carbohydrate. *New Zealand Journal of Agricultural Research* 21, 355–364.

Metson, A.J., Gibson, E.J., Hunt, J.L. and Saunders, W.M.H. (1979) Seasonal variations in chemical composition of pasture III. Silicon, aluminium, iron, zinc, copper and manganese. *New Zealand Journal of Agricultural Research*, 22, 309–318.

Michell, A.R. (1985) Sodium in health and disease: a comparative review with emphasis on herbivores. *Veterinary Record* 116, 653–657.

Mika, V. (1975) [Levels of nutrients in roots and in above-ground mass of a grass stand]. *Die Bodenkultur* 26, 146–162.

Mikkelsen, R.L. and Camberato, J.J. (1995) Potassium, sulphur, lime and micronutrient fertilizers. In: Rechcigl, J.E. (ed.) *Soil Amendments and Environmental Quality*. Lewis Publishers, Boca Racon, pp. 109–137.

Mikkelsen, R.L., Page, A.L. and Bingham, F.T. (1989) Factors affecting selenium accumulation by agricultural crops. In: Jacobs, L.W. (ed.) *Selenium in Agriculture and the Environment*. American Society of Agronomy, Madison, pp. 65–94.

Miles, D.G., ap Griffith, G. and Walters, R.J.K. (1964) Variation in the chemical composition of four grasses. In: *Annual Report, 1963.* Welsh Plant Breeding Station, Aberystwyth, pp. 110–114.

Millar, K.R., Craig, J. and Dawe, L. (1973) α-Tocopherol and selenium levels in pasteurised cows' milk from different areas of New Zealand. *New Zealand Journal of Agricultural Research* 16, 301–303.

Millard, P., Sharp, G.S. and Scott, N.M. (1985) The effect of sulphur deficiency on the uptake and incorporation of nitrogen in ryegrass. *Journal of Agricultural Science, Cambridge* 105, 501–504.

Millard, P., Gordon, A.H., Richardson, A.J. and Chesson, A. (1987) Reduced ruminal degradation of ryegrass caused by sulphur limitation. *Journal of the Science of Food and Agriculture* 40, 305–314.

Miller, E.R., Lei, X. and Ullrey, D.E. (1991) Trace elements in animal nutrition. In: Mortvedt, J.J. (ed.) *Micronutrients in Agriculture*, 2nd edn. Soil Science Society of America, Madison, pp. 593–662.

Miller, J.K., Swanson, E.W. and Spalding, G.E. (1975) Iodine absorption, excretion, recycling and tissue distribution in the dairy cow. *Journal of Dairy Science* 58, 1578–1593.

Miller, W.J. (1970) Zinc nutrition of cattle: a review. *Journal of Dairy Science* 53, 1123–1135.

Miller, W.J. (1979) *Dairy Cattle Feeding and Nutrition.* Academic Press, New York, 411 pp.

Miller, W.J., Adams, W.E., Nussbaumer, R., McCreery, R.A. and Perkins, H.F. (1964) Zinc content of Coastal Bermudagrass as influenced by frequency and season of harvest, location and level of N and lime. *Agronomy Journal* 56, 198–201.

Mills, C.F. (1992) Trace element investigations in man and animals. In: Widdowson, E.M. and Mathers, J.C. (eds) *The Contribution of Nutrition to Human and Animal Health.* Cambridge University Press, Cambridge, pp. 162–173.

Mills, C.F. and Davis, G.K. (1987) Molybdenum. In: Mertz, W. (ed.) *Trace Elements in Human and Animal Nutrition*, 5th edn, Vol. 1. Academic Press, San Diego, pp. 429–463.

Miltimore, J.E., Mason, J.L. and Ashby, D.L. (1970) Copper, zinc, manganese and iron variation in five feeds for ruminants. *Canadian Journal of Animal Science* 50, 293–300.

Ministry of Agriculture, Fisheries and Food (1975) *ADAS Science Service Annual Report, 1974.* HMSO, London, 116 pp.

Ministry of Agriculture, Fisheries and Food (1993) *Code of Good Agricultural Practice for the Protection of Soil.* MAFF, London, 55 pp.

Ministry of Agriculture, Fisheries and Food (1994) *Fertiliser Recommendations for Agricultural and Horticultural Crops*, 6th edn. Reference Book 209, HMSO, London.

Minson, D.J. (1990) *Forage in Ruminant Nutrition.* Academic Press, London, 483 pp.

Mitchell, R.L. (1963) Soil aspects of trace element problems in plants and animals. *Journal of the Royal Agricultural Society* 124, 75–86.

Mitchell, R.L., Reith, J.W.S. and Johnston, I.M. (1957a) Soil copper status and plant uptake. In: *Plant Analysis and Fertilizer Problems.* IRHO, Paris, pp. 249–259.

Mitchell, R.L., Reith, J.W.S. and Johnston, I.M. (1957b) Trace element uptake in relation to soil content. *Journal of the Science of Food and Agriculture* 8, S51–S59.

Moraghan, J.T. and Mascagni, H.J. (1991) Environmental and soil factors affecting micronutrient deficiencies and toxicities. In: Mortvedt, J.J. (ed.) *Micronutrients in Agriculture*, 2nd edn. Soil Science Society of America, Madison, pp. 371–425.

Moré, E. and Coppenet, M. (1980) [Selenium contents of forage plants: effects of fertilization and selenite applications]. *Annales Agronomiques* 31, 297–317.

Morris, E.R. (1987) Iron. In: Mertz, W. (ed.) *Trace Elements in Human and Animal Nutrition*, 5th edn, Vol. 1. Academic Press, San Diego, pp. 79–142.

Morris, J.G. (1980) Assessment of sodium requirement of grazing beef cattle: a review. *Journal of Animal Science* 50, 145–152.

Morrison, I.N. and Cohen, A.S. (1980) Plant uptake, transport and metabolism. In: Hutzinger, O. (ed.) *The Handbook of Environmental Chemistry*, Vol. 2A. Springer Verlag, Berlin, pp. 193–219.

Morrison, J., Jackson, M.V. and Sparrow, P.E. (1980) *The Response of Perennial Ryegrass to Fertilizer Nitrogen in Relation to Climate and Soil.* Technical Report 27, Grassland Research Institute, Hurley.

Morse, D., Head, H.H., Wilcox, C.J., Van Horn, H.H., Hissen, C.D. and Harris, B. (1992) Effects of concentration of dietary phosphorus on amount and route of excretion. *Journal of Dairy Science* 75, 3039–3049.

Mortensen, W.P., Baker, A.S. and Dermanis, P. (1964) Effects of cutting frequency of orchardgrass and nitrogen rate on yield, plant nutrient composition and removal. *Agronomy Journal* 56, 316–320.

Morton, J.D. and Baird, D.B. (1990) Spatial distribution of dung patches under sheep grazing. *New Zealand Journal of Agricultural Research* 33, 285–294.

Morton, J.D., Sinclair, A.G., Johnstone, P.D., Smith, L.C., O'Connor, M.B., Roberts, A.H.C., Risk, W.H., Nguyen, L. and Shannon, P.W. (1995) Effects of frequency of application of triple superphosphate amd Sechura phosphate rock on pasture DM production, herbage P concentration and Olsen P soil tests. *New Zealand Journal of Agricultural Research* 38, 543–552.

Morton, J.D., Smith, L.C. and Metherell, A.K. (1999) Pasture yield responses to phosphorus, sulphur and potassium applications in North Otago soils. *New Zealand Journal of Agricultural Research* 42, 133–146.

Moseley, G. and Baker, D.H. (1991) The efficacy of a high magnesium grass cultivar in controlling hypomagnesaemia in grazing animals. *Grass and Forage Science* 46, 375–380.

Mouat, M.C.H. (1983) Competitive adaptation by plants to nutrient shortage through modification of root growth and surface charge. *New Zealand Journal of Agricultural Research* 26, 327–332.

Mouat, M.C.H. and Nes, P. (1983) Effect of the interaction of nitrogen and phosphorus on the growth of ryegrass. *New Zealand Journal of Agricultural Research* 26, 333–336.

Mudd, A.J. (1970a) The influence of heavily fertilized grass on mineral metabolism of dairy cows. *Journal of Agricultural Science, Cambridge* 74, 11–21.

Mudd, A.J. (1970b) Trace mineral composition of heavily fertilized grass in relation to ARC standards for the requirements of dairy cows. *British Veterinary Journal* 126, 38–44.

Mundy, G.N. (1984) Effects of potassium and sodium application to soil on growth and cation accumulation of herbage. *Australian Journal of Agricultural Research* 35, 85–97.

Murphy, L.S. and Smith, G.E. (1967) Nitrate accumulation in forage crops. *Agronomy Journal* 59, 171–174.

Murphy, M.D. and Boggan, J.M. (1988) Sulphur deficiency in herbage in Ireland. 1. Causes and extent. *Irish Journal of Agricultural Research* 27, 83–90.

Murphy, M.D. and Quirke, W.A. (1997) The effect of sulphur/nitrogen/selenium interactions on herbage yield and quality. *Irish Journal of Agricultural and Food Research* 36, 31–38.

National Research Council, USA (1984) *Nutrient Requirements of Domestic Animals: Nutrient Requirements of Beef Cattle*, 6th edn. National Academy Press, Washington, DC, 90 pp.

National Research Council, USA. (1985) *Nutrient Requirements of Domestic Animals: Nutrient Requirements of Sheep*, 6th edn. National Academy Press, Washington, DC, 99 pp.

National Research Council, USA (Panel on Nitrates) (1978) *Nitrates: An Environmental Assessment*. National Academy of Sciences, Washington DC, 723 pp.

National Research Council, USA (Subcommittee on Dairy Cattle Nutrition) (1989) *Nutrient Requirements of Dairy Cattle*, 6th edn, updated. National Academy Press, Washington, DC, 157 pp.

National Research Council, USA (Subcommittee on Nitrogen Usage in Ruminants) (1985) *Ruminant N Usage*. National Academy of Sciences, Washington, DC.

Naylor, J.M. (1991) The major minerals. In: Naylor, J.M. and Ralston, S.L. (eds) *Large Animal Clinical Nutrition*. Mosby-Year Book, St. Louis, Missouri, pp. 35–54.

Nelson, P.N., Cotsaris, E. and Oades, J.M. (1996) Nitrogen, phosphorus and organic carbon in streams draining two grazed catchments. *Journal of Environmental Quality* 25, 1221–1229.

Newman, E.I. (1995) Phosphorus inputs to terrestrial ecosystems. *Journal of Ecology* 83, 713–726.

Nguyen, M.L. and Goh, K.M. (1992a) Nutrient cycling and losses based on a mass-balance model in grazed pastures receiving long-term superphosphate applications in New Zealand. 1. Phosphorus. *Journal of Agricultural Science, Cambridge* 119, 89–106.

Nguyen, M.L. and Goh, K.M. (1992b) Nutrient cycling and losses based on a mass-balance model in grazed pastures receiving long-term superphosphate applications in New Zealand. 2. Sulphur. *Journal of Agricultural Science, Cambridge* 119, 107–122.

Nguyen, M.L. and Goh, K.M. (1993) An overview of the New Zealand S-cycling model for recommending S requirements in grazed pastures. *New Zealand Journal of Agricultural Research* 36, 474–491.

Nguyen, M.L. and Goh, K.M. (1994a) Sulphur cycling and its implications on sulphur fertilizer requirements of grazed grassland ecosystems. *Agriculture, Ecosystems and Environment* 49, 173–206.

Nguyen, M.L. and Goh, K.M. (1994b) Distribution, transformations and recovery of urinary sulphur and sources of plant-available soil sulphur in irrigated pasture soil-plant systems treated with 35sulphur-labelled urine. *Journal of Agricultural Science, Cambridge* 122, 91–105.

Nguyen, M.L., Tillman, R.W. and Gregg, P.E.H. (1989) Evaluation of sulphur status of East Coast pastures of the North Island, New Zealand. 2. Field trials. *New Zealand Journal of Agricultural Research* 32, 273–280.

Nielsen, F.H. (1986) Other elements. In: Mertz, W. (ed.) *Trace Elements in Human and Animal Nutrition*, Vol. 2, 5th edn. Academic Press, Orlando, pp. 415–463.

Nielsen, K.F. and Cunningham, R.K. (1964) The effects of soil temperature and form and level of nitrogen on the growth and chemical composition of Italian ryegrass. *Proceedings of the Soil Science Society of America* 28, 213–218.

Nielsen, K.F., Halstead, R.L., Maclean, A.J., Holmes, R.M. and Bourget, S.J. (1960) Effects of soil temperature on the growth and chemical composition of lucerne. In: *Proceedings of the 8th International Grassland Congress, Reading*, British Grassland Society, Hurley, 287–292.

Nobel, P.S. (1974) *Biophysical Plant Physiology*. W.H. Freeman, San Francisco, 488 pp.

Norman, B.R. and Mackey, E.A. (1997) Concentrations of iodine determined by pre-irradiation combustion and neutron activation analysis in powdered grass as a function of particle size. *Science of the Total Environment* 205, 151–158.

Norman, M.J.T. (1956) Intervals of superphosphate application to downland permanent pasture. *Journal of Agricultural Science* 47, 157–171.

Norrish, K. (1975) Geochemistry and mineralogy of trace elements. In: Nicholas, D.J.D. and Egan, A.R. (eds) *Trace Elements in Soil–Plant–Animal Systems*. Academic Press, New York, pp. 55–81.

Nowakowski, T.Z. (1961) The effect of different nitrogenous fertilizers, applied as solids or solutions, on the yield and nitrate-N content of established grass and newly-sown ryegrass. *Journal of Agricultural Science, Cambridge* 56, 287–292.

Nowakowski, T.Z. and Cunningham, R.K. (1966) Nitrogen fractions and soluble carbohydrates in Italian ryegrass. 2. Effects of light intensity, form and level of nitrogen. *Journal of the Science of Food and Agriculture* 17, 145–150.

Nowakowski, T.Z., Cunningham, R.K. and Nielsen, K.F. (1965) Nitrogen fractions and soluble carbohydrates in Italian ryegrass. 1. Effects of soil temperature, form and level of nitrogen. *Journal of the Science of Food and Agriculture* 16, 124–134.

Oberle, S.L. and Keeney, D.R. (1994) Interactions of sewage sludge with soil–crop–water systems. In: Clapp, C.E., Larson, W.E. and Dowdy, R.H. (eds) *Sewage Sludge: Land Utilization and the Environment*. American Society of Agronomy, Madison, pp. 17–20.

Oenema, O. and Heinen, M. (1999) Uncertainties in nutrient budgets due to biases and errors. In: Smaling, E.M.A., Oenema, O. and Fresco, L.O. (eds) *Nutrient Disequilibria in Agroecosystems: Concepts and Case Studies*. CAB International, Wallingford, pp. 75–97.

O'Riordan, E.G., Dodd, V.A. and Fleming, G.A. (1994) Spreading a low-metal sludge on grassland: effects on soil and herbage heavy metal concentrations. *Irish Journal of Agricultural and Food Research* 33, 61–69.

Ørskov, E.R. (1992) *Protein Nutrition in Ruminants*, 2nd edn. Academic Press, London, 175 pp.

Owens, L.B., Van Keuren, R.W. and Edwards, W.M. (1998) Budgets of non-nitrogen nutrients in a high fertility pasture system. *Agriculture, Ecosystems and Environment* 70, 7–18.

Ozanne, P.G. (1980) Phosphate nutrition of plants – a general treatise. In: Khasawneh, F.E., Sample, E.C. and Kamprath, E.J. (eds) *The Role of Phosphorus in Agriculture*. American Society of Agronomy, Madison, pp. 559–589.

Ozanne, P.G. and Howes, K.M.W. (1971a) The effect of grazing on the phosphorus requirement of an annual pasture. *Australian Journal of Agricultural Research* 22, 81–92.

Ozanne, P.G. and Howes, K.M.W. (1971b) Preference of grazing sheep for pasture of high phosphate content. *Australian Journal of Agricultural Research* 22, 941–950.

Paasikallio, A. (1978) The mineral element content of timothy (*Phleum pratense*, L.) in Finland. II. The elements aluminium, boron, molybdenum, strontium, lead and nickel. *Acta Agriculturae Scandinavica*, Suppl. 20, 40–52.

Pais, I. and Benton Jones, J. (1997) *The Handbook of Trace Elements*. St Lucie Press, Boca Raton, 223 pp.

Parente, G. (1996) Grassland and land use systems. In: Parente, G., Frame, J. and Orsi, S. (eds) *Grassland Science in Europe. Proceedings of 16th General Meeting of the European Grassland Federation, 1996*, pp. 23–34.

Parfitt, R.L. (1980) A note on the losses from a phosphate cycle under grazed pasture. *New Zealand Journal of Experimental Agriculture* 8, 215–217.

Parks, W.L. and Fisher, W.B. (1958) Influence of soil temperature and nitrogen on ryegrass growth and chemical composition. *Proceedings of the Soil Science Society of America* 22, 257–259.

Parsons, A.J. (1988) The effects of season and management on the growth of grass swards. In: Jones, M.B. and Lazenby, A. (eds) *The Grass Crop*. Chapman and Hall, London, pp. 129–177.

Parsons, A.J., Orr, R.J., Penning, P.D. and Lockyer, D.R. (1991) Uptake, cycling and fate of nitrogen in grass-clover swards continuously grazed by sheep. *Journal of Agricultural Science, Cambridge* 116, 47–61.

Patil, B.D. and Jones, D.I.H. (1970) The mineral status of some temperate herbage varieties in relation to animal performance. In: *Proceedings of the 11th International Grassland Congress, Surfer's Paradise*, University of Queensland Press, St. Lucia, pp. 726–730.

Paul, E.A. and Clark, F.E. (1996) *Soil Microbiology and Biochemistry*, 2nd edn. Academic Press, San Diego.

Paul, E.A. and Huang, P.M. (1980) Chemical aspects of soil. In: Hutzinger, O. (ed.) *Handbook of Environmental Chemistry*, Vol. 1, Part A. Springer Verlag, Berlin, pp. 69–86.

Payne, J.M. (1989) *Metabolic and Nutritional Diseases of Cattle*. Blackwell Scientific Publishers, Oxford, 149 pp.

Paynter, R.M. and Dampney, P.M.R. (1991) The effect of rate and timing of phosphate fertilizer on the yield and phosphate offtake of grass grown for silage at moderate to high levels of soil phosphorus. *Grass and Forage Science* 46, 131–137.

Pegtel, D.M., Bakker, J.P., Verweij, G.L. and Fresco, L.F.M. (1996) N, P and K deficiency in chronosequential cut summer-dry grasslands on gley podsol after the cessation of fertilizer application. *Plant and Soil* 178, 121–131.

Peoples, M.B., Ladha, J.K. and Herridge, D.F. (1995) Enhancing legume N_2 fixation through plant and soil management. *Plant and Soil* 174, 83–101.

Perrott, K.W. and Mansell, G.P. (1989) Effect of fertilizer phosphorus and liming on inorganic and organic soil phosphorus fractions. *New Zealand Journal of Agricultural Research* 32, 63–70.

Perrott, K.W. and Sarathchandra, S.U. (1987) Nutrient and organic matter levels in a range of New Zealand soils under established pasture. *New Zealand Journal of Agricultural Research* 30, 249–259.

Perrott, K.W. and Sarathchandra, S.U. (1989) Phosphorus in the microbial biomass of New Zealand soils under established pasture. *New Zealand Journal of Agricultural Research* 32, 409–413.

Petersen, S.O., Lind, A.-M. and Sommer, S.G. (1998a) Nitrogen and organic matter losses during storage of cattle and pig manure. *Journal of Agricultural Science, Cambridge* 130, 69–79.

Petersen, S.O., Sommer, S.G., Aaes, O. and Soegaard, K. (1998b) Ammonia losses from urine and dung of grazing cattle: effect of N intake. *Atmospheric Environment* 32, 295–300.

Peterson, L.A. and Newman, R.C. (1976) Influence of soil pH on the availability of added boron. *Soil Science Society of America Journal* 40, 280–282.

Peterson, P.J. (1969) The distribution of zinc-65 in *Agrostis tenuis* Sibth and *A. stolonifera* L. tissues. *Journal of Experimental Botany* 20, 863–875.

Petrie, S.E. and Jackson, T.L. (1982) Effects of lime, P and Mo application on Mo concentration in subclover. *Agronomy Journal* 74, 1077–1081.

Philipson, T. (1953) Boron in plant and soil. *Acta Agriculturae Scandinavica* 3, 121–242.

Phipps, R.H. (1975) The effects on dairy cows of grazing pasture containing high levels of nitrate-nitrogen. *Journal of the British Grassland Society* 30, 45–49.

Pinkerton, A. and Randall, P.J. (1994) Internal phosphorus requirements of six legumes and two grasses. *Australian Journal of Experimental Agriculture* 34, 373–379.

Pinkerton, B.W. and Brown, K.W. (1985) Plant accumulation and soil sorption of cobalt from Co-amended soils. *Agronomy Journal* 77, 634–638.

Powell, K., Reid, R.L. and Balasko, J.A. (1978) Performance of lambs on perennial ryegrass, smooth bromegrass, orchardgrass and tall fescue pastures. II. Mineral utilization, *in vitro* digestibility and chemical composition of herbage. *Journal of Animal Science* 46, 1503–1514.

Power, J.F. (1968) Mineralization of nitrogen in grass roots. *Proceedings of the Soil Science Society of America* 32, 673–674.

Power, J.F. (1981) Long-term recovery of fertilizer nitrogen applied to a native mixed pasture. *Soil Science Society of America Journal* 45, 782–786.

Power, J.F. (1986) Nitrogen cycling in seven cool-season perennial grass species. *Agronomy Journal* 78, 681–687.

Powlson, D.S. (1994) Quantification of nutrient cycles using long-term experiments. In: Leigh, R.A. and Johnston, A.E. (eds) *Long-term Experiments in Agricultural and Ecological Sciences*. CAB International, Wallingford, pp. 97–115.

Pratley, J.E. and McFarlane, J.D. (1974) The effect of sulphate on the selenium content of pasture plants. *Australian Journal of Experimental Agriculture and Animal Husbandry* 14, 533–538.

Price, J. (1989) *The Nutritive Value of Grass in Relation to Mineral Deficiencies and Imbalances in the Ruminant*. Proceedings of the Fertilizer Society no. 289, London.

Price, N.O. and Moschler, W.W. (1965) Effect of residual lime in soil on minor elements in plants. *Journal of Agricultural and Food Chemistry* 13, 163–165.

Price, N.O. and Moschler, W.W. (1970) Residual lime effects in soils on certain mineral elements in barley, fescue and oats. *Journal of Agricultural and Food Chemistry* 18, 5–8.

Prins, W.H., Den Boer, D.J. and Van Burg, P.F.J. (1986) Requirements for phosphorus, potassium and other nutrients for grassland in relation to nitrogen usage. In: Cooper, J.P. and Raymond, W.F. (eds) *Grassland Manuring*. Occasional Symposium no. 20, British Grassland Society, London, pp. 28–45.

Pritchard, G.I., Pigden, W.J. and Folkins, L.P. (1964) Distribution of potassium, calcium, magnesium and sodium in grasses at progressive stages of maturity. *Canadian Journal of Plant Science* 44, 318–324.

Pumphrey, F.V. and Moore, D.P. (1965a) Sulphur and nitrogen content of alfalfa herbage during growth. *Agronomy Journal* 57, 237–239.

Pumphrey, F.V. and Moore, D.P. (1965b) Diagnosing sulphur deficiency of alfalfa (*Medicago sativa* L.) from plant analysis. *Agronomy Journal* 57, 364–366.

Rahman, H., McDonald, P. and Simpson, K. (1960) Effects of nitrogen and potassium fertilizers on the mineral status of perennial ryegrass. *Journal of the Science of Food and Agriculture* 11, 422–428.

Rahman, A.A.A., Shalaby, A.F. and El Monayeri, M.O. (1971) Effect of moisture stress on metabolic products and ion accumulation. *Plant and Soil* 34, 65–90.

Rangeley, A. (1989) The suitability of plant tissue tests to indicate P deficiency in perennial ryegrass (*L. perenne* L.). *Grass and Forage Science* 44, 91–95.

Rangeley, A. and Newbould, P. (1985) Growth responses to lime and fertilizers and critical concentrations in herbage of white clover in Scottish hill soils. *Grass and Forage Science* 40, 265–277.

Raven, K.P. and Loeppert, R.H. (1997) Trace element composition of fertilizers and soil amendments. *Journal of Environmental Quality* 26, 551–557.

Reay, P.F. and Marsh, B. (1976) Element composition of ryegrass and red clover leaves during a growing season. *New Zealand Journal of Agricultural Research* 19, 469–472.

Reay, P.F. and Waugh, C. (1983) Element composition of ryegrass and white clover leafblades – seasonal variation in a continuously stocked pasture. *New Zealand Journal of Agricultural Research* 26, 341–348.

Reddy, G.D., Alston, A.M. and Tiller, K.G. (1981) Seasonal changes in the concentrations of copper, molybdenum and sulphur in pasture plants. *Australian Journal of Experimental Agriculture and Animal Husbandry* 21, 498–505.

Redmann, R.E. (1992) Primary productivity. In: Coupland, R.T. (ed.) *Ecosystems of the World. 8A. Natural Grasslands*. Elsevier, Amsterdam, pp. 75–93.

Redshaw, E.S., Martin, P.J. and Laverty, D.H. (1978) Iron, manganese, copper, zinc and selenium concentrations in Alberta grains and roughages. *Canadian Journal of Animal Science* 58, 553–558.

Reid, D. (1966) The response of herbage yields and quality to a wide range of nitrogen application rates. In: *Proceedings, 10th International Grassland Congress, Helsinki, 1966*, Valtioneuvoston Kirjapaino, Helsinki, pp. 209–213.

Reid, D. and Strachan, N.H. (1974) The effects of a wide range of nitrogen rates on some chemical constituents of herbage from perennial ryegrass swards with and without white clover. *Journal of Agricultural Science, Cambridge* 83, 393–401.

Reid, R.L. (1980) Relationship between phosphorus nutrition of plants and the phosphorus nutrition of animals and man. In: Khasawneh, F.E., Sample, E.C.

and Kamprath, E.J. (eds) *The Role of Phosphorus in Agriculture*. American Society of Agronomy, Madison, pp. 847–886.

Reid, R.L. (1983) Biological availability of magnesium from natural and supplemental sources. In: Fontenot, J.P., Bunce, J.E., Webb, K.E. and Allen, V.G. (eds) *Role of Magnesium in Animal Nutrition*. Virginia Polytechnic Institute, Blacksburg, pp. 121–142.

Reid, R.L. and Jung, G.A. (1965) Influence of fertilizer treatment on the intake, digestibility and palatability of tall fescue hay. *Journal of Animal Science* 24, 615–625.

Reid, R.L. and Jung, G.A. (1974) Effects of nutrients other than nitrogen on the nutritive value of forage. In: Mays, D.A. (ed.) *Forage Fertilization*. American Society of Agronomy, Madison, pp. 395–436.

Reid, R.L., Odhuba, E.K. and Jung, G.A. (1967a) Evaluation of tall fescue pasture under different fertilization treatments. *Agronomy Journal* 59, 265–271.

Reid, R.L., Jung, G.A. and Kinsey, C.M. (1967b) Nutritive value of nitrogen-fertilized orchardgrass pasture at different periods of the year. *Agronomy Journal* 59, 519–525.

Reinbott, T.M. and Blevins, D.G. (1994) Phosphorus and temperature effects on magnesium, calcium and potassium in wheat and tall fescue leaves. *Agronomy Journal* 86, 523–529.

Reith, J.W.S. (1963) The magnesium contents of soils and crops. *Journal of the Science of Food and Agriculture* 14, 417–426.

Reith, J.W.S. (1970) Soil factors influencing the trace element content of herbage. In: Mills, C.F. (ed.) *Trace Element Metabolism in Animals*. E. & S. Livingstone, Edinburgh, pp. 410–412.

Reith, J.W.S. and Mitchell, R.L. (1964) The effect of soil treatment on trace element uptake by plants. In: *Plant Analysis and Fertilizer Problems IV*. American Society of Horticultural Science, Beltsville, Maryland, pp. 241–254.

Reith, J.W.S., Inkson, R.H.E., Holmes, W., MacLusky, D.A., Reid, D., Heddle, R.G. and Copeman, G.J.F. (1964) The effect of fertilizers on herbage production. 2. The effect of nitrogen, phosphorus and potassium on botanical and chemical composition. *Journal of Agricultural Science, Cambridge* 63, 209–219.

Reith, J.W.S., Burridge, J.C., Berrow, M.L. and Caldwell, K.S. (1983) Effects of the application of fertilizers and trace elements on the cobalt content of herbage cut for conservation. *Journal of the Science of Food and Agriculture* 34, 1163–1170.

Reith, J.W.S., Burridge, J.C., Berrow, M.L. and Caldwell, K.S. (1984) Effects of fertilizers on the contents of copper and molybdenum in herbage cut for conservation. *Journal of the Science of Food and Agriculture* 35, 245–256.

Renner, E., Schaafsma, G. and Scott, K.J. (1989) Micronutrients in milk. In: Renner, E. (ed.) *Micronutrients in Milk and Milk-based Food Products*. Elsevier, London, pp. 1–70.

Reuter, D.J., Loneragan, J.F., Robson, A.D. and Plaskett, D. (1982) Zinc in sub-terranean clover. 1. Effects of zinc supply on distribution of Zn and dry weight among plant parts. *Australian Journal of Agricultural Research* 33, 989–999.

Richards, I.R. (1999) Options for change. In: Corrall, A.J. (ed.) *Accounting for Nutrients*. Occasional Symposium no. 33, British Grassland Society, Reading, pp. 49–56.

Richards, I.R. and Wolton, K.M. (1976) The spatial distribution of excreta under intensive cattle grazing. *Journal of the British Grassland Society* 31, 89–92.

Rickaby, C.D. (1981) The selenium requirement of ruminants. In: Haresign, W. (ed.) *Recent Advances in Animal Nutrition – 1981*. Butterworths, London, pp. 121–128.

Riggs, K.S., Syers, J.K., Rimmer, D.L. and Sumner, M.E. (1995) Effect of liming on calcium and magnesium concentrations in herbage. *Journal of the Science of Food and Agriculture* 69, 169–174.

Rinne, S.-L., Sillanpää, M., Huokuna, E. and Hiivola, S.-L. (1974s) Effects of heavy nitrogen fertilization on potassium, calcium, magnesium and phosphorus contents in ley grasses. *Annales Agriculturae Fenniae* 13, 96–108.

Rinne, S.-L., Sillanpää, M., Huokuna, E. and Hiivloa, S.-L. (1974b) Effects of heavy nitrogen fertilization on iron, manganese, sodium, zinc, copper, strontium, molybdenum and cobalt contents in ley grasses. *Annales Agriculturae Fenniae* 13, 109–118.

Robinson, D. (1994) The response of plants to non-uniform supplies of nutrients. *New Phytologist* 127, 635–674.

Robinson, D.L., Kappel, L.C. and Boling, J.A. (1989) Management practices to overcome the incidence of grass tetany. *Journal of Animal Science* 67, 3470–3484.

Robson, A.D. (1988) Manganese in soils and plants – an overview. In: Graham, R.D., Hannam, R.J. and Uren, N.C. (eds) *Manganese in Soils and Plants*. Kluwer Academic Publishers, Dordrecht, pp. 329–333.

Roche, J.R., Dalley, D., Moate, P., Grainger, C., Hannah, M., O'Mara, F. and Rath, M. (2000) Variations in the dietary cation-anion difference and the acid-base balance of dairy cows on a pasture-based diet in south-eastern Australia. *Grass and Forage Science* 55, 26–36.

Rodger, J.B.A. (1982) The effect of fertilizer nitrogen, phosphorus and potassium on the calcium, magnesium and phosphorus status of pasture cut for silage. *Journal of Agricultural Science, Cambridge* 99, 199–205.

Rogers, J.A. and Davis, G.E. (1973) Growth and chemical composition of four grass species in relation to soil moisture and aeration factors. *Journal of Ecology* 61, 455–472.

Romheld, V. and Marschner, H. (1991) Function of micronutrients in plants. In: Mortvedt, J.J. (ed.) *Micronutrients in Agriculture*, 2nd edn. Soil Science Society of America, Madison, pp. 297–328.

Rominger, R.S., Smith, D. and Peterson, L.A. (1975) Yields and elemental composition of alfalfa plant parts at late bud under two fertility levels. *Canadian Journal of Plant Science* 55, 69–75.

Rosenfeld, I. and Beath, O.A. (1964) *Selenium: Geobotany, Biochemistry, Toxicity and Nutrition*. Academic Press, New York, 411 pp.

Ross, S.M. (1994) Retention, transformation and mobility of toxic metals in soils. In: Ross, S.M. (ed.) *Toxic Metals in Soil–Plant Systems*. John Wiley, Chichester, pp. 63–152.

Rowarth, J.S., Gillingham, A.G., Tillman, R.W. and Syers, J.K. (1985) Release of phosphorus from sheep faeces on grazed hill country pastures. *New Zealand Journal of Agricultural Research* 28, 497–504.

Rowarth, J.S., Gillingham, A.G., Tillman, R.W. and Syers, J.K. (1988) Effects of season and fertilizer rate on phosphorus concentrations in pasture and sheep faeces in hill country. *New Zealand Journal of Agricultural Research* 31, 187–193.

Rowarth, J.S., Tillman, R.W. and Gillingham, A.G. (1990) Comparison of release of phosphorus from monocalcium phosphate and sheep faeces. *New Zealand Journal of Agricultural Research* 33, 473–477.

Rumball, W., Butler, G.W. and Jackman, R.H. (1972) Variation in nitrogen and mineral composition in populations of prairie grass (*Bromus uniloides* H.B.K.). *New Zealand Journal of Agricultural Research* 15, 33–42.

Ryan, M. and Finn, T. (1976) Grassland productivity. 3. Effect of phosphorus on the yield of herbage at 26 sites. *Irish Journal of Agricultural Research* 15, 11–23.

Ryden, J.C. (1986) Gaseous losses of nitrogen from grassland. In: Van der Meer, H.G., Ryden, J.C. amd Ennik, G.C. (eds) *Nitrogen Fluxes in Intensive Grassland Systems*. Martinus Nijhoff, Dordrecht, pp. 59–73.

Ryser, P. and Lambers, H. (1995) Root and leaf attributes accounting for the performance of fast- and slow-growing grasses at different nutrient supply. *Plant and Soil* 170, 251–265.

Safley, L.M., Barker, J.C. and Westermann, P.W. (1984) Characteristics of fresh dairy manure. *Transactions of the American Society of Agricultural Engineers* 27, 1150–1153.

Saggar, S., Hedley, M.J. and Phimsarn, S. (1998) Dynamics of sulfur transformations in grazed pastures. In: Maynard, D.G. (ed.) *Sulfur in the Environment*. Marcel Dekker, New York, pp. 45–94.

Saini, H.S., Attieh, J.M. and Hanson, D. (1995) Biosynthesis of halomethanes and methanethiol by higher plants via a novel methyltransferase reaction. *Plant, Cell and Environment* 18, 1027–1033.

Sakadeven, K., Hedley, M.J. and Mackay, A.D. (1993a) Mineralisation and fate of soil sulphur and nitrogen in hill pasture. *New Zealand Journal of Agricultural Research* 36, 271–281.

Sakadeven, K., Mackay, A.D. and Hedley, M.J. (1993b) Influence of sheep excreta on pasture uptake and leaching losses of sulphur, nitrogen and potassium from grazed pastures. *Australian Journal of Soil Research* 31, 151–162.

Salette, J. (1978) Sulphur content in grasses during primary growth. In: Brogan, J.C. (ed.) *Sulphur in Forages*. An Foras Taluntais, Dublin, pp. 142–152.

Salo, M.L. and Sormunen, R. (1976) Investigations of grass silage prepared in ordinary farm conditions. 2. Changes in mineral composition and mineral losses. *Journal of the Scientific Agricultural Society of Finland* 48, 128–137.

Sanders, I.R. and Fitter, A.H. (1992) The ecology and functioning of vesicular-arbuscular mycorrhizas in co-existing grassland species. II. Nutrient uptake and growth of vesicular-arbuscular mycorrhizal plants in a semi-natural grassland. *New Phytologist* 120, 525–533.

Sanyal, S.K. and De Datta, S.K. (1991) Chemistry of phosphorus transformations in soil. *Advances in Soil Science* 16, 1–120.

Saul, G.R., Kearney, G.A., Flinn, P.C. and Lescun, C.L. (1999) Effects of superphosphate fertilizer and stocking rate on the nutritive value of perennial ryegrass and subterranean clover herbage. *Australian Journal of Agricultural Research* 50, 537–545.

Saunders, W.M.H. (1984) Mineral composition of soil and pasture from areas of grazed paddocks, affected and unaffected by dung and urine. *New Zealand Journal of Agricultural Research* 27, 405–412.

Saunders, W.M.H. and Metson, A.J. (1971) Seasonal variation of phosphorus in soil and pasture. *New Zealand Journal of Agricultural Research* 14, 307–328.

Saunders, W.M.H., Cooper, D.M. and Gravett, I.M. (1988) Response to S and comparison of soil tests for predicting S status of soils for grazed grass–clover pasture in New Zealand. *New Zealand Journal of Agricultural Research* 31, 195–204.

Schellberg, J., Möseler, B.M., Kühbauch, W. and Rademacher, I.F. (1999) Long-term effects of fertilizer on soil nutrient concentration, yield, forage quality and floristic composition of a hay meadow in the Eifel mountains, Germany. *Grass and Forage Science* 54, 195–207.

Schoenau, J.J. and Bettany, J.R. (1987) Organic matter leaching as a component of carbon, nitrogen, phosphorus and sulphur cycles in a forest, grassland and gleyed soil. *Soil Science Society of America Journal* 51, 646–651.

Schoenau, J.J. and Karamanos, R.E. (1993) Sodium bicarbonate-extractable P, K and N. In: Carter, M.R. (ed.) *Soil Sampling and Methods of Analysis*. Lewis Publishers, Boca Raton, pp. 51–58.

Scholefield, D., Garwood, E.A. and Titchen, N.M. (1988) The potential of management practices for reducing losses of nitrogen from grazed pastures. In: Jenkinson, D.S. and Smith, K.A. (eds) *Nitrogen Efficiency in Agricultural Soils*. Elsevier, London, pp. 220–230.

Schroder, J. and Dilz, K. (1987) Cattle slurry and farmyard manure as fertilizers for forage maize. In: Van der Meer, H.G., Unwin, R.J., Van Dijk, T.A. and Ennick, G.C. (eds) *Animal Manures on Grassland and Fodder Crops: Fertilizer or Waste*. Martinus Nijhoff, Dordrecht, pp. 137–156.

Schuster, J.L. (1964) Root development of native plants under three grazing intensities. *Ecology* 45, 63–70.

Schutte, K.H. (1964) *The Biology of Trace Elements*. Crosby Lockwood, London, 228 pp.

Scott, B.J. and Robson, A.D. (1990) Distribution of magnesium in subterranean clover (*Trifolium subterraneum* L.) in relation to supply. *Australian Journal of Agricultural Research* 41, 499–510.

Scott, D. (1999) Sustainability of New Zealand high-country pastures under contrasting development inputs. 1. Site and shoot nutrients. *New Zealand Journal of Agricultural Research* 42, 365–383.

Scott, N.M. (1985) Sulphur in soils and plants. In: Vaughan, D. and Malcolm, R.E. (eds) *Soil Organic Matter and Biological Activity*. Martinus Nijhoff/Dr W Junk, Dordrecht, pp. 379–401.

Scott, R.O. (1970) Problems in trace element analysis. In: Mills, C.F. (ed.) *Trace Element Metabolism in Animals*. E. & S. Livingstone, Edinburgh, pp. 497–503.

Scottish Agricultural Colleges. (1982) *Trace Element Deficiency in Ruminants – Report of a Study Group*. East Scotland College, Edinburgh.

Senesi, N. and Polemio, M. (1981) Trace element addition to soil by application of NPK fertilizers. *Fertilizer Research* 2, 289–302.

Shamberger, R.J. (1983) *Biochemistry of Selenium*. Plenum Press, New York, 334 pp.

Shamberger, R.J. (1984) Selenium. In: Frieden, E. (ed.) *Biochemistry of the Essential Ultratrace Elements*. Plenum Press, New York, pp. 201–237.

Sharpley, A., Foy, B. and Withers, P. (2000) Practical and innovative measures for the control of agricultural phosphorus losses to water: an overview. *Journal of Environmental Quality* 29, 1–9.

Sharpley, A.N., Hedley, M.J., Sibbesen, E., Hillbricht-Ilkowska, A., House, W.A. and Ryszkowski, L. (1995) Phosphorus transfers from terrestrial to aquatic ecosystems. In: Tiessen, H. (ed.) *Phosphorus in the Global Environment.* John Wiley and Sons, Chichester, pp. 171–199.

Sheppard, M.I., Hawkins, J.L. and Smith, P.A. (1996) Linearity of iodine sorption and sorption capacities for seven soils. *Journal of Environmental Quality* 25, 1261–1267.

Sheppard, S.C., Evenden, W.G. and Amiro, B.D. (1993) Investigations of the soil-to-plant pathway for I, Br, Cl and F. *Journal of Environmental Radioactivity* 21, 9–32.

Sherrell, C.G. (1966) Boron deficiency in white clover seedlings grown in an organic soil. *New Zealand Journal of Agricultural Research* 9, 1025–1031.

Sherrell, C.G. (1978) A note on sodium concentrations in New Zealand pasture species. *New Zealand Journal of Experimental Agriculture* 6, 189–190.

Sherrell, C.G. (1983) Boron nutrition of perennial ryegrass, cocksfoot and timothy. *New Zealand Journal of Agricultural Research* 26, 205–208.

Sherrell, C.G. (1990a) Effect of cobalt application on the cobalt status of pastures. 2. Pastures without previous cobalt application. *New Zealand Journal of Agricultural Research* 33, 305–311.

Sherrell, C.G. (1990b) Effect of cobalt application on the cobalt status of pastures. 3. Comparison of chelate and sulphate as cobalt sources for topdressing deficient pastures. *New Zealand Journal of Agricultural Research* 33, 313–317.

Sherrell, C.G. and Rawnsley, J.S. (1982) Effect of copper application on copper concentration in white clover and perennial ryegrass on some Northland soils and a yellow-brown pumice soil. *New Zealand Journal of Agricultural Research* 25, 363–368.

Sherrell, C.G., Percival, N.S. and Gee, T.M. (1990) Effect of cobalt application on the cobalt status of pasture. 1. Pastures with history of regular cobalt application. *New Zealand Journal of Agricultural Research* 33, 295–304.

Shuman, L.M. (1991) Chemical forms of micronutrients in soils. In: Mortvedt, J.J. (ed.) *Micronutrients in Agriculture*, 2nd edn. Soil Science Society of America, Madison, Wisconsin, pp. 113–144.

Sillanpää, M. and Rinne, S.-L. (1975) The effect of heavy nitrogen fertilization on the uptake of nutrients and on some properties of soils cropped with grasses. *Annales Agriculturae Fenniae* 14, 210–226.

Simán, G. and Jansson, S.L. (1976) Sulphur exchange betweeen soil and atmosphere with special attention to sulphur release directly to the atmosphere. 2. *Swedish Journal of Agricultural Research* 6, 135–144.

Simpson, D., Wilman, D. and Adams, W.A. (1988) Response of white clover and grass to applications of potassium and nitrogen on a potassium-deficient hill soil. *Journal of Agricultural Science, Cambridge* 110, 159–167.

Simpson, J.R. and Stobbs, J.H. (1981) Nitrogen supply and annual production from pasture. In: Morley, F.H.W. (ed.) *Grazing Animals.* Elsevier, Amsterdam, pp. 261–287.

Sims, J.T. and Johnson, G.V. (1991) Micronutrient soil tests. In: Mortvedt, J.J. (ed.) *Micronutrients in Agriculture*, 2nd edn. Soil Science Society of America, Madison, pp. 427–476.

Sinclair, A.G., Smith, L.C., Morrison, J.D. and Dodds, K.G. (1996a) Effects and interactions of phosphorus and sulphur on a mown white clover/ryegrass sward. 1. Herbage dry matter production and balanced nutrition. *New Zealand Journal of Agricultural Research* 39, 421–433.

Sinclair, A.G., Morrison, J.D., Smith, L.C. and Dodds, K.G. (1996b) Effects and interactions of phosphorus and sulphur on a mown white clover/ryegrass sward. 2. Concentrations and ratios of phosphorus, sulphur and nitrogen in clover herbage in relation to balanced plant nutrition. *New Zealand Journal of Agricultural Research* 39, 435–445.

Sinclair, A.G., Morrison, J.D., Smith, L.C. and Dodds, K.G. (1997) Effects and interactions of phosphorus and sulphur on a mown white clover/ryegrass sward. 3. Indices of nutrient adequacy. *New Zealand Journal of Agricultural Research* 40, 297–307.

Sinclair, A.H., MacDonald, D.C., Ferrier, R. and Edwards, A.C. (1993) The use of minor elements for grassland. In: Hopkins, A. and Younie, D. (eds) *Forward with Grass into Europe*. Occasional Symposium no. 27, British Grassland Society, Reading, 73–84.

Sleper, D.A., Garner, G.B., Nelson, C.J. and Sebaugh, J.L. (1980) Mineral composition of tall fescue genotypes grown under controlled conditions. *Agronomy Journal* 72, 720–722.

Smith, C.M., Gregg, P.E.H. and Tillman, R.W. (1983) Drainage losses of sulphur from a yellow-grey earth soil. *New Zealand Journal of Agricultural Research* 26, 363–371.

Smith, D. (1970) Influence of temperature on the yield and chemical composition of five forage legume species. *Agronomy Journal* 62, 520–523.

Smith, D. (1975) Effects of potassium topdressing a low fertility silt loam soil on alfalfa herbage yields and composition and on soil K values. *Agronomy Journal* 67, 60–64.

Smith, D. and Rominger, R.S. (1974) Distribution of elements among individual parts of the orchardgrass shoot and influence of two fertility levels. *Canadian Journal of Plant Science* 54, 485–492.

Smith, D., Rohweder, D.A. and Jorgensen, N.A. (1974) Chemical composition of three legumes and four grass herbages harvested at early flower during three years. *Agronomy Journal* 66, 817–819.

Smith, G.S. and Cornforth, I.S. (1982) Concentrations of nitrogen, phosphorus, sulphur, magnesium and calcium in North Island pastures in relation to plant and animal nutrition. *New Zealand Journal of Agricultural Research* 25, 373–387.

Smith, G.S. and Edmeades, D.C. (1983) Manganese status of New Zealand pastures 2. Pasture concentrations. *New Zealand Journal of Agricultural Research* 26, 223–225.

Smith, G.S. and Middleton, K.R. (1978) Sodium and potassium content of top-dressed pastures in New Zealand in relation to plant and animal nutrition. *New Zealand Journal of Experimental Agriculture* 6, 217–225.

Smith, G.S., Middleton, K.R. and Edmonds, A.S. (1978) A classification of pasture and fodder plants according to their ability to translocate sodium from their roots into aerial parts. *New Zealand Journal of Experimental Agriculture* 6, 183–188.

Smith, G.S., Edmeades, D.C. and Upsdell, M. (1983) Manganese status of New Zealand pastures 1. Toxicity in ryegrass, white clover and lucerne. *New Zealand Journal of Agricultural Research* 26, 215–221.

Smith, G.S., Cornforth, I.S. and Henderson, H.V. (1985) Critical leaf concentrations for deficiencies of nitrogen, potassium, phosphorus, sulphur and magnesium in perennial ryegrass. *New Phytologist* 101, 393–409.

Smith, J.H., Carter, D.L., Brown, M.J. and Douglas, C.L. (1968) Differences in chemical composition of plant sample fractions resulting from grinding and screening. *Agronomy Journal* 60, 149–151.

Smith, K.A. and Paterson, J.E. (1995) Manganese and cobalt. In: Alloway, B.J. (ed.) *Heavy Metals in Soils*, 2nd edn. Blackie Academic and Professional, London, pp. 224–244.

Smith, K.A. and van Dijk, T.A. (1987) Utilization of phosphorus and potassium from animal manures on grassland and forage crops. In: Van der Meer, H.G., Unwin, R.J., van Dijk, T.A. and Ennik, G.C. (eds) *Animal Manures on Grassland and Fodder Crops: Fertilizer or Waste*. Martinus Nijhoff, Dordrecht, pp. 87–102.

Smith, L.C., Morton, J.D. and Catto, W.D. (1999) The effects of fertilizer iodine application on herbage iodine concentration and animal blood levels. *New Zealand Journal of Agricultural Research* 42, 433–440.

Smith, R.M. (1987) Cobalt. In: Mertz, W. (ed.) *Trace Elements in Human and Animal Nutrition*, 5th edn, Vol. 1. Academic Press, San Diego, pp. 143–183.

Smith, R.S. (1994) Effects of fertilizers on plant species composition and conservation interest of UK grassland. In: Hagggar, R.J. and Peel, S. (eds) *Grassland Management and Nature Conservation*. Occasional Symposium no. 28, British Grassland Society, Reading, pp. 64–73.

Smith, S.R. (1991) Effects of sewage sludge application on soil microbial processes and soil fertility. *Advances in Soil Science* 16, 191–212.

Smith, S.R. (1996) *Agricultural Recycling of Sewage Sludge and the Environment*. CAB International, Wallingford, 382 pp.

Sommer, S.G. and Ersboll, A.K. (1996) Effect of air flow rate, lime amendments and chemical soil properties on the volatilization of ammonia from fertilizers applied to sandy soils. *Biology and Fertility of Soils* 21, 53–60.

Sommer, S.G. and Husted, S. (1995) The chemical buffer system in raw and digested animal slurry. *Journal of Agricultural Science, Cambridge* 124, 45–53.

Sommers, L.E. (1977) Chemical composition of sewage sludge and analysis of their potential use as fertilizers. *Journal of Environmental Quality* 6, 225–232.

Sommers, L.E. and Sutton, A.L. (1980) Use of waste materials as sources of phosphorus. In: Khasawneh, F.E., Sample, E.C. and Kamprath, E.J. (eds) *The Role of Phosphorus in Agriculture*. American Society of Agronomy, Madison, pp. 515–544.

Soon, Y.K., Bates, T.E. and Moyer, J.R. (1980) Land application of chemically treated sewage sludge. III. Effects on soil and plant heavy metal content. *Journal of Environmental Quality* 9, 497–504.

Spears, J.W. (1991) Advances in mineral nutrition in grazing ruminants. In: *Proceedings 2nd Grazing Livestock Nutrition Conference, Oklahoma State University*, Stillwater, pp. 138–149.

Spears, J.W., Burns, J.C. and Hatch, P.A. (1985) Sulfur fertilization of cool season grasses and effect on utilization of minerals, nitrogen and fibre by steers. *Journal of Dairy Science* 68, 347–355.

Spencer, K. (1978) Sulphur nutrition of clover: effect of plant age on the composition–yield relationship. *Communications in Soil Science and Plant Analysis* 9, 883–895.

Spencer, K., Jones, M.B. and Freney, J.R. (1977) Diagnostic indices for sulphur status of subterranean clover. *Australian Journal of Agricultural Research* 28, 401–412.

Sposito, G. and Page, A.L. (1984) Cycling of metal ions in the soil environment. In: Sigel, H. (ed.) *Metal Ions in Biological Systems*. Marcel Dekker, New York, pp. 287–332.

Stamm, C., Flühler, H., Gächter, R., Leuenberger, J. and Wunderli, H. (1998) Preferential transport of phosphorus in drained grassland soils. *Journal of Environmental Quality* 27, 515–522.

Statham, M. and Bray, A.C. (1975) Congenital goitre in sheep in southern Tasmania. *Australian Journal of Agricultural Research* 26, 751–768.

Steele, K.W., Judd, M.J. and Shannon, P.W. (1984) Leaching of nitrate and other nutrients from a grazed pasture. *New Zealand Journal of Agricultural Research* 27, 5–12.

Steenvoorden, J.H.A.M. (1986) Nutrient leaching losses following application of farm slurry and water quality considerations in the Netherlands. In: Dam Kofoed, A., Williams, J.H. and L'Hermite, P. (eds) *Efficient Land Use of Sludge and Manure*. Elsevier, London, pp. 168–176.

Stevens, R.J. (1985) Evaluation of the sulphur status of some grasses for silage in Northern Ireland. *Journal of Agricultural Science, Cambridge* 105, 581–585.

Stevens, R.J. and Laughlin, R.J. (1996) Effects of lime and nitrogen fertilizer on two sward types over a 10-year period. *Journal of Agricultural Science, Cambridge* 127, 451–461.

Stevens, R.J. and Laughlin, R.J. (1997) The impact of cattle slurries and their management on ammonia and nitrous oxide emissions. In: Jarvis, S.C. and Pain, B.F. (eds) *Gaseous Nitrogen Emissions from Grasslands*. CAB International, Wallingford, pp. 233–256.

Stevens, R.J., O'Bric, C.J. and Carton, O.T. (1995) Estimating nutrient content of animal slurries using electrical conductivity. *Journal of Agricultural Science, Cambridge* 125, 233–238.

Stevenson, F.J. (1986) *Cycles of Soil: Carbon, Nitrogen, Phosphorus, Sulfur, Micronutrients*. John Wiley and Sons, New York, 380 pp.

Stevenson, F.J. (1991) Organic matter–micronutrient reactions in soil. In: Mortvedt, J.J. (ed.) *Micronutrients in Agriculture*. Soil Science Society of America, Madison, pp. 145–186.

Stevenson, F.J. (1994) *Humus Chemistry*, 2nd edn. John Wiley and Sons, New York, 496 pp.

Stewart, A.B. and Holmes, W. (1953) Nitrogenous manuring of grassland. 1. Some effects of heavy dressings of nitrogen on the mineral composition. *Journal of the Science of Food and Agriculture* 4, 401–408.

Stewart, A.G. and Pharoah, P.O.D. (1996) Clinical and epidemiological correlates of iodine deficiency disorders. In: Appleton, J.D., Fuge, R. and McCall, G.J.H. (eds) *Environmental Geochemistry and Health*. Geological Society, London, pp. 223–230.

Stewart, F.B. and Axley, J.H. (1956) Seasonal variation in the boron content of alfalfa. *Agronomy Journal* 48, 259–262.

Stewart, J.W.B. and McConaghy, S. (1963) Some effects of reaction (pH) changes in a basaltic soil on the mineral composition of growing crops. *Journal of the Science of Food and Agriculture* 14, 613–621.

Stockdale, C.R. (1999) Effects of season and time since defoliation on the nutritive characteristics of three irrigated perennial pasture species in northern Victoria. 2. Macro-minerals. *Australian Journal of Experimental Agriculture* 39, 567–577.

Stout, W.L., Gburek, W.J., Schnabel, R.R., Folmar, G.J. and Weaver, S.R. (1998) Soil-climate effects on nitrate leaching from cattle excreta. *Journal of Environmental Quality* 27, 992–998.

Stratton, S.D. and Sleper, D.A. (1979) Genetic variation and interrelationships of several minerals in orchardgrass herbage. *Crop Science* 19, 477–481.

Suttle, N.F. (1979) Copper, iron, manganese and zinc concentrations in the carcasses of lambs and calves and the relationship to trace element requirements for growth. *British Journal of Nutrition* 42, 89–96.

Suttle, N.F. (1983) Effects of molybdenum concentration in fresh herbage, hay and semi-purified diets on the copper metabolism of sheep. *Journal of Agricultural Science, Cambridge* 100, 651–656.

Suttle, N.F. (1986) Problems in the diagnosis and anticipation of trace element deficiencies in grazing livestock. *Veterinary Record* 119, 148–152.

Suttle, N.F. (1987) The absorption, retention and function of minor nutrients. In: Hacker, J.B. and Ternouth, J.H. (eds) *The Nutrition of Herbivores*. Academic Press, Sydney, pp. 333–361.

Suttle, N.F. (1988) Relationships between the trace element status of soils, pasture and animals in relation to growth rate in lambs. In: Thornton, I. (ed.) *Geochemistry and Health*. Science Reviews, Northwood, pp. 69–79.

Suttle, N.F. (1994) Meeting the copper requirements of ruminants. In: Garnsworthy, P.C. and Cole, D.J.A. *Recent Advances in Animal Nutrition, 1994*, Nottingham University Press, Nottingham, pp. 173–187.

Suttle, N.F. and Jones, D.G. (1989) Recent developments in trace element metabolism and function: trace elements, disease resistance and immune responsiveness in ruminants. *Journal of Nutrition* 119, 1055–1061.

Suttle, N., Mills, C. and Allen, M. (1984) Preventing trace element deficiencies in cattle and sheep. In: *Grassland Research Today*. AFRC, London, pp. 20–21.

Sweeney, D.W., Moyer, J.L. and Havlin, J.L. (1996) Multinutrient fertilization and placement to improve yield and nutrient concentration in tall fescue. *Agronomy Journal* 88, 982–986.

Swift, R.S. (1983) Mineralisation and immobilisation of sulphur in soil. In: Gregg, P.E.H. and Syers, J.K. (eds) *Sulphur in New Zealand Agriculture*. Proceedings Technical Workshop, Massey University, Palmerston North, pp. 18–27.

Syers, J.K., Skinner, R.J. and Curtin, D. (1987) *Soil and Fertilizer Sulphur in UK Agriculture*. Proceedings of the Fertilizer Society no. 264, London, 43 pp.

Ta, T.C. and Faris, M.A. (1987) Effects of alfalfa proportions and clipping frequencies on timothy–alfalfa mixtures. 1. Competition and yield advantages. *Agronomy Journal* 79, 817–820.

Tallowin, J.R.B. and Brookman, S.K.E. (1988) Herbage potassium levels in a permanent pasture under grazing. *Grass and Forage Science* 43, 209–212.

Tamminga, S. (1992) Nutrition management of dairy cows as a contribution to pollution control. *Journal of Dairy Science* 75, 345–357.

Tang, C., Barton, L. and Raphael, C. (1998) Pasture legume species differ in their capacity to acidify soil. *Australian Journal of Agricultural Research* 49, 53–58.

Tate, K.R. (1984) The biological transformation of P in soil. *Plant and Soil* 76, 245–256.

Tate, K.R. (1985) Soil phosphorus. In: Vaughan, D. and Malcolm, R.E. (eds) *Soil Organic Matter and Biological Activity*. Martinus Nijhoff, Dordrecht, pp. 329–377.

Tate, R.L. (1995) *Soil Microbiology*. John Wiley, New York, 398 pp.

Taylor, K., Woof, C., Ineson, P., Scott, W.A., Rigg, E. and Tipping, E. (1999) Variation in seasonal precipitation chemistry with altitude in the northern Pennines, UK. *Environmental Pollution* 104, 1–9.

Ternouth, J.H., Bortolussi, S., Coates, D.B., Hendricksen, R.E. and McLean, R.W. (1996) The phosphorus requirements of growing cattle consuming forage diets. *Journal of Agricultural Science, Cambridge* 126, 503–510.

Thill, J.L. and George, J.R. (1975) Cation concentrations and K to Ca + Mg ratio of nine cool-season grasses and implications with hypomagnesaemia. *Agronomy Journal* 67, 89–91.

Thomas, B., Thompson, A., Oyenuga, V.A. and Armstrong, R.H. (1952) The ash constituents of some herbage plants at different stages of maturity. *Empire Journal of Experimental Agriculture* 20, 10–22.

Thomas, J.G. (1967) The effect of time of cutting on the mineral content of a single-species sward. *Journal of the British Grassland Society* 22, 282–288.

Thompson, J.F., Smith, I.K. and Madison, J.T. (1986) Sulphur metabolism in plants. In: Tabatabai, M.A. (ed.) *Sulfur in Agriculture*. American Society of Agronomy, Madison, pp. 57–121.

Thompson, J.K. (1980) *The Prevention and Treatment of Trace Element Deficiencies in Sheep and Cattle*. North of Scotland College of Agriculture Bulletin no. 22, Aberdeen.

Thompson, J.K. and Warren, R.W. (1979) Variations in composition of pasture herbage. *Grass and Forage Science* 34, 83–88.

Thornton, I. (1977) Biogeochemical studies on molybdenum in the United Kingdom. In: Chappell, W.R. and Petersen, K.K. (eds) *Molybdenum in the Environment*, Vol. 2. Marcel Dekker, New York, pp. 341–369.

Thornton, I. and Abrahams, P. (1981) The role of soil ingestion in the intake of metals by livestock. In: Hemphill, D.E. (ed.) *Trace Substances in Environmental Health XV*. University of Missouri, Columbia, pp. 366–371.

Till, A.R. (1975) Sulphur cycling in grazed pastures. In: McLachlan, K.D. (ed.) *Sulphur in Australasian Agriculture*. Sydney University Press, Sydney, pp. 68–75.

Till, A.R. (1981) Cycling of plant nutrients in pasture. In: Morley, F.H.W. (ed.) *Grazing Animals*. Elsevier, Amsterdam, pp. 35–53.

Till, A.R. and Blair, G.J. (1978) The utilization by grass of sulphur and phosphorus from clover litter. *Australian Journal of Agricultural Research* 29, 235–242.

Tilman, D., Dodd, M.E., Silvertown, J., Poulton, P.R., Johnston, A.E. and Crawley, M.J. (1994) The Park Grass Experiment: insights from the most long-term ecological study. In: Leigh, R.A. and Johnston, A.E. (eds) *Long-term Experiments in Agricultural and Ecological Studies*. CAB International, Wallingford, pp. 287–303.

Tilman, D., Wedin, D. and Knops, J. (1996) Productivity and sustainability influenced by biodiversity in grassland ecosystems. *Nature* 379, 718–720.

Timmons, D.R. and Holt, R.F. (1977) Nutrient losses in surface runoff from a native pasture. *Journal of Environmental Quality* 6, 369–373.

Tinker, P.B. and Gildon, A. (1983) Mycorrhizal fungi and ion uptake. In: Robb, D.A. and Pierpoint, W.S. (eds) *Metals and Micronutrients: Uptake and Utilization by Plants*. Academic Press, London, pp. 21–32.

Tisdale, S.L., Nelson, W.L. and Beaton, J.D. (1985) *Soil Fertility and Fertilizers*, 4th edn. Macmillan, New York.

Treeby, M., Marschner, H. and Romheld, V. (1989) Mobilization of iron and other micronutrient cations from a calcareous soil by plant-borne, microbial and synthetic metal chelators. *Plant and Soil* 114, 217–226.

Troughton, A. (1981) Length of life of grass roots. *Grass and Forage Science* 36, 117–120.

Tunney, H., Breeuwsma, A., Withers, P.J.A. and Ehlert, P.A.I. (1997) Phosphorus fertilizer strategies: present and future. In: Tunney, H., Carton, O.T., Brookes, P.C. and Johnston, A.E. (eds) *Phosphorus Loss from Soil to Water*. CAB International, Wallingford, pp. 177–203.

Turner, R.G. (1970) The subcellular distribution of zinc and copper within the roots of metal-tolerant clones of *Agrostis tenuis* Sibth. *New Phytologist* 69, 725–731.

Tyler, G. and Ström, L. (1995) Differing organic acid exudation pattern explains calcifuge and acidfuge behaviour of plants. *Annals of Botany* 75, 75–78.

Tyler, G. and Zohlen, A. (1998) Plant seeds as mineral nutrient resource for seedlings – a comparison of plants from calcareous and silicate soils. *Annals of Botany* 81, 455–459.

Tyndale-Biscoe, M. (1994) Dung burial by native and introduced dung beetles (Scarabaeidae). *Australian Journal of Agricultural Research* 45, 1799–1808.

Tyson, K.C., Scholefield, D., Jarvis, S.C. and Stone, A.C. (1997) A comparison of animal output and nitrogen leaching losses recorded from drained fertilized grass and grass/clover pasture. *Journal of Agricultural Science, Cambridge* 129, 315–325.

Úlehlová, B. (1992) Micro-organisms. In: Coupland, R.T. (ed.) *Ecosystems of the World. 8A. Natural Grasslands*. Elsevier, Amsterdam, pp. 95–119.

Ullrey, D.E. (1987) Biochemical and physiological indicators of selenium status in animals. *Journal of Animal Science* 65, 1712–1726.

Ulrich, A., Tabatabai, M.A., Ohki, K. and Johnson, C.M. (1967) Sulphur content of alfalfa in relation to growth in filtered and unfiltered air. *Plant and Soil* 26, 235–252.

Ulrich, B. (1991) Deposition of acids and metal compounds. In: Merian, E. (ed.) *Metals and their Compounds in the Environment*. VCH, Weinheim, pp. 369–378.

Umaly, R.C. and Poel, L.W. (1970) Effects of various concentrations of iodine as potassium iodide on the growth of barley, tomato and pea in nutrient solution culture. *Annals of Botany* 34, 919–926.

Umoh, J.E., Harbers, L.H. and Smith, E.F. (1982) The effects of burning on mineral contents of Flint Hill range forages. *Journal of Range Management* 35, 231–234.

Underhay, V.H.S. and Dickinson, C.H. (1978) Water, mineral and energy fluctuations in decomposing cattle dung pats. *Journal of the British Grassland Society* 33, 189–196.

Underwood, E.J. (1981) *The Mineral Nutrition of Livestock*, 2nd edn. Commonwealth Agricultural Bureaux, Farnham Royal, 180 pp.

Underwood, E.J. and Mertz, W. (1987) Introduction. In: Mertz, W. (ed.) *Trace Elements in Human and Animal Nutrition*, 5th edn. Academic Press, London, pp. 1–19.

Underwood, E.J. and Suttle, N.F. (1999) *The Mineral Nutrition of Livestock*, 3rd edn. CAB International, Wallingford, 600 pp.

Ure, A.M., Bacon, J.R., Berrow, M.L. and Watt, J.J. (1979) The total trace element content of some Scottish soils by spark source mass spectrometry. *Geoderma* 22, 1–23.

Valk, H., Metcalf, J.A. and Withers, P.J.A. (2000) Prospects for minimizing phosphorus excretion in ruminants by dietary manipulation. *Journal of Environmental Quality* 29, 28–36.

Van Breemen, N. and van Dijk, H.F.G. (1988) Ecosystem effects of atmospheric deposition of nitrogen in the Netherlands. *Environmental Pollution* 54, 249–274.

Van Burg, P.F.J. (1966) Nitrate as an indicator of the nitrogen-nutrition of grass. In: *Proceedings of the 10th International Grassland Congress, Helsinki*, Valtioneuvoston Kirjapaino, Helsinki, pp. 267–272.

Van Burg, P.F.J. and van Brakel, G.D. (1965) The fertilizer value of anhydrous ammonia on permanent grassland. *Stikstof* (English edition) 9, 28–36.

Van der Meer, H.G. and Van der Putten, A.H.J. (1995) Reduction of nutrient emissions from ruminant livestock farms. In: Pollott, R.J.L. (ed.) *Grassland into the 21st Century*. Occasional Symposium no. 29, British Grassland Society, Reading, pp. 118–134.

Van der Weerden, T.J. and Jarvis, S.C. (1997) Ammonia emission factors for N fertilizers applied to two contrasting grassland soils. *Environmental Pollution* 95, 205–211.

Van Dijk, T.A. and Sturm, H. (1983) *Fertiliser Value of Animal Manures on the Continent*. Proceedings of the Fertilizer Society no. 220, London, 45 pp.

Van Faassen, H.G. and van Dijk, H. (1987) Manure as a source of nitrogen and phosphorus in soils. In: Van der Meer, H.G., Unwin, R.J., van Dijk, T.A. and Ennik, G.C. (eds) *Animal Manure on Grassland and Fodder Crops: Fertilizer or Waste?* Martinus Nijhoff, Dordrecht, pp. 27–43.

Van Koetsveld, E.E. and Lehr, J.J. (1961) [The zinc content of soil and grass in the Netherlands and the significance of this for the feeding of cattle.] *Landbouwk Tijdshrift* (Wageningen) 73, 371–382 (cited in *Herbage Abstracts* 32, 1370).

Van Riper, G.E. and Smith, D. (1959) *Changes in the Chemical Composition of the Herbage of Alfalfa, Medium Red Clover, Ladino Clover and Bromegrass with Advance in Maturity*. Research Report no. 4, Madison, Wisconsin, Wisconsin Agricultural Experiment Station, 25 pp.

Van Steveninck, R.F.M., Babare, A., Fernando, D.R. and Van Steveninck, M.E. (1993) The binding of zinc in root cells of crop plants by phytic acid. *Plant and Soil* 155/156, 525–528.

Van Straalen, W.M. and Tamminga, S. (1990) Protein degradation of ruminant diets. In: Wiseman, J. and Cole, D.J.A. (eds) *Feedstuff Evaluation*. Butterworths, London, pp. 55–72.

Van Vuuren, A.M. and Meijs, J.A.C. (1987) Effects of herbage composition and supplement feeding on the excretion of nitrogen in dung and urine by grazing dairy cows. In: Van der Meer, H.G., Unwin, R.J., Van Dijk, T.A. and Ennik, G.C. (eds) *Animal Manure on Grassland and Forage Crops: Fertilizer or Waste*. Martinus Nijhof, Dordrecht, pp. 17–26.

Van Vuuren, A.M. and Smits, M.C.T. (1997) Effect of nitrogen and sodium chloride intake on production and composition of urine in dairy cows. In: Jarvis, S.C. and Pain, B.F. (eds) *Gaseous Nitrogen Emissions from Grassland.* CAB International, Wallingford, pp. 95–96.

Verloo, M. and Willaert, G. (1990) Direct and indirect effects of fertilization practices on heavy metals in plants and soils. In: Merckz, R., Vereecken, H. and Vlassak, K. (eds) *Fertilization and the Environment.* Leuven University Press, Leuven, pp. 79–87.

Vlamis, J. and Williams, D.E. (1967) Manganese and silicon interactions in the gramineae. *Plant and Soil* 27, 131–140.

Volenec, J.J. and Nelson, C.J. (1995) Forage crop management: application of emerging technologies. In: Barnes, R.F., Miller, D.A. and Nelson, C.J. (eds) *Forages,* Vol. 2, *The Science of Grassland Agriculture.* Iowa State University Press, Ames, pp. 3–20.

Vought, R.L., London, W.T. and Brown, F.A. (1964) A note on atmospheric iodine and its absorption in man. *Journal of Clinical Endocrinology* 24, 414–416.

Wacquant, J.P. (1977) Physicochemical selectivity for cations and CEC of grass roots. *Plant and Soil* 47, 257–262.

Wadsworth, G.A. and Webber, J. (1980) Deposition of minerals and trace elements in rainfall. In: *Inorganic Pollution and Agriculture.* MAFF Reference Book 326, HMSO, London, pp. 47–55.

Waite, R. (1965) The chemical composition of grasses in relation to agronomical practice. *Proceedings of the Nutrition Society* 24, 38–46.

Walker, T.W. and Adams, A.F.R. (1958) Competition for sulphur in a grass–clover association. *Plant and Soil* 9, 353–366.

Walker, W.M. and Pesek, J. (1967) Yield of Kentucky bluegrass (*Poa pratensis*) as a function of its percentage of nitrogen, phosphorus and potassium. *Agronomy Journal* 59, 44–47.

Walworth, J.E. and Sumner, M.E. (1987) The diagnosis and recommendation integrated system (DRIS). *Advances in Soil Science* 6, 149–188.

Ward, G., Harbers, L.H. and Blaha, J.J. (1979) Calcium-containing crystals in alfalfa: their fate in cattle. *Journal of Dairy Science* 62, 715–722.

Watson, C.J. and Stevens, R.J. (1986) The sulphur content of slurries and fertilizers. In: *Record of Agricultural Research, Northern Ireland, 1986,* HMSO, Belfast, pp. 5–7.

Waughman, G.J. and Bellamy, D.J. (1981) Movement of cations in some plant species prior to leaf senescence. *Annals of Botany* 47, 141–145.

Weaver, J.E. (1961) The living network in prairie soils. *Botanical Gazette* 123, 16–28.

Webb, J., Whinham, N. and Unwin, R.J. (1990) Response of cut grass to potassium and the mineral sylvinite. *Journal of the Science of Food and Agriculture* 53, 297–312.

Webb, K.E. and Bergman, E.N. (1991) Amino acid and peptide absorption and transport across the intestine. In: Tsuda, T., Sasaki, Y. and Kawashima, R. (eds) *Physiological Aspects of Digestion and Metabolism in Ruminants.* Academic Press, San Diego, pp. 111–128.

Webster, A.J.F. (1996) The metabolizable protein system for ruminants. In: Garnsworthy, P.C. and Cole, D.J.A. (eds) *Recent Developments in Ruminant Nutrition – 3.* Nottingham University Press, Nottingham, pp. 55–70.

Wedin, D.A. and Tilman, D. (1990) Species effects on nitrogen cycling: a test with perennial grasses. *Oecologia* 84, 433–441.

Wedin, D.A. and Tilman, D. (1996) Influence of nitrogen loading and species composition on the carbon balance of grassland. *Science* 274, 1720–1723.

Weeda, W.C. (1977) Effects of cattle dung patches on soil tests and botanical and chemical composition of herbage. *New Zealand Journal of Agricultural Research* 20, 471–478.

Weinmann, H. (1940) Seasonal chemical changes in the roots of some South African highveld grasses. *Journal of South African Botany* 6, 131–145.

Welch, R.M. (1993) Zinc concentrations and forms in plants for humans and animals. In: Robson, A.D. (ed.) *Zinc in Soils and Plants.* Kluwer Academic Publishers, Dordrecht, pp. 183–195.

Welch, R.M., Allaway, W.H., House, W.A. and Kubota, J. (1991) Geographic distribution of trace element problems. In: Mortvedt, J.J. (ed.) *Micronutrients in Agriculture*, 2nd edn. Soil Science Society of America, Madison, pp. 31–57.

West, C.P., Mallarino, A.P., Wedin, W.F. and Marx, D.B. (1989) Spatial variability of soil chemical properties in grazed pastures. *Soil Science Society of America Journal* 53, 784–789.

West, T.S. (1981) Soil as the source of trace elements. *Philosophical Transactions of the Royal Society of London, B* 294, 19–39.

Westermann, D.T. (1975) Indexes of sulphur deficiency in alfalfa. II. Plant analysis. *Agronomy Journal* 67, 265–268.

Westermann, D.T. and Robbins, C.W. (1974) Effect of SO_4-S fertilization on Se concentration of alfalfa (*Medicago sativa*, L.). *Agronomy Journal* 66, 207–208.

Westerman, P.W., Safley, L.M., Barker, J.C. and Cheschier, G.M. (1985) Available nutrients in livestock waste. In: *Proceedings of the 5th International Symposium on Agricultural Wastes, Chicago, 1985.* American Society of Agricultural Engineers, St Joseph, pp. 295–307.

Wheeler, D.M. (1998) Investigation into the mechanisms causing lime responses in a grass/clover pasture on a clay loam soil. *New Zealand Journal of Agricultural Research* 41, 497–515.

White, C.L. (1993) The zinc requirement of grazing ruminants. In: Robson, A.D. (ed.) *Zinc in Soils and Plants.* Kluwer Academic Publishers, Dordrecht, pp. 197–206.

White, J.C.D. and Davies, D.T. (1958) The relation between the chemical composition of milk and the stability of the caseinate complex. *Journal of Dairy Research* 25, 236–255.

Whitehead, D.C. (1966) *Data on the Mineral Composition of Grassland Herbage from the Grassland Research Institute, Hurley, and the Welsh Plant Breeding Station, Aberystwyth.* Technical Report, 4, Grassland Research Institute, Hurley, 55 pp.

Whitehead, D.C. (1970) Carbon, nitrogen, phosphorus and sulphur in herbage plant roots. *Journal of the British Grassland Society* 25, 236–241.

Whitehead, D.C. (1973a) Studies on iodine in British soils. *Journal of Soil Science* 24, 260–270.

Whitehead, D.C. (1973b) Uptake and distribution of iodine in grass and white clover plants grown in solution culture. *Journal of the Science of Food and Agriculture* 24, 43–50.

Whitehead, D.C. (1978) Iodine in soil profiles in relation to iron and aluminium oxides and organic matter. *Journal of Soil Science* 29, 88–94.

Whitehead, D.C. (1979) Iodine in the UK environment with particular reference to agriculture. *Journal of Applied Ecology* 16, 269–279.

Whitehead, D.C. (1981) The volatilization from soils and mixtures of soil components of iodine added as potassium iodide. *Journal of Soil Science* 32, 97–102.

Whitehead, D.C. (1982) Yield of white clover and its fixation of nitrogen as influenced by nutritional and soil factors under controlled environment conditions. *Journal of the Science of Food and Agriculture* 33, 1227–1234.

Whitehead, D.C. (1984) The distribution and transformations of iodine in the environment. *Environment International* 10, 321–339.

Whitehead, D.C. (1986) Sources and transformations of organic nitrogen in intensively managed grassland soils. In: Van der Meer, H.G., Ryden, J.C. and Ennik, G.C. (eds) *Nitrogen Fluxes in Intensive Grassland Systems*. Martinus Nijhoff, Dordrecht, pp. 47–58.

Whitehead, D.C. (1987) Some soil–plant and root–shoot relationships of copper, zinc and manganese in white clover and perennial ryegrass. *Plant and Soil* 97, 47–56.

Whitehead, D.C. (1995) *Grassland Nitrogen*. CAB International, Wallingford, 397 pp.

Whitehead, D.C. and Jones, E.C. (1969) Nutrient elements in the herbage of white clover, red clover, lucerne and sainfoin. *Journal of the Science of Food and Agriculture* 20, 584–591.

Whitehead, D.C. and Jones, L.H.P. (1978) Nitrogen/sulphur relationships in grass and legumes. In: Brogan, J.C. (ed.) *Sulphur in Forages*. An Foras Taluntais, Dublin, pp. 126–141.

Whitehead, D.C., Barnes, R.J. and Jones, L.H.P. (1983) Nitrogen, sulphur and other mineral elements in white clover and perennial ryegrass grown in mixed swards, with and without fertilizer N, at a range of sites in the UK. *Journal of the Science of Food and Agriculture* 34, 901–909.

Whitehead, D.C., Goulden, K.M. and Hartley, R.D. (1985) The distribution of nutrient elements in cell wall and other fractions of some grasses and legumes. *Journal of the Science of Food and Agriculture* 36, 311–318.

Whitehead, D.C., Jones, L.H.P. and Barnes, R.J. (1978) The influence of fertilizer N plus K on N, S and other mineral elements in perennial ryegrass at a range of sites. *Journal of the Science of Food and Agriculture* 21, 1–11.

Wild, A. and Jones, L.H.P. (1988) Mineral nutrition of crop plants. In: Wild, A. (ed.) *Russell's Soil Conditions and Plant Growth*. Longman, Harlow, pp. 69–112.

Wilkins, P.W., Allen, D.K. and Mytton, L.R. (2000) Differences in the nitrogen use efficiency of perennial ryegrass varieties under simulated rotational grazing and their effects on nitrogen recovery and herbage nitrogen content. *Grass and Forage Science* 55, 69–76.

Wilkins, R.J. (1987) Grassland into the twenty-first century. In: Norsk Hydro Fertilizers (ed.) *Farming into the Twenty-First Century*. Norsk Hydro Fertilizers, Levington, pp. 93–115.

Wilkins, R.J. (1995) Improved technology for production efficiency and utilization efficiency. In: Pollott, G.E. (ed.) *Grassland into the 21st Century: Challenges and Opportunities*. Occasional Symposium 29, British Grassland Society, Reading, pp. 210–224.

Wilkins, R.J. and Harvey, H.J. (1994) Management options to achieve agricultural and nature conservation objectives. In: Haggar, R.J. and Peel, S. (eds) *Grassland*

Management and Nature Conservation. Occasional Symposium 28, British Grassland Society, Reading, pp. 86–94.

Wilkinson, S.R. and Gross, C.F. (1967) Macro- and micronutrient distribution within Ladino clover (*Trifolium repens* L.). *Agronomy Journal* 59, 372–374.

Wilkinson, S.R. and Lowrey, R.W. (1973) Cycling of mineral nutrients in pasture ecosystems. In: Butler, G.W. and Bailey, R.W. (eds) *Chemistry and Biochemistry of Herbage*, Vol. 2. Academic Press, London, pp. 247–315.

Wilkinson, S.R. and Mays, D.A. (1979) Mineral nutrition. In: Buckner, R.C. and Bush, L.P. (eds) *Tall Fescue.* American Society of Agronomy, Madison, pp. 41–73.

Wilkinson, S.R., Stuedemann, J.A. and Belesky, D.P. (1989) Soil potassium distribution in grazed K-31 tall fescue pastures as affected by fertilization and endophytic fungus infection level. *Agronomy Journal* 81, 508–512.

Williams, B.L., Shand, C.A., Sellers, S. and Young, M.E. (1999) Impact of synthetic sheep urine on N and P in two pastures in the Scottish uplands. *Plant and Soil* 214, 93–103.

Williams, C. and Thornton, I. (1972) The effect of soil additives on the uptake of molybdenum and selenium from different environments. *Plant and Soil* 36, 395–406.

Williams, C. and Thornton, I. (1973) The use of soil extractants to estimate plant-available molybdenum and selenium in potentially toxic soils. *Plant and Soil* 39, 149–159.

Williams, P.H. and Haynes, R.J. (1993a) Fate of ^{35}S-labelled urine sulphate in urine affected areas of pasture soil under field conditions. *Journal of Agricultural Science, Cambridge* 121, 83–89.

Williams, P.H. and Haynes, R.J. (1993b) Forms of sulphur in sheep excreta and their fate after application to pasture soil. *Journal of the Science of Food and Agriculture* 62, 323–329.

Williams, P.H. and Haynes, R.J. (1994) Comparison of initial wetting pattern, nutrient concentrations in soil solution and the fate of ^{15}N-labelled urine in sheep and cattle urine patch areas of pasture soil. *Plant and Soil* 162, 49–59.

Williams, P.H. and Haynes, R.J. (1995) Effect of sheep, deer and cattle dung on herbage production and soil nutrient content. *Grass and Forage Science* 50, 263–271.

Williams, P.H., Hedley, M.J. and Gregg, P.E.H. (1989) Uptake of potassium and nitrogen by pasture from urine-affected soil. *New Zealand Journal of Agricultural Research* 32, 415–421.

Williams, P.H., Gregg, P.E.H. and Hedley, M.J. (1990) Fate of potassium in dairy cow urine applied to intact soil cores. *New Zealand Journal of Agricultural Research* 33, 151–158.

Williamson, P. (1976) Above-ground primary production of chalk grassland allowing for leaf death. *Journal of Ecology* 64, 1059–1075.

Wilman, D. (1965) The effect of nitrogenous fertilizer on the rate of growth of Italian ryegrass. *Journal of the British Grassland Society* 20, 248–254.

Wilman, D. and Derrick, R.W. (1994) Concentration and availability to sheep of N, P, K, Ca, Mg and Na in chickweed, dandelion, dock, ribwort and spurrey compared with perennial ryegrass. *Journal of Agricultural Science, Cambridge* 122, 217–223.

Wilman, D. and Hollington, P.A. (1985) Effects of white clover and fertilizer nitrogen on herbage production and chemical composition and soil water. *Journal of Agricultural Science, Cambridge* 104, 453–467.

Wilman, D. and Mzamane, N. (1982) The content and yield of six elements in four grasses as affected by nitrogen application and interval between harvests. *Fertilizer Research* 3, 97–110.

Wilman, D. and Mzamane, N. (1986) The effect of field drying on the concentrations of some major elements in herbage. *Journal of Agricultural Science, Cambridge* 107, 9–13.

Wilman, D., Acuna, P.G.H. and Michaud, P.J. (1994) Concentrations of N, P, K, Ca, Mg and Na in perennial ryegrass and white clover leaves of different ages. *Grass and Forage Science* 49, 422–428.

Wilson, R.K. and McCarrick, R.B. (1967) A nutritional study of grass swards at progressive stages of maturity. *Irish Journal of Agricultural Research* 6, 267–279.

Wilson, S.B. and Hallsworth, E.G. (1965) The distribution of cobalt in *Trifolium subterraneum*. *Plant and Soil* 23, 60–78.

Withers, P.J.A. and Jarvis, S.C. (1998) Mitigation options for diffuse phosphorus loss to water. *Soil Use and Management* 14, 186–192.

Withers, P.J. and Sharpley, A.N. (1995) Phosphorus fertilizers. In: Rechcigl, J.E. (ed.) *Soil Amendments and Environmental Quality*. Lewis Publishers, Boca Raton, pp. 65–107.

Wolton, K.M. (1979) Dung and urine as agents of sward change: a review. In: Charles, H.A. and Haggar, R.J. (eds) *Changes in Sward Composition and Productivity*. Occasional Symposium 10, British Grassland Society, Hurley, pp. 131–135.

Wolton, K.M., Brockman, J.S., Brough, D.W.T. and Shaw, P.G. (1968) The effect of nitrogen, phosphate and potash fertilizers on three grass species. *Journal of Agricultural Science, Cambridge* 70, 195–202.

Wood, M. (1995) *Environmental Soil Biology*, 2nd edn. Blackie Academic and Professional, London.

Wu, L. and Huang, Z.-Z. (1992) Selenium assimilation and nutrient element uptake in white clover and tall fescue under the influence of sulphate concentration and selenium tolerance of the plants. *Journal of Experimental Botany* 43, 549–555.

Wu, L., Huang, Z.-Z. and Burau, R.G. (1988) Selenium accumulation and selenium salt co-tolerance in five grass species. *Crop Science* 28, 517–522.

Yano, F., Yano, H. and Breves, G. (1991) Calcium and phosphorus metabolism in ruminants. In: Tsuda, T., Sasaki, Y. and Kawashima, R. (eds) *Physiological Aspects of Digestion and Metabolism in Ruminants*. Academic Press, San Diego, pp. 277–295.

Yokoyama, K., Kai, H., Koga, T. and Aibe, T. (1991) Nitrogen mineralization and microbial populations in cow dung, dung balls and underlying soil affected by paracoprid dung beetles. *Soil Biology and Biochemistry* 23, 649–653.

Young, D.J.B. (1958) A study of the influence of nitrogen on the root weight and nodulation of white clover in a mixed sward. *Journal of the British Grassland Society* 13, 106–114.

Zyrin, N.G. and Zborishchuk, Y.N. (1975) Iodine content in the plow layer of soils in the European USSR. *Soviet Soil Science* 7, 558–563.

Appendix

Botanical names of plant species mentioned by their common name in the text and References

Alfalfa	*Medicago sativa*
Bent-grass	*Agrostis* spp. (esp. *tenuis* and *stolonifera*)
Birdsfoot trefoil	*Lotus corniculatus*
Bluegrass	*Poa pratensis*
Brome-grass	*Bromus* spp. (esp. *inermis*)
Browntop	*Agrostis tenuis*
Burnet	*Sanguisorba minor*
Chickweed	*Stellaria media*
Chicory	*Cichorium intybus*
Coastal Bermudagrass	*Cynodon dactylon*
Cocksfoot	*Dactylis glomerata*
Couch grass	*Agropyron* spp.
Crested dogstail	*Cynosurus cristatus*
Dandelion	*Taraxacum officinale*
Italian ryegrass	*Lolium multiflorum*
Kentucky bluegrass	*Poa pratensis*
Ladino clover	*Trifolium repens*
Lucerne	*Medicago sativa*
Maize	*Zea mays*
Meadow fescue	*Festuca pratensis*
Meadow foxtail	*Alopecurus pratensis*
Medick	*Medicago* spp. (other than *sativa*)
Orchardgrass	*Dactylis glomerata*
Perennial ryegrass	*Lolium perenne*
Plantain	*Plantago* spp.
Prairie-grass	*Bromus uniloides* (and *B. willdenowi*)
Red clover	*Trifolium pratense*
Red fescue	*Festuca rubra*

Ribwort plantain	*Plantago lanceolata*
Reed canary-grass	*Phalaris arundinacea*
Sainfoin	*Onobrychis viciifolia*
Smooth brome-grass	*Bromus inermis*
Smooth-stalked meadow-grass	*Poa pratensis*
Subterranean clover	*Trifolium subterraneum*
Tall fescue	*Festuca arundinacea*
Timothy	*Phleum pratense*
White clover	*Trifolium repens*
Yarrow	*Achillea millefolium*
Yorkshire fog	*Holcus lanatus*

Index

INTERNATIONAL FORENSIC SCIENCE
AND INVESTIGATION SERIES

FIREARMS, THE LAW, AND FORENSIC BALLISTICS

Second Edition

INTERNATIONAL FORENSIC SCIENCE AND INVESTIGATION SERIES

Edited by James Robertson
Forensic Sciences Division, Australian Federal Police

Firearms, the Law and Forensic Ballistics
T A Warlow
ISBN 0 7484 0432 5
1996

Scientific Examination of Documents:
methods and techniques, 2nd edition
D Ellen
ISBN 0 7484 0580 1
1997

Forensic Investigation of Explosions
A Beveridge
ISBN 0 7484 0565 8
1998

Forensic Examination of Human Hair
J Robertson
ISBN 0 7484 0567 4
1999

Forensic Examination of Fibres,
2nd edition
J Robertson and M Grieve
ISBN 0 7484 0816 9
1999

Forensic Examination of Glass and Paint:
analysis and interpretation
B Caddy
ISBN 0 7484 0579 9
2001

Forensic Speaker Identification
P Rose
ISBN 0 415 27182 7
2002

The Practice of Crime Scene Investigation
J Horswell
ISBN 0 7484 0609 3
2004

Fire Investigation
N Nic Daéid
ISBN 0 415 24891 4
2004

Fingerprints and Other Ridge Skin
Impressions
*C Champod, C J Lennard, P Margot,
and M Stoilovic*
ISBN 0 415 27175 4
2004

Firearms, the Law, and Forensic Ballistics
Second Edition
Tom Warlow
ISBN 0 415 316014
2004